CONCISE PHYSICAL CHEMISTRY

CONCISE PHYSICAL CHEMISTRY

DONALD W. ROGERS
Department of Chemistry and Biochemistry
The Brooklyn Center
Long Island University
Brooklyn, NY

A JOHN WILEY & SONS, INC., PUBLICATION

Published by John Wiley & Sons, Inc., Hoboken, New Jersey.
Published simultaneously in Canada.

Don Rogers is an amateur jazz musician and painter who lives in Greenwich Village, NY.

Library of Congress Cataloging-in-Publication Data:

Rogers, Donald W.
Concise physical chemistry / by Donald W. Rogers.
p. cm.
Includes index.
Summary: "This book is a physical chemistry textbook that presents the essentials of physical chemistry as a logical sequence from its most modest beginning to contemporary research topics. Many books currently on the market focus on the problem sets with a cursory treatment of the conceptual background and theoretical material, whereas this book is concerned only with the conceptual development of the subject. It contains mathematical background, worked examples and problemsets. Comprised of 21 chapters, the book addresses ideal gas laws, real gases, the thermodynamics of simple systems, thermochemistry, entropy and the second law, the Gibbs free energy, equilibrium, statistical approaches to thermodynamics, the phase rule, chemical kinetics, liquids and solids, solution chemistry, conductivity, electrochemical cells, atomic theory, wave mechanics of simple systems, molecular orbital theory, experimental determination of molecular structure, and photochemistry and the theory of chemical kinetics"– Provided by publisher.
ISBN 978-0-470-52264-6 (pbk.)
1. Chemistry, Physical and theoretical–Textbooks. I. Title.
QD453.3.R63 2010
541–dc22

2010018380

10 9 8 7 6 5 4 3 2 1

CONTENTS

FOREWORD

Among many advantages of being a professional researcher and teacher is the pleasure of reading a new and good textbook that concisely summarizes the fundamentals and progress in your research area. This reading not only gives you the enjoyment of looking once more at the whole picture of the edifice that many generations of your colleagues have meticulously build but, most importantly, also enhances your confidence that your choice to spend your entire life to promote and contribute to this structure is worthwhile. Clearly, the perception of the textbook by an expert in the field is quite different, to say the least, from the perception of a junior or senior undergraduate student who is about to register for a class. A simple look at a textbook that is jam-packed with complex integrals and differential equations may scare any prospective students to death. On the other hand, eliminating the mathematics entirely will inevitably eliminate the rigor of scientific statements. In this respect, the right compromise between simplicity and rigor in explaining complex scientific topics is an extremely rare talent. The task is especially large given the fact that the textbook is addressed to students for whom a particular area of science is not among their primary interests. In this respect, Professor Rogers's *Concise Physical Chemistry* is a textbook that ideally suits all of the above-formulated criteria of a new and good textbook.

Although the fundamental laws and basic principles of physical chemistry were formulated long ago, research in the area is continuously widening and deepening. As a result, the original boundaries of physical chemistry as a science become more and more vague and difficult to determine. During the last two decades, physical chemistry has made a tremendous progress mainly boosted by a spectacular increase in our computational capabilities. This is especially visible in quantum molecular modeling. For instance, on my first acquaintance with physical chemistry about 30 years ago, the only molecule that could be quantitatively treated with an accuracy close to experimental data by wave mechanics was the hydrogen molecule. In a lifetime, I have witnessed a complete change of the research picture in which thermodynamic and kinetic data are theoretically obtained routinely with an accuracy often exceeding the experimental one. Quite obviously, to keep the pace with the progress in research, textbooks should be permanently updated and revised. In his textbook Professor Rogers sticks to the classical topics that are conventionally considered as part of physical chemistry. However, these classical topics are deciphered from a modern point of view, and here lies the main strength of this textbook as well as what actually makes this textbook different from many other similar textbooks.

Traditionally, physical chemistry is viewed as an application of physical principles in explaining and rationalizing chemical phenomena. As such, the powerful principles and theories that physical chemistry borrows from physics are accompanied by an advanced and mandatory set of mathematical tools. This makes the process of learning physical chemistry very difficult albeit challenging, exciting, and rewarding. The level of mathematics used by Professor Rogers to formulate and prove the physicochemical principles is remarkably consistent throughout the whole text. Thus, only the most general algebra and calculus concepts are required to understand the essence of the topics discussed. Professor Rogers's way of reasoning is succinct and easy to follow while the examples used to illustrate the theoretical developments are carefully selected and always make a good point. There is no doubt that this textbook is a work of great value, and I heartily recommend it for everybody who wants to enter the wonderful world of physical chemistry.

Worcester Polytechnic Institute ILIE FISHTIK
Worcester, MA
July 2010

PREFACE

Shall I call that wise or foolish, now; if it be really wise it has a foolish look to it; yet, if it be really foolish, then has it a sort of wiseish look to it.
Moby-Dick (Chapter 99) —Herman Melville

Physical chemistry stands at the intersection of the power and generality of classical and quantum physics with the minute molecular complexity of chemistry and biology. Any molecular process that can be envisioned as a flow from a higher energy state to a lower state is subject to analysis by the methods of classical thermodynamics. Chemical thermodynamics tells us where a process is going. Chemical kinetics tells us how long it will take to get there.

Evidence for and application of many of the most subtle and abstract principles of quantum mechanics are to be found in the physical interpretation of chemical phenomena. The vast expansion of spectroscopy from line spectra of atoms well known in the nineteenth century to the magnetic resonance imaging (MRI) of today's diagnostic procedures is a result of our gradually enhanced understanding of the quantum mechanical interactions of energy with simple atomic or complex molecular systems.

Mathematical methods developed in the domain of physical chemistry can be successfully applied to very different phenomena. In the study of seemingly unrelated phenomena, we are astonished to find that electrical potential across a capacitor, the rate of isomerization of cyclopentene, and the growth of marine larvae either as individuals or as populations have been successfully modeled by the same first-order differential equation.

Many people in diverse fields use physical chemistry but do not have the opportunity to take a rigorous three-semester course or to master one of the several ~1000-page texts in this large and diverse field. *Concise Physical Chemistry* is intended to meet (a) the needs of professionals in fields other than physical chemistry who need to be able to master or review a limited portion of physical chemistry or (b) the need of instructors who require a manageable text for teaching a one-semester course in the essentials of the subject. The present text is not, however, a diluted form of physical chemistry. Topics are treated as brief, self-contained units, graded in difficulty from a reintroduction to some of the concepts of general chemistry in the first few chapters to research-level computer applications in the later chapters.

I wish to acknowledge my obligations to Anita Lekhwani and Rebekah Amos of John Wiley and Sons, Inc. and to Tony Li of Scientific Computing, Long Island University. I also thank the National Center for Supercomputing Applications and the National Science Foundation for generous allocations of computer time, and the H. R. Whiteley Foundation of the University of Washington for summer research fellowships during which part of this book was written.

Finally, though many people have helped me in my attempts to better appreciate the beauty of this vast and variegated subject, this book is dedicated to the memory of my first teacher of physical chemistry, Walter Kauzmann.

DONALD W. ROGERS

1

IDEAL GAS LAWS

In the seventeenth and eighteenth centuries, thoughtful people, influenced by the success of early scientists like Galileo and Newton in the fields of mechanics and astronomy, began to look more carefully for quantitative connections among the phenomena around them. Among these people were the chemist Robert Boyle and the famous French balloonist Jacques Alexandre César Charles.

1.1 EMPIRICAL GAS LAWS

Many physical chemistry textbooks begin, quite properly, with a statement of Boyle's and Charles's laws of ideal gases:

$$pV = k_1 \qquad \text{(Boyle, 1662)}$$

and

$$V = k_2 T \qquad \text{(Charles, 1787)}$$

The constants k_1 and k_2 can be approximated simply by averaging a series of experimental measurements, first of pV at constant temperature T for the Boyle equation, then of V/T at constant pressure p for Charles's law. All this can be done using simple manometers and thermometers.

Concise Physical Chemistry, by Donald W. Rogers

1.1.1 The Combined Gas Law

These two laws can be combined to give a new constant

$$\frac{pV}{T} = k_3$$

Subsequently, it was found that if the quantity of gas taken is the number of grams equal to the atomic or molecular weight of the gas, the constant k_3, now written R under the new stipulations, is given by

$$pV = RT$$

For the number of *moles* of a gas, n, we have

$$pV = nRT$$

The constant R is called the *universal gas constant*.

1.1.2 Units

The pressure of a confined gas is the sum of the force exerted by all of the gas molecules as they impact with the container walls of area A in unit time:

$$p = \frac{f \text{ in units of N}}{A \text{ in units of m}^2}$$

The summed force f is given in units of newtons (N), and the area is in square meters (m^2). The N m^{-2} is also called the pascal (Pa). The pascal is about five or six orders of magnitude smaller than pressures encountered in normal laboratory practice, so the convenient unit 1 bar $\equiv 10^5$ Pa was defined.

The logical unit of volume in the MKS (meter, kilogram, second) system is the m^3, but this also is not commensurate with routine laboratory practice where the liter is used. One thousand liters equals 1 m^3, so the MKS name for this cubic measure is the cubic decimeter—that is, one-tenth of a meter cubed (1 dm^3). Because there are 1000 cubic decimeters in a cubic meter and 1000 liters in a cubic meter, it is evident that 1 L = 1 dm^3.

The unit of temperature is the kelvin (K), and the unit of weight is the kilogram (kg). Formally, there is a difference between weight and mass, which we shall ignore for the most part. Chemists are fond of expressing the *amount* of a pure substance in terms of the number of moles n (a pure, unitless number), which is the mass in kg divided by an experimentally determined unit molar mass M, also in kg:[1]

$$n = \frac{\text{kg}}{\text{M}}$$

If the pressure is expressed as N m^{-2} and volume is in m^3, then pV has the unit N m, which is a unit of energy called the *joule* (J). From this, the expression

$$R = \frac{pV}{nT}$$

gives the unit of R as J K^{-1} mol^{-1}. Experiment revealed that

$$R = 8.314 \text{ J K}^{-1} \text{ mol}^{-1} = 0.08206 \text{ L atm K}^{-1} \text{ mol}^{-1}$$

which also defines the *atmosphere*, an older unit of pressure that still pervades the literature.

1.2 THE MOLE

The concept of the *mole* (gram molecular weight in early literature) arises from the deduction by Avogadro in 1811 that equal volumes of gas at the same pressure and temperature contain the same number of particles. This somewhat intuitive conclusion was drawn from a picture of the gaseous state as being characterized by repulsive forces between gaseous particles whereby doubling, tripling, and so on, the weight of the sample taken will double, triple, and so on, its number of particles, hence its volume. It was also known at the time that electrolysis of water produced *two* volumes of hydrogen for every volume of oxygen, so Avogadro deduced the formula H_2O for water on the basis of his hypothesis of equal volume for equal numbers of particles in the gaseous state.

By Avogadro's time, it was also known that the number of grams of oxygen obtained by electrolysis of water is 8 times the number of grams of hydrogen. By his 2-for-1 hypothesis, Avogadro reasoned that the less numerous oxygen atoms must be $2(8) = 16$ times as heavy as the more numerous hydrogen atoms. This theoretical vision led directly to the concept of atomic and molecular weight and to the mass of pure material equal to its atomic weight or molecular weight, which we now call the mole.[2] Various experimental methods have been used to determine the number of particles comprising one mole of a pure substance with the result

[1] General practice is to write experimentally determined quantities in italics and units in Roman letters, but there is some overlap and we shall not be strict in this observance.
[2] The word is mole, but the unit is mol.

6.022×10^{23}, which is now appropriately called Avogadro's number, N_A. One mole of an ideal gas contains N_A particles and occupies 24.79 dm^3 at 1 bar pressure and 298.15 K.

1.3 EQUATIONS OF STATE

The equation $pV = RT$ with the stipulation of one mole of a pure gas is an *equation of state*. Given that R is a constant, the combined gas law equation can be written in a more general way:

$$p = f(V, T)$$

which suggests that there are other ways of writing an equation of state. Indeed, many equations of state are used in various applications (Metiu, 2006). The common feature of these equations is that only two *independent variables* are combined with constants in such a way as to produce a third *dependent* variable. We can write the general form as $p = f(V, T)$, or

$$V = f(p, T)$$

or

$$T = f(p, V)$$

so long as there are two independent variables and one dependent variable. One mole of a pure substance always has two *degrees of freedom.* Other observable properties of the sample can be expressed in the most general form:

$$z = f(x_1, x_2)$$

The variables in the general equation may seem unconnected to p and V, but there always exists, in principle, an equation of state, with two and only two independent variables, connecting them.

An infinitesimal change in a state function z for a system with two degrees of freedom is the sum of the infinitesimal changes in the two dependent variables, each multiplied by a sensitivity coefficient $(\partial z/\partial x_1)_{x_2}$ or $(\partial z/\partial x_2)_{x_1}$ which may be large if the dependent variable is very sensitive to independent variable x_i or small if dz is insensitive to x_i:

$$dz = \left(\frac{\partial z}{\partial x_1}\right)_{x_2} dx_1 + \left(\frac{\partial z}{\partial x_2}\right)_{x_1} dx_2$$

The subscripts x_1 and x_2 on the parenthesized derivatives indicate that when one degree of freedom is varied, the other is held constant. We shall investigate state functions in more detail in the chapters that are to come.

1.4 DALTON'S LAW

At constant temperature and pressure, by Avogadro's principle, the volume of an ideal gas is directly proportional to the number of particles of the gas measured in moles:

$$V = n\left[\frac{RT}{p}\right] = nN_A$$

This principle holds *regardless of the nature of the particles*:

$$p = n\left[\frac{RT}{V}\right]_{\text{const}}$$

Since the nature of the particles plays no role in determining the pressure, the total pressure of a mixture of ideal gases[3] is determined by the total number of moles of gas present:

$$p = n_1\left[\frac{RT}{V}\right]_{\text{const}} + n_2\left[\frac{RT}{V}\right]_{\text{const}} + \cdots = \sum_i n_i\left[\frac{RT}{V}\right]_{\text{const}} = \left[\frac{RT}{V}\right]_{\text{const}}\sum_i n_i$$

Each gas acts as though it were alone in the container, which leads to the concept of a *partial pressure* p_i exerted by one component of a mixture relative to the total pressure. This idea is embodied in Dalton's law for the total pressure of a mixture as the sum of its partial pressures:

$$p_{\text{total}} = \sum_i p_i$$

Apart from emphasizing Avogadro's idea that the ideal gaseous state is characterized by the number of particles, not by their individual nature, Dalton's law also leads to the idea of a *pressure fraction* of one component of a mixture relative to the total pressure exerted by all the components of the mixture:

$$X_{p_i} = \frac{p_i}{\sum_i p_i}$$

[3]Many real gases are nearly ideal under normal room conditions.

1.5 THE MOLE FRACTION

Recognizing that the pressure of each gas is directly proportional to the number of moles through the same constant, we may write the pressure fraction as a *mole fraction*:

$$X_i = \frac{n_i}{\sum_i n_i}$$

The pressure of a real gas follows Dalton's law only as an approximation, but the number of particles (measured in moles) is not dependent upon ideal behavior; hence the summation of mole fractions

$$X_{\text{total}} = \sum_i X_i$$

is exact for ideal or nonideal gases and for other states of matter such as liquid and solid mixtures and solutions.

1.6 EXTENSIVE AND INTENSIVE VARIABLES

Mass m is an *extensive* variable. Density ρ is an *intensive* variable. If you take twice the amount of a sample, you have twice as many grams, but the density remains the same at constant p and T. Molar quantities are intensive. For example, if you double the amount of sample under at constant p and T, the molar volume (volume *per mole*) V_m remains the same just as the density did.

1.7 GRAHAM'S LAW OF EFFUSION

Knowing the molar gas constant $R = 8.314\,\text{J K}^{-1}\ \text{mol}^{-1} = 0.08206\,\text{L atm K}^{-1}\ \text{mol}^{-1}$, which follows directly from measurements of p and V on known amounts of a gas at specific values of T, one can determine the atomic or molecular weight of an independent sample within the limits of the ideal gas approximation. Another way of finding the molecular weight of a gas is through Graham's law of effusion, which states that the rate of escape of a confined gas through a very small hole is inversely proportional to its particle weight—that is, its atomic or molecular weight. This being the case, measuring the rate of effusion of two gases—one of known molecular weight and the other of unknown molecular weight—gives the ratio $\text{MW}_{\text{known}}/\text{MW}_{\text{unknown}}$ and hence easy calculation of $\text{MW}_{\text{unknown}}$.

Aside from important medical applications (dialysis), Graham's work also focused attention on the random motions of gaseous particles and the speeds with which they move. We can rationalize Graham's law as the result of a very large *ensemble* of particles colliding with the wall of a constraining container, supposing that the wall has a hole in it. Only a few particles escape the container because the hole is small. Escape probability is determined by how fast the particle is moving. Fast particles collide with the walls of the container more often than do slow ones.

By a standard derivation (Exercise 1.2), one finds

$$pV = \tfrac{1}{3}N_A m\bar{u}_x^2$$

where $\bar{u}_x$ is the average speed of an ensemble consisting of one mole of an ideal gas. Notice that because $pV = RT$ has the units J K^{-1} mol^{-1} K = J mol^{-1}, pV is a *molar energy*. Increasing the temperature of a gas requires an input of energy. We usually write the kinetic energy E_{kin} of a single moving mass such as a baseball as $E_{\text{kin}} = \frac{1}{2}mv^2$, where v is its speed and m is its mass. Consider a hypothetical one-dimensional x-space along which point particles can move without interference. If the kinetic energy of molecular particles follows the same kind of law as more massive particles, we obtain

$$\tfrac{1}{2}m\bar{u}_x^2 = \bar{E}_{\text{kin}}$$

where $\bar{E}_{\text{kin}}$ is the average *kinetic energy* because kinetic energy is the only kind an ensemble of point particles can have. Substitute $2\bar{E}_{\text{kin}}$ for $m\bar{u}_x^2$ in

$$pV = \tfrac{2}{3}N_A\bar{E}_{kin}$$

but pV also equals RT for one mole of a gas, so

$$pV = RT = \tfrac{1}{3}N_A m\bar{u}_x^2$$

This enables us to calculate $\bar{u}_x$ at any specified temperature. The calculation gives high speeds. For example, nitrogen molecules move at about 400 m s^{-1} (meters per second) at room temperature and hydrogen molecules move at an astonishing speed of nearly 2000 m s^{-1}. There are different ways of calculating averages (mean, mode, root mean square), which give slightly different results for molecular speeds.

1.8 THE MAXWELL–BOLTZMANN DISTRIBUTION

All particles of a confined gas do not move with the same velocity even if T is constant. Rather, they move with a velocity *probability density* ρ_v which is randomly distributed about $v = 0$ and which follows the familiar Gaussian distribution e^{-v^2}.

The probability density function drops off at large values of $\pm v$ because the probability of finding particles with velocities very much different from the mean is small. The curve is symmetrical because, picking an arbitrary axis, the particle may be going either to the left or to the right, having a velocity v or $-v$. The peak at $v = 0$ is somewhat

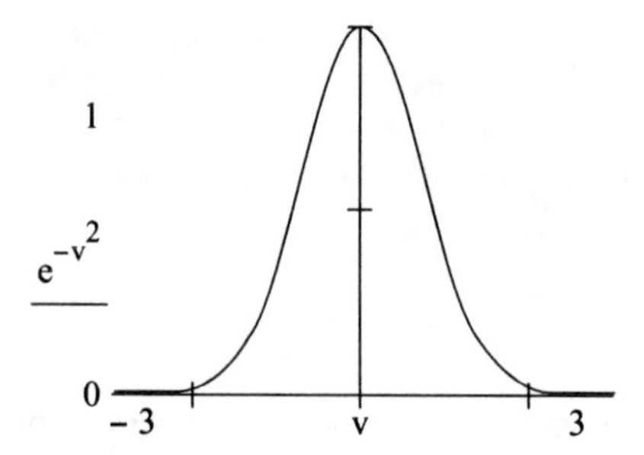

FIGURE 1.1 The probability density for velocities of ideal gas particles at $T \neq 0$.

misleading because it may suggest that the most probable velocity is zero. Not so. The particles are not standing still at any temperature above absolute zero. The peak at $v = 0$ arises because we don't know which direction any particle is going, left or right. In our ignorance, assuming a random distribution, the best bet is to guess zero. We will always be wrong, but the sum of squares of our error over many trials will be minimized. This is an example of the principle of *least squares*.

The Maxwell–Boltzmann distribution of molecular speeds was originally derived assuming that particle velocities are distributed along a continuous spectrum like Fig. 1.1. This implies that E_{kin} can take any value in a continuum as well. The laws of quantum mechanics, however, deny this possibility. They require a distribution over a *discontinuous* energy *spectrum* or manifold of energy levels like that in Fig. 1.2. The connection between Figs. 1.1 and 1.2 can be seen by tilting the page 90° to the left. The number of particles at higher energies tails off according to a Gaussian distribution. The Maxwell–Boltzmann distribution over nondegenerate, discontinuous energy levels is

$$\frac{N_i}{N_0} = e^{-E_i/k_B T}$$

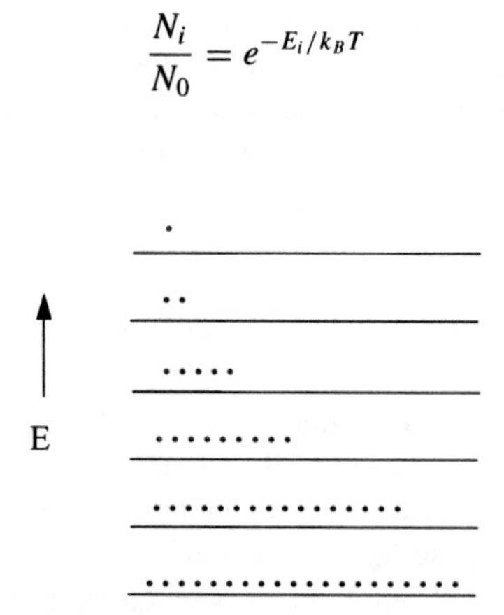

FIGURE 1.2 A Maxwell–Boltzmann distribution over discontinuous energy levels. Particles are not static; they exchange energy levels rapidly. The levels need not be equally spaced.

where N_i is the number of particles at the level having energy E_i. In this expression, N_0 is the number at the lowest energy, usually designated zero $E_0 = 0$ in the absence of a reason to do otherwise.[4] Energy, being a scalar, is proportional to the square of the *speed* of an ensemble of molecules. The *population* of the energy levels in Fig. 1.2 drops off rapidly at higher energies.

The term *degeneracy* is used when two or more levels exist at the same energy, which sometimes happens under the laws of quantum mechanics. Now the number of particles at level E_i is multiplied by the number of levels g_i having that energy

$$\frac{N_i}{N_0} = g_i e^{-E_i/k_B T}$$

The degeneracy is always an integer and it is usually small. Also, from $E_{\text{kin}} = \frac{3}{2}RT$ for one mole, we can find the *expectation value* $\langle \varepsilon_{\text{kin}} \rangle$ of the kinetic energy per representative or average particle

$$\langle \varepsilon_{\text{kin}} \rangle = \frac{3}{2} \frac{R}{N_A} T$$

This leads to the important constant

$$\frac{R}{N_A} = k_B = \frac{8.3145}{6.022 \times 10^{23}} = 1.381 \times 10^{-23} \text{J K}^{-1}$$

and

$$\langle \varepsilon_{\text{kin}} \rangle = \tfrac{3}{2} k_B T$$

where k_B, the universal gas constant *per particle*, is called the *Boltzmann constant*. It should be evident that $k_B T$ must have the units of energy because N_i/N_0 is a unitless (pure) number, $\ln(N_i/N_0) = -E_i/k_B T$, which is also unitless, hence the units of $k_B T$ must be the same as E_i. We are taking advantage of the fact that if $y = e^x$, then $\ln y = x$.

1.9 A DIGRESSION ON "SPACE"

The terms density and probability density were used in Section 1.8. These are different but analogous uses of the word density. In the first case, density was used in the usual sense of weight or mass per unit volume, *m/V*. In the second case, the *probability density* is defined as the probability in a specified space. Any variable measured

[4] Like all energies, this zero point is arbitrary.

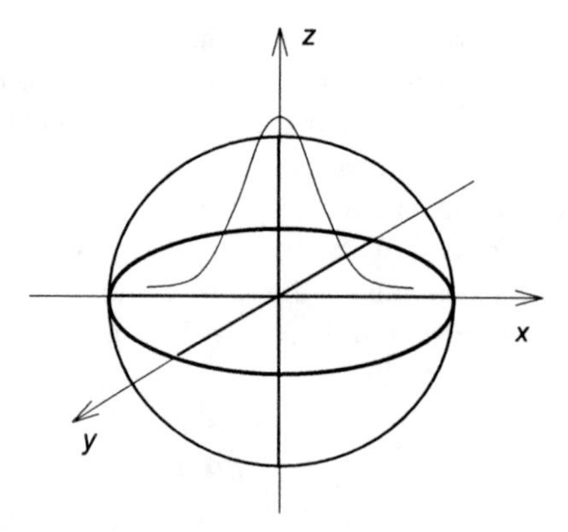

FIGURE 1.3 The Gaussian Probability Density Distribution in 3-Space. The distribution curve is in the fourth dimension of the space. The probability maximum is at the center of the sphere.

along an axis defines a *space*. For example, plotting x along a horizontal axis defines a one-dimensional x-space. Space in the x, y, and z dimensions is the familiar *3-space* often called a Cartesian space in honor of the seventeenth-century mathematician and philosopher René Descartes. We usually plot functions along mutually perpendicular or *orthogonal* axes for mathematical convenience. If velocity is plotted along a $\mathbf{v}$ axis, we have a one-dimensional velocity space. If probability density ρ is plotted along one axis, and velocity is plotted along another axis, the result is a *probability density–velocity space* of two dimensions. If $\rho(\mathbf{v})$ is plotted in $\mathbf{v}_x,\mathbf{v}_y$ space, the result is a function in 3-space; or if it is thought of as a function of all three Cartesian coordinates, the resulting function is in 4-space. That is, $\rho(\mathbf{v})$ in 1-, 2-, or 3-space gives a function in 2-, 3-, or 4-space, one dimension more. There should be nothing terrifying about many-dimensional *space* or *hyperspace*; it is merely an algebraic generalization of the more commonplace use of the term.

The four-dimensional surface of the Gaussian distribution in Cartesian 3-space cannot be precisely drawn but it can be imagined as a figure with spherical symmetry, having a maximum at the center of the sphere. Imagine that the sphere in Fig. 1.3 can be rotated any amount in any angular direction, leaving the distribution curve unchanged.

1.10 THE SUM-OVER-STATES OR PARTITION FUNCTION

Adding up all the particles in all the states of a system gives the total number of particles in the system:

$$\sum N_i = N$$

or

$$\sum N_i = \sum N_0 g_i e^{-\varepsilon_i/k_BT} = N$$

Since N_0 is a number appearing in each term of the sum, it can be factored out:

$$N_0 \sum g_i e^{-\varepsilon_i/k_BT} = N$$

Dividing $N_i = N_0 g_i e^{-\varepsilon_i/k_BT}$ by $N_0 \sum g_i e^{-\varepsilon_i/k_BT} = N$, gives

$$\frac{N_i}{N} = \frac{N_0 g_i e^{-\varepsilon_i/k_BT}}{N_0 \sum g_i e^{-\varepsilon_i/k_BT}} = \frac{g_i e^{-\varepsilon_i/k_BT}}{Q}$$

where we have given the symbol Q to the summation $\sum g_i e^{-\varepsilon_i/k_BT}$. This important summation appears frequently and is given the name *sum-over-states* or *partition function*. Rewriting the ratio N_i/N, we have

$$N_i Q = \underbrace{N g_i e^{-\varepsilon_i/k_BT}}_{\text{fixed at } T = \text{const}}$$

We see that, for a given number of molecules N at temperature T, the right hand side of the equation is fixed for any specific energy state, that is, $N_i Q = \text{const}$. The sum-over-states is then a scaling factor, determining the relative population of a state, $N_i = \text{const}/Q$. If Q is large, the state is sparsely populated. If Q is small, it is densely populated. Another way of looking at Q is that it is an indicator of the number of quantum states available to a system. For many available states (large Q) a given state is less densely populated than it would be if only a few states were available (small Q).

The *occupation number* of a quantum state relative to the total number of particles N_i/N is, strictly speaking, a probability; however, given the immense number of particles in a mole of gas, we may treat it as a certainty. The summation of all possible fractions N_i/N must be 1:

$$\sum_i \frac{N_i}{N} = \frac{\sum_i N_i}{N} = \frac{N}{N} = 1$$

The probability distribution of molecular velocities in 3-space is a collection of randomly oriented vectors away from an origin of the zero (least squares) estimate. Plotted in velocity-space, the probability density approaches zero near the origin. This comes about when we use spherical polar coordinates to express our probability densities because, near the origin, the volume of the spherical space becomes very small. As the radial distance is increased, the space becomes larger but the probability

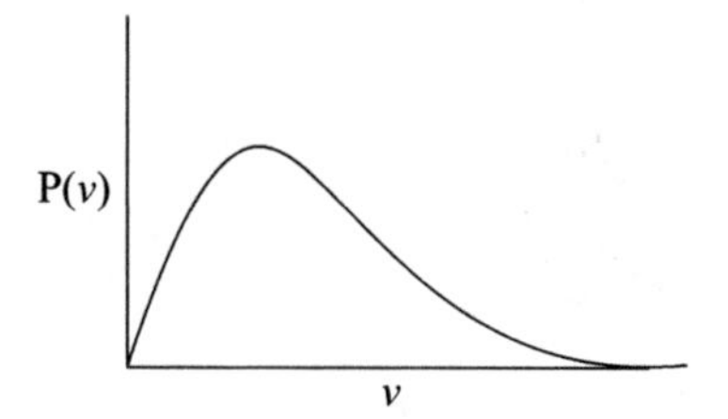

FIGURE 1.4 The probability density of molecular velocities in a spherical velocity space.

density drops off as a Gaussian function. In between these two approaches to zero, the probability density must go through a maximum as shown in Fig. 1.4.

PROBLEMS AND EXERCISES

Exercise 1.1 The Combined Gas Law

Combine Boyle's law and Charles's law to obtain the combined gas law.

Solution 1.1 Take an ideal gas under the arbitrary conditions $p_1V_1T_1$ and convert it to $p_2V_2T_2$ by a two-step process, varying the pressure first and the temperature second. After the pressure is changed from p_1 to p_2, according to Boyle's law, the volume, still at T_1, is at an intermediate value V_x

$$p_1V_1 = p_2V_x \qquad T_1 = \text{const}$$

$$V_x = \frac{p_1V_1}{p_2}$$

Now change the temperature to T_2 at constant p_2. By Charles's law, the volume goes from V_x to V_2

$$\frac{V_x}{T_1} = \frac{V_2}{T_2} \qquad p_2 = \text{const}$$

$$V_x = T_1\frac{V_2}{T_2}$$

Now equate V_x from the equations above:

$$\frac{p_1V_1}{p_2} = T_1\frac{V_2}{T_2}$$

$$\frac{p_1V_1}{T_1} = \frac{p_2V_2}{T_2}$$

Since the *pV/T* quotients are equal to each other for any arbitrary variations in *p*, *V*, and *T*, they must be equal to the same constant *k*:

$$\frac{pV}{T} = \text{const} = k$$

which is the combined gas law.

If we accept the combined gas law, there is a gas constant for any specified quantity of each individual gas, subject only to the restriction of ideal behavior. If we demand that *V* be the *molar volume* V_m

$$\frac{pV_m}{T} = \frac{p\left(\frac{V}{n}\right)}{T} = k$$

where *n* is the number of moles in the gas sample, then this equation becomes

$$\frac{pV}{T} = nR$$

where the symbol *R* is used to denote a *universal* gas constant applicable to one mole of any gas in the approximation of ideal behavior. One can obtain a value for *R* by arbitrarily assigning a pressure of 1 bar at $T = 298.15$ K to precisely 1.0 mole of an ideal gas. We know the molar volume to be 24.790 dm^3 at this pressure and temperature, so

$$R = \frac{p\left(\frac{V}{n}\right)}{T} = \frac{1.000(24.790)}{298.15} = 0.083146$$

with units of bar $\text{dm}^3\ \text{K}^{-1}\ \text{mol}^{-1}$. If the volume is expressed in m^3, then *R* is in bar $\text{m}^3\ \text{K}^{-1}\ \text{mol}^{-1} = 100(0.0831) = 8.310\ \text{J K}^{-1}\ \text{mol}^{-1}$. Remember that the unit J K^{-1} is for a molar gas constant. Notice that the numerator is an energy. The tabulated value is 8.3144725 $\text{J K}^{-1}\ \text{mol}^{-1}$ (*CRC Handbook of Chemistry and Physics, 2008–2009*, 89th ed.)

Exercise 1.2 The Maxwell–Boltzmann Distribution

Derive

$$pV = \tfrac{1}{3}N_A m v_x^2$$

where N_A, m, and v_x are Avogadro's number, the mass, and the average *x*-component of the velocity of a collection of ideal gas particles confined to a cubic box *l* dm on an edge.

Solution 1.2 Consider a particle (molecule or atom) moving in the x direction in a cubic box l units on an edge, oriented so that the particle path is perpendicular to one of its faces. Consider for now only collisions with one wall. The momentum of the particle is mv on the way into the collision, and $-mv$ (in the opposite direction) on the way out of the collision. The change of momentum is $2mv$. The number of collisions per second (collision frequency) is $v/2l$ because the particle must travel the length of the box l to collide with the opposite wall and l once again on the return trip. Force is the change in momentum with respect to time, $d\mathbf{p}/dt$ (Newton's second law). Force per collision is $F = m\mathbf{a} = d\mathbf{p}/dt$, where $\mathbf{a}$ is the acceleration, a change in speed over unit time (in seconds). The total force exerted by the particle on the wall for many collisions is the force per collision times the number of collisions per unit time (a frequency):

$$F = 2mv\frac{v}{2l} = \frac{mv^2}{l}$$

Pressure is force per unit area $A = l^2$:

$$p = \frac{mv^2}{l}\left(\frac{1}{A}\right) = \frac{mv^2}{l}\left(\frac{1}{l^2}\right) = \frac{mv^2}{l^3} = \frac{mv^2}{V}$$

where the volume of the box is the cube of one of its edges. Given that there are very many molecules in the box, on average, only one-third of them are moving in the x direction; or, better said, only one-third of all components of all velocity vectors are oriented in the x direction. (The other two-thirds are oriented in the y and z directions.) For Avogadro's number, N_A, of particles we find,

$$p = \frac{1}{3}N_A\frac{mv^2}{V}$$

where v is an unbiased or average particle speed. This is the equation we sought.

We already know that pV is an energy and that $\frac{1}{2}mv^2$ is the kinetic energy of a moving mass m at velocity v. One sees the proportionality in the preceding equation. More explicitly, multiply and divide the preceding equation by 2. Now,

$$pV = \frac{2}{3}N_A\frac{1}{2}m\bar{v}_x^2 = \frac{2}{3}\bar{E}_{\text{kinetic}}$$

where the overbar notation for $\bar{v}_x$ and $\bar{E}_{\text{kinetic}}$ is added to stress that the speed (scalar magnitude of the velocity component) is an average value over all of the N_A particles.

Problem 1.1

Calculate the volume of 50.0 gr of methane at 400 K and $p = 12.0$ bar. $R = 8.314\,\mathrm{J\,K^{-1}\,mol^{-1}}$. Discuss units.

Problem 1.2

A quantity of an ideal gas occupies 37.5 L at 1.00 atm pressure. How many liters will it occupy when compressed to 4.50 atm pressure at constant temperature? What is the Boyle's law constant k? Give units. 1 bar = 0.986923 atm.

Problem 1.3

Plot the pV curves for 1.00 dm^3 of three ideal gas samples, each expanded from a volume of 1.00 dm^3 in increments of 10.0 dm^3 to 100 dm^3, where the three samples are maintained at energies of 500, 1500, and 2500 joules, respectively.

Problem 1.4

Four grams of an organic liquid vaporizes to produce 1.00 dm^3 of vapor at 298.15 K and 1.00 atmosphere. Find an approximate molar mass of the liquid.

Problem 1.5 Mathcad© Computer Exercise

(a) The volume of a fixed quantity of a real gas at 298.15 K was measured at five different pressures, $p = 0.160, 0.219, 0.310, 0.498,$ and 0.652 atm. The experimental results were $V = 3.42,\ 2.48,\ 1.71,\ 1.03,\ 0.75$ dm^3. These pressures and volumes were tabulated as column vectors $\mathbf{p}$ and $\mathbf{V}$. Calculate five approximate Boyle's law constants from these measurements.

(b) What would the Boyle's law constant have been if the gas had been ideal?

(c) If the amount of sample, identified as carbon dioxide, is 1.000 g, what is its molar volume?

Problem 1.6

The volume of a 0.5333-g sample of gas was measured at 298 K and pressures of 0.0590, 0.143, 0.288, 0.341, and 0.489 bar with the results $V = 14.8, 6.07, 2.99, 2.54,$ and 1.75 dm^3. What is the molar mass of the gas?

Problem 1.7

A mixture of 8.00 g of H_2 and 2.00 g of D_2 was allowed to effuse through a minute orifice, and the composition of the effusing gas mixture was monitored by glc. What was the percent composition of the first trace of gas mixture so monitored?

Problem 1.8

We have two expressions for the molar volume of an ideal gas: (1) $V_m = 22.414$ L at 0°C and 1 atm and (2) $V_m = 24.790$ dm^3 at 25°C and 1 bar. (The atm and bar are taken to have an indefinite number of significant digits.) Use this information to obtain absolute zero ($T = 0$ K) on the Centigrade scale.

Problem 1.9

If a gas occupies 47.6 dm^3 at NSTP = 298.15 K, what is its volume at 1.00 bar and 500 K? The acronym for *new standard temperature and pressure* NSTP replaces the old STP for standard temperature and pressure.

Problem 1.10

(a) Suppose that 18.44 g of N_2 occupies a container at *new standard temperature and pressure* NSTP and 24.35 g of a sample of a different gas is introduced into the container, keeping the temperature constant. If the pressure after addition is 3.20 bar, what is the average molar mass M_{av} of the mixture?

(b) What is the molar mass of the introduced gas?

Problem 1.11

A pure gas takes twice as long as helium to effuse through a porous membrane. What is its molar mass?

Problem 1.12

Compute the root mean square speed of H_2 molecules at 1000 K.

Problem 1.13

What is the translational energy of 1 mol of an ideal gas at $T = 298.15$ K?

Computer Exercise 1.14

Using a standard plotting package, plot a graph of $pV = k$ where $k = 1$, taking p as the vertical axis and V as the horizontal axis. Play with your plotting package so as to produce many different plots, thereby learning the idiosyncrasies of your particular package.

Problem 1.15

What is the expectation value of the molecular speed among an ensemble of nitrogen molecules at 298 K?

Problem 1.16

(a) Calculate the expectation value of the speed of hydrogen molecules among an ensemble at 298 K. Give units.

(b) At the same temperature, $\langle v \rangle = 515\ \mathrm{m\,s^{-1}}$ for nitrogen molecules. What is the ratio $\langle v_{H_2} \rangle / \langle v_{N_2} \rangle$? Explain this ratio. Give units.

Problem 1.17

A sample of 2.50 mol of a gas was confined to a certain volume at 1.00 atm pressure and 298 K. Assuming ideal behavior, what volume did it occupy?

Problem 1.18

If the molar volume is 24.79 dm^3 at 298.15 K and 1 bar pressure, what is the universal gas constant. Give units.

2

REAL GASES: EMPIRICAL EQUATIONS

The ideal gas laws are based on two assumptions, neither of which is true. First, atoms or molecules comprising the ideal gas are assumed to have no volume. They are treated as mathematical point masses for convenience. Second, they are assumed to have no interactions with each other. Attractive and repulsive forces are ignored by setting them to zero.

2.1 THE VAN DER WAALS EQUATION

The Dutch physicist van der Waals remedied both of these failures. He treated the first of them by simply subtracting an empirical parameter taken to represent the volume of the particles, called the *excluded volume* b, from the total volume V of the gas to leave an *effective volume* of $(V - b)$. There is less space for each molecule to move in because of the space taken up by its neighbors.

Attractive and repulsive forces often operate through an inverse square law. For example, gravitational force on masses m_1 and m_2 at a separation of r is

$$F = G\frac{m_1 m_2}{r^2}$$

where G is the gravitational constant. The constants G, m_1, and m_2 can be grouped as a single constant k in the numerator to give an attractive or repulsive force F that varies as $F = k/r^2$. Van der Waals reasoned that the *distance* between gas particles

Concise Physical Chemistry, by Donald W. Rogers

increases directly with the volume occupied by the gas; thus, for the attractive force between particles, he wrote

$$F = \frac{a}{V^2}$$

The conversion factor from distance to volume is not needed because it is included in the parameter a, which is determined experimentally. Pressure is force per unit area on the walls of a container, so van der Waals added the force term to the pressure and rewrote the ideal gas equation for one mole of a real gas as the corrected pressure times the volume remaining after subtracting the volume of the particles:

$$\left(p + \frac{a}{V^2}\right)(V - b) = RT$$

The van der Waals equation is a *semiempirical* equation because the ideal gas law on which it is based can be derived from pure theory (see below), but a and b are empirical parameters found by trial and error. One can start with a pair of plausible estimates for a and b, vary them, compare the results with measured p, V, T behavior, and select the values of a and b that give the best agreement with experimental measurements. Computer routines are available that make many thousands of estimates and give the best curve fit in a matter of seconds.

A knowledge of b permits one to calculate order-of-magnitude radii of molecules. For example, the van der Waals radius of methane is 190 pm (picometers, 10^{-9} m) as compared to the spectroscopic value (obtained many years later) of 109 pm for the C–H bond length of methane. Numerous similar calculations give comparable results. This rough agreement supports van der Waals's qualitative picture of the excluded volume of real gases.

2.2 THE VIRIAL EQUATION: A PARAMETRIC CURVE FIT

The *virial* equation is an example of a more general and frequently more accurate curve fitting routine than the van der Waals equation, but it that gives less insight into possible causes of nonideal behavior than the van der Waals equation does. Any data set can be graphed and fit by an analytical equation (an equation that can be written out in terms of a limited number of variables and some accompanying parameters). A parameter is a number entering into an equation that takes on a fixed value for one system but may change to some other fixed value for another system. For example, what is usually called the Boyle's law "constant" $pV = k$ is, in fact, a parameter because it is valid only for a specified, fixed temperature. Change the temperature and the parameter takes on a different value, but it acts like a constant as long as you maintain the temperature fixed. The gas law constant R is a true constant; for one mole of an ideal gas it is always the same.

Of any two equations, one will be a better fit to a given data set than the other. Of three equations, one will be a better fit than either of the other two, and so on.

Computer speed makes it possible to try very many equations and select the best one, but there is a point of diminishing returns. An equation may have so many parameters that no one will ever use it or it may follow random fluctuations in the data set that tell us nothing about the physics of the actual system. If you are not prudent in your use of curve-fitting programs, you may be calculating the characteristics of a unicorn to many significant figures. These caveats apply to any curve-fitting problem, not just those of real gases.

A very nice balance that avoids daunting complexity but achieves good accuracy is the series equation

$$y = a + bx + cx^2 + dx^3 + \cdots$$

which has an infinite number of terms but which is cut off or *truncated* at some reasonable number of terms, usually 3 or 4. As applied to real gases, this series is the *virial equation of state*:

$$pV_m = RT + B_2[T]\left(\frac{RT}{V_m}\right) + B_3[T]\left(\frac{RT}{V_m}\right)^2 + B_4[T]\left(\frac{RT}{V_m}\right)^3 + \cdots$$

The parameters $B_2[T]$, $B_3[T]$, and $B_4[T]$ are called the second, third, fourth, ... *virial coefficients* and the notation $V_m = V/n$ is used to remind us that the volume taken is a *molar* quantity. The notation $B_2[T]$, $B_3[T]$, $B_4[T]$, ... is used to indicate temperature dependence of the virial coefficients. The square brackets do not indicate multiplication. By a simple algebraic manipulation, it is possible to express the virial coefficients in terms of the van der Waals constants and find $B_2[T] = b - a/RT$. By another simple manipulation, one obtains the *compressibility factor*.

2.3 THE COMPRESSIBILITY FACTOR

The difference between ideal and real gaseous behavior can be made clearer if we define a *compressibility factor* Z, a way of indicating the degree of nonideality of a gas

$$Z = \frac{pV_m}{RT} = \frac{pV_m}{(pV_m)_{\text{ideal}}}$$

If Z is less than one, nonideality is largely due to attractive forces between molecules. If Z is greater than one, the nonideal behavior can be ascribed to the volume taken up by individual molecules treated as hard spheres or to repulsive forces, or both. An ideal gas would show a compressibility factor of 1.00 at all pressures. At high temperatures, the total volume is large for any selected pressure. Molecular crowding becomes less significant, and attractive or repulsive forces are weaker because they act over longer distances. The gas approaches ideal behavior and Z approaches the constant value of 1.00 as p approaches zero.

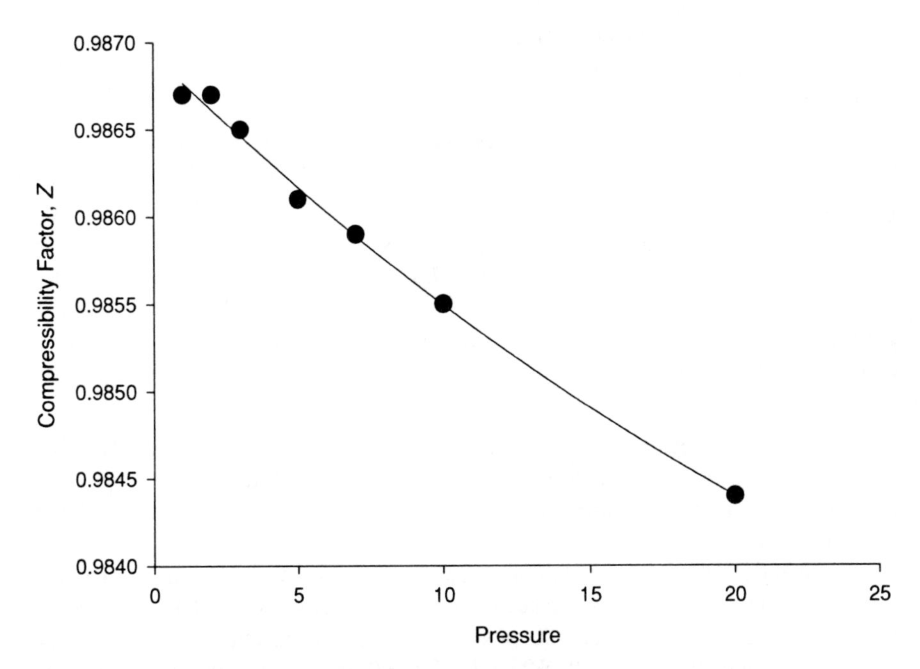

FIGURE 2.1 A quadratic least-squares fit to an experimental data set for the compressibility factor of nitrogen at 300 K and low pressures (sigmaplot 11.0©).

Dividing pV_m by RT over a set of different pressures at a fixed temperature gives a series of Z values that can then be plotted against p. This has been done for nitrogen at 300 K to give Fig. 2.1. Commercial curve-fitting software can be used to give the least-squares expression for a polynomial fit to the data points. One needs to select the degree of the polynomial to be fit to the points. The data set shown in Fig. 2.1 shows a little experimental scatter at the lower pressures and (perhaps) some slight curvature. Therefore, we selected a simple quadratic fit to the points.

Real gas law calculations like this one have considerable practical value. The engineering literature contains data sets of a much more complicated nature, over a much larger range than Fig. 2.1. The curve-fitting technique is the same, although one might choose a cubic or quartic curve fit. The output for the simple nitrogen curve fit is given in File 2.1. The **Rsqr** (square of the residual), being close to 1.0, indicates a good fit, although the extrapolated intercept y0 is not as close to 1.0 as we would like to see it.

The two virial coefficients are quite small, -0.0002 and 1.69×10^{-6}; nitrogen is nearly ideal at 300 K over the short pressure range 1–10 bar. The second virial coefficient is negative, reflecting the gentle downward slope away from ideal behavior. Note that in File 2.1 the notation f=y0+a*x+b*x^2 is used so that the somewhat overworked parameters a and b appear in a new and different context. Now,

Data Source: Data 1 in N2 molar density
Equation: Polynomial, Quadratic
f=y0+a*x+b*x^2

R	Rsqr	Adj Rsqr	Standard Error of Estimate
0.9977	0.9954	0.9932	6.7995E-005

Coefficient	Std. Error	t	P	
y0	0.9869	6.0263E-005	16377.0857	<0.0001
a	−0.0002	1.6559E-005	−9.6859	0.0006
b	1.6977E-006	7.5741E-007	2.2415	0.0885

FILE 2.1 Partial output from a quadratic least-squares curve fit to the compressibility factor of nitrogen at 300 K (SigmaPlot 11.0©).

$a = B_2[T]$ is the second virial coefficient and $b = B_3[T]$ is the third. The first virial coefficient (term rarely used) is 1.0 by definition. In File 2.1, the third virial coefficient is positive, indicating a slight upward curvature. (Experiments at higher pressures confirm the curvature.)

Nonideal gas behavior is nearly linear at low pressures. That is why the slope of the linear function is a measure of the second virial coefficient $B_2[T]$. The temperature variation of the second virial coefficients of helium, nitrogen, and carbon dioxide are shown schematically in Fig. 2.2. When $B_2[T] = 0$, the slope of the virial equation for Z is zero and $Z = 1$ over the range. If this is true, the gas is ideal. Helium shows ideal or nearly ideal behavior over most of the temperature range. Carbon dioxide is very nonideal over the range, and N_2 is in between. This order is pretty much what we would expect from our qualitative knowledge of the three gases. Helium is a "noble" gas, CO_2 is commonly available in the condensed state as "dry ice," and atmospheric N_2 is in between. Nitrogen is not as easily driven into the condensed state as CO_2, but liquid nitrogen is far easier to produce than liquid helium.

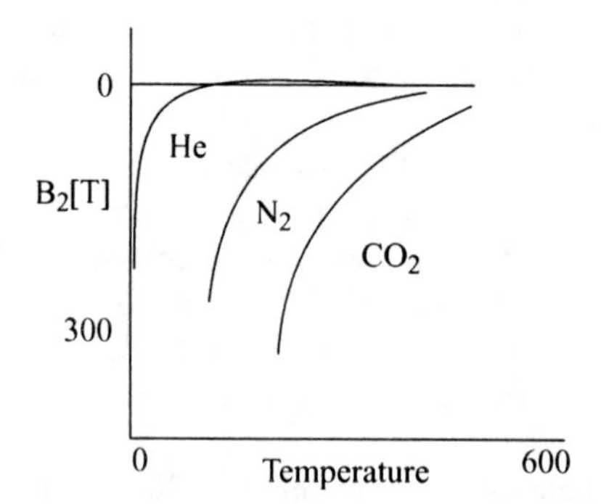

FIGURE 2.2 The second virial coefficient of three gases as a function of temperature. Notice the slight maximum in the curve for helium. It is not a computational error, helium really does that. Intermolecular repulsion brings about a small positive deviation of Z from Z_{ideal} over part of the temperature range.

2.3.1 Corresponding States

An interesting comparison from among Z factors is shown schematically in Figs. 2.3 and 2.4 which might be $Z = f(p)$ for two different gases or $Z = f(p)$ for *the same gas at two different temperatures*. Simply by choosing the right two temperatures, one can make any two gases identical to each other in their degree of nonideality, that is, their Z factors. When different gases at different temperatures behave in the same way, they are said to be in *corresponding states*. To define the *state* of a real gas, we must describe it in such detail that a colleague can reproduce it in *all* of its physical properties from the description. Because of the essentially infinite number of physical properties one can measure, this would seem to be a tall order; but, if it is pure, the number of degrees of freedom for one mole of a real gas is 2 regardless of its degree of nonideality, so specifying any two properties is equivalent to specifying all of them. The equation of state is written with two independent variables, but it doesn't matter which two.

To summarize up to this point: We are left with a van der Waals qualitative picture of nonideal gas behavior that is quite reasonable but gives an equation that doesn't work very well outside of common laboratory conditions. Our alternative is to rely upon empirical equations that work quite well in most cases but are hard to interpret. The term "empirical" as applied to the virial equation in this context has become somewhat of a misnomer over the years, however, because considerable progress has been made in theoretical interpretation of the virial equation and Fig. 2.3. Indeed, the statistical mechanics of these curves and others like them is an active research topic.

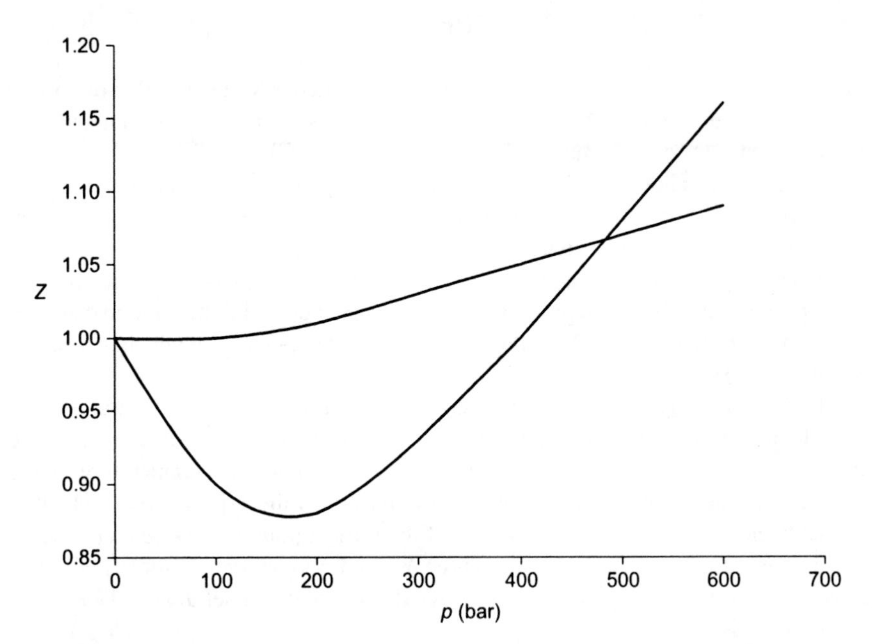

FIGURE 2.3 The $Z = f(p)$ curve for two different gases or for the same gas at two different temperatures. The unit—bar—is approximately one atmosphere.

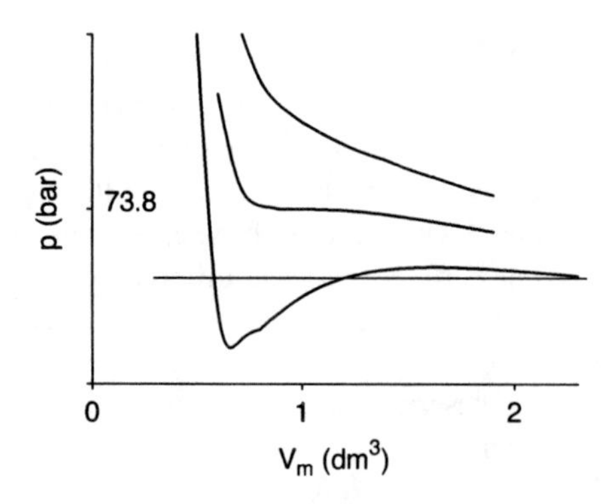

FIGURE 2.4 Three isotherms of a van der Waals gas. The top isotherm is above T_c, the middle isotherm is at the critical temperature T_c and the bottom curve is below T_c. The critical pressure is 73.8 bar.

The apparent paradox that there are only two degrees of freedom in the equation of state of a pure substance which may have an infinite number of terms in an equation of state is removed by noting that each term contains only the pressure, p, and an adjustable parameter (not a variable) that is a function of the temperature. Hence the only true variables in the equation are p and T.

2.4 THE CRITICAL TEMPERATURE

At lower temperatures, gas molecules occupy a smaller volume and move more slowly than they do at higher temperatures. Attractive forces among molecules or atoms become more important at lower temperatures. Ultimately, they become so strong that the gas liquefies. Thus, a useful physical picture of the liquid state is that liquefaction is the *limiting behavior* of an extremely nonideal gas and it results from large interparticle attractive forces.

As the temperature of a real gas is lowered, its deviation from hyperbolic (Boyle's law) behavior becomes more pronounced until the p–V curve has become so distorted that it goes through a horizontal inflection point. The temperature at which this occurs is called the *critical temperature*, T_c.

The curves in Fig. 2.4 arise from plotting the van der Waals equation at each of three temperatures above, at, and below T_c. The locus of p–V points at the same temperature is called an *isotherm*. The pressure, volume, and temperature at the inflection of the critical isotherm define a point called the *critical point*, which is unique to each real gas. The coordinates of the critical point (the *critical constants*) are the critical pressure p_c, critical volume V_c, and critical temperature T_c. Critical constants vary widely. For example, the critical temperature of helium is 5.19 K while that for CO_2 is 304 K.

Below the critical temperature, the system may be in the liquid or gaseous state, or it may exist as an equilibrium between liquid and vapor. The term "vapor" means

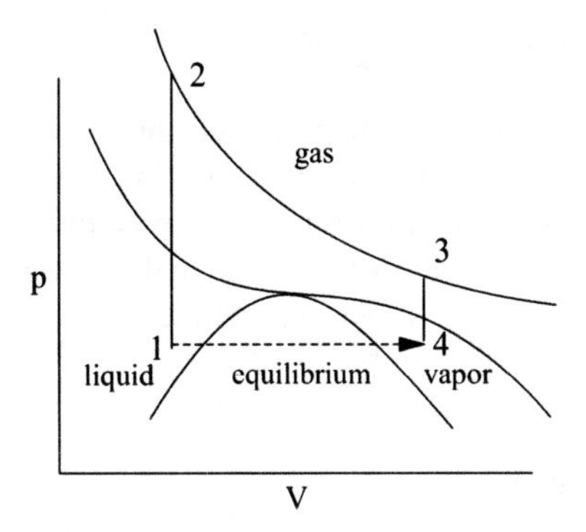

FIGURE 2.5 Conversion of a liquid to its vapor without boiling (1–4).

gas, but it is usually applied to a gas in equilibrium with its liquid form. When a liquid is in equilibrium with its vapor, heat can be applied with no change in temperature but with conversion of some or all of the liquid to its vapor. When spheroids of vapor rise from the bottom of a heated liquid to the top, we say that the liquid boils. It is sometimes said that "no gas can be liquefied above the critical point." This is true, but it is a little misleading because *there is no distinction* between liquid and gas above the critical isotherm. Above the critical isotherm, the system is a *supercritical fluid*.

The segment of the critical isotherm forming the upper boundary of the liquid region is particularly interesting. The system passes from liquid to gas (or back again) with no discontinuity. That is, it goes from liquid to gas but it does not boil. Points just below the isotherm represent liquids of low density. Those above it represent gases of high density. On the isotherm, the liquid and gaseous states become one and the same.

To get a better feeling for the meaning of the critical isotherm, let us heat a subcritical liquid (1) to one of its supercritical isotherms (2), expand it (3), and cool it to its original temperature (4) as in Fig. 2.5. At the end of the process, the liquid has been transformed to a state that is clearly in the gaseous region, but there is no discernible phase change (boiling or vaporization → gas) during this process.

2.4.1 Subcritical Fluids

A subcritical curve in Fig. 2.5 has three real roots, predicting three different volumes for the same fluid below T_c. This seems absurd. How can a gas have three different volumes at the same time? The answer is that the term "fluid," meaning *that which flows*, is more general than "gas." The term fluid includes both liquids and gases. The volume of a subcritical liquid is small and is given by the leftmost intersection (root) of the subcritical isotherm with the horizontal. The volume of the vapor is large and is given by the rightmost intersection (root). The middle root of the subcritical isotherm is not observed experimentally but has theoretical significance.

The constant pressure horizontal that intersects the subcritical isotherm represents vaporization going from left to right or condensation going from right to left. It is called an *isobar*. In vaporization, the volume of the system gets much larger but the pressure stays the same. Along an isobar, liquid and vapor are in equilibrium. If the pressure indicated by the isobar is atmospheric pressure, the subcritical isotherm is at the normal boiling point T_b.

2.4.2 The Critical Density

Below the critical temperature, one can measure the densities of a gas and a liquid at equilibrium in a closed container. Upon raising the temperature slightly at constant pressure, the density of the gas increases because liquid vaporizes. The density of the liquid decreases due to expansion. These densities, one increasing and the other decreasing, must approach each other as in Fig. 2.6. At the point where they meet, the densities of the gas and liquid *and all other properties* become identical. This is the critical temperature. Determination of the critical temperature of O_2 by density measurements is shown in Fig. 2.6.

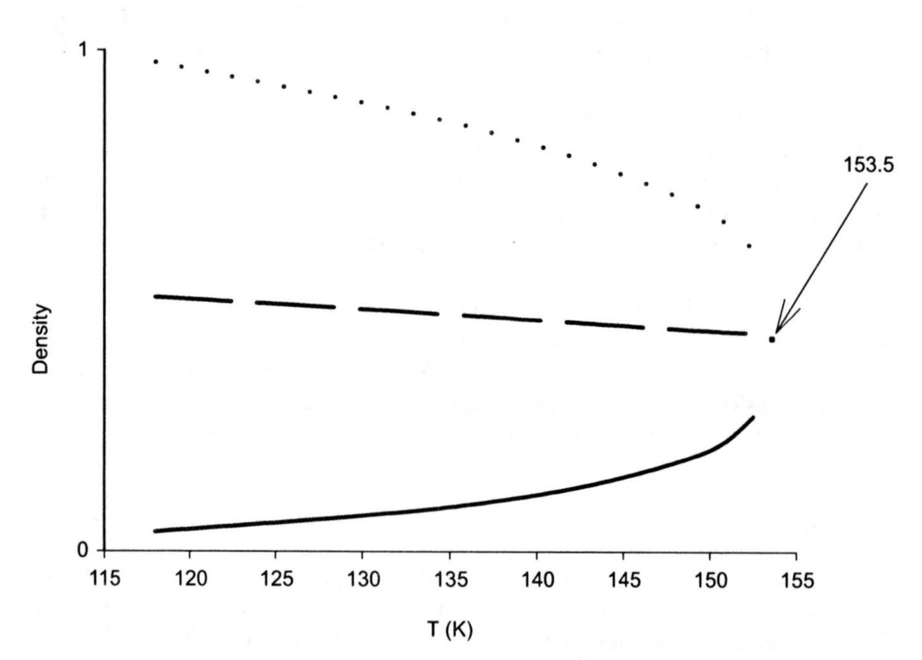

FIGURE 2.6 Density ρ curves for Liquid and Gaseous Oxygen. The straight line represents the arithmetic mean density of the liquid and gas $(\rho_l + \rho_g)/2$. The three curves meet at $T =$ 153.5 K. The tabulated value for T_c is 154.6 K (*CRC Handbook of Chemistry and Physics 2008–2009*, 89th ed.).

2.5 REDUCED VARIABLES

The degree of nonideality for real gases is determined by how near the temperature is to the critical temperature. We can express the "nearness" of the temperature of a gas to its critical temperature as the unitless ratio T/T_c. This ratio is called the *reduced temperature*

$$T_R = \frac{T}{T_c}$$

The other reduced variables are defined in the same way. The (unitless) *reduced pressure* and *reduced volume* are, for one mole,

$$p_R = \frac{p}{p_c} \quad \text{and} \quad V_R = \frac{V}{V_c}$$

These new variables are scaling factors by which we take an entire family of isotherms suggested by the three representative isotherms in Fig. 2.4 and move, stretch, or compress them until their critical isotherms coincide. Having done that, at the same T_R, the p_R and, V_R behavior of all gases fall approximately on the same family of curves. The gas data have been manipulated into *corresponding states* (Section 2.3). Knowing the behavior of *one* gas in terms of its reduced variables p_R, V_R, and T_R, we know the behavior of *any* real gas, provided only that we know its critical constants p_c, V_c, and T_c. Needless to say, industrial chemists and chemical engineers are delighted by this, and they have devoted considerable effort to construct standard tables and Z-curves in terms of the reduced variables.

2.6 THE LAW OF CORRESPONDING STATES, ANOTHER VIEW

Replacing the parameters a and b and the constant R in the van der Waals equation for one mole

$$p = \frac{RT}{(V - b)} - \frac{a}{V^2}$$

with

$$b = \frac{V_c}{3}, \qquad a = 3p_c V_c^2, \qquad \text{and} \qquad R = \frac{8p_c V_c}{3T_c}$$

we get, with some algebraic manipulation (see Problems and Exercise 2.1),

$$\left[p_R + \frac{3}{V_R^2}\right]\left(V_R - \frac{1}{3}\right) = \frac{8}{3}T_R$$

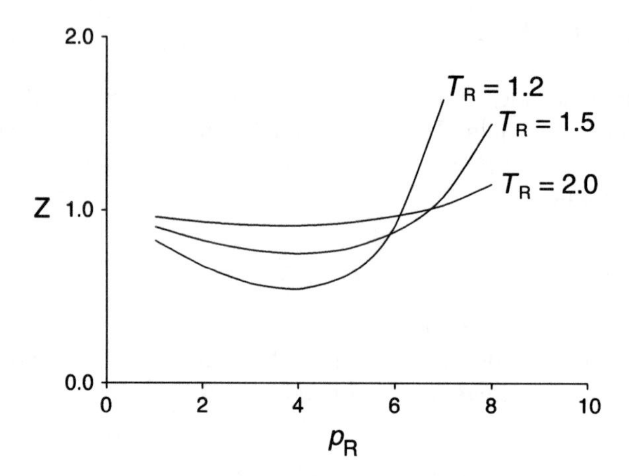

FIGURE 2.7 Compressibility factors calculated from the van der Waals constants.

The van der Waals constants, which are characteristic of individual gases, have been eliminated in the more general equation, which holds for any gas. Z can be calculated as a function of p_R from the equation of corresponding states as shown in Fig. 2.7. Portrayal of extensive families of curves like those in Fig 2.7 can be found in the chemical engineering literature. The equation of corresponding states is approximate and holds only insofar as the van der Waals equation, on which it is based, holds.

2.7 DETERMINING THE MOLAR MASS OF A NONIDEAL GAS

A nonideal gas, even if it is pure, will generally not occupy a molar volume of 24.789 dm^3 at $p = 1.000$ bar and $T = 298.15$ K; hence any molar mass computed on the basis of this molar volume will be wrong. Depending on the temperature, the error may be 50% or more. If, however, the weight and the volume of a pure gas sample are known along with the pressure and temperature, the (incorrect) molar mass, often called the *effective molecular weight* (EMW), can be calculated from the ideal gas law. Historically, EMWs were measured at several different pressures, treating a real gas as though it were ideal. The EMWs were then extrapolated as a function of p to $p = 0$ to obtain the true molecular weight. It is fruitless to ask about the meaning of the properties of a gas at zero pressure; one has simply "extrapolated out" the error due to nonideality. Extrapolating out is a common applied mathematical device. A number of more accurate methods for determining molar mass now exist.

PROBLEMS AND EXERCISES

Exercise 2.1 The van der Waals Cubic

Show that the van der Waals equation is a cubic.

Solution 2.1 Expanding the van der Waals equation and collecting terms, we get

$$\left(p+\frac{a}{V^2}\right)(V-b) = RT$$

$$pV - pb + \frac{a}{V} - \frac{ba}{V^2} = RT$$

$$pV^3 - pbV^2 + aV - ba = RTV^2$$

$$pV^3 - (pb+RT)\,V^2 + aV - ba = 0$$

Comment: Nonideal Behavior—Another View We know that molecules are not dimensionless points. Figure 2.8 shows that Boyle's law, the lower curve, is not obeyed above about 200 bars pressure for nitrogen. Positive deviation from Boyle's law is a high-pressure phenomenon shown by all real gases because all real molecules occupy a nonzero volume. The real sample volume is larger than it "ought to be" on the basis of Boyle's law. Molecules, which may be thought of roughly as hard spheres, are bumping into each other and refusing to invade each other's space. The volume in which the particles can move is the total volume minus the volume actually taken up by the particles. As higher pressures are imposed, the particle volume becomes a larger proportion of the total volume; hence the deviation from the Boyle's law curve is larger. The aggregate of molecular volume, as distinct from total volume, for all molecules in the sample is the *excluded volume.*

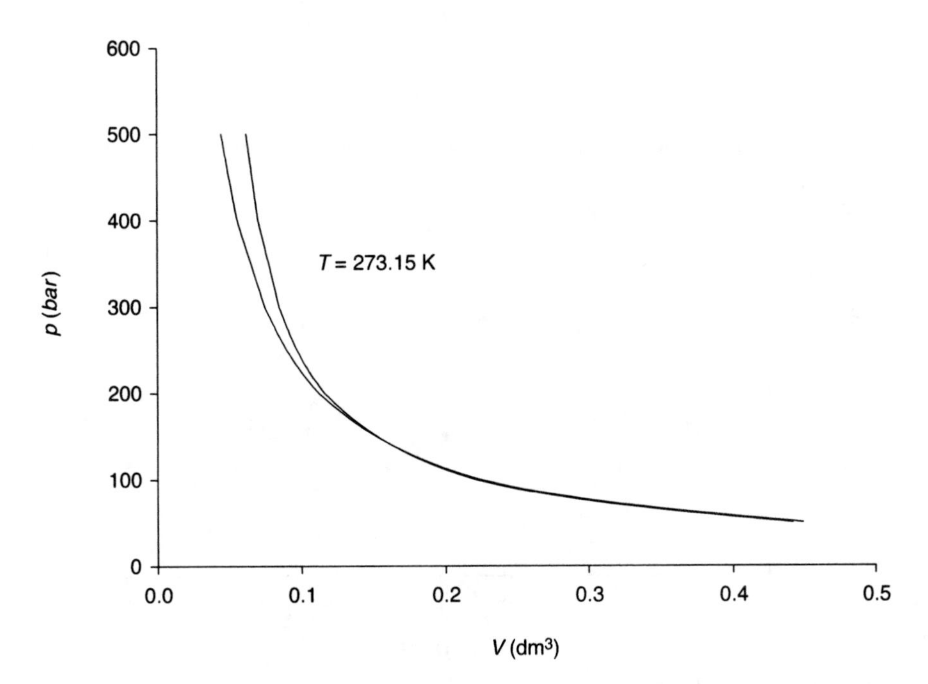

FIGURE 2.8 Boyle's law plot for an ideal gas (lower curve) and for nitrogen (upper curve).

TABLE 2.1 Observed Real Gas Behavior from 10 to 100 bar Expressed as (p, pV_m).

p (bar)	pV_m (dm^3 bar)
10.0000	24.6940
20.0000	24.6100
30.0000	24.5400
40.0000	24.4820
50.0000	24.4380
60.0000	24.4070
70.0000	24.3880
80.0000	24.3830
90.0000	24.3910
100.0000	24.4120

Exercise 2.2

Given the problem that experimental values of pV_m between 10 and 100 bar have been measured at intervals of 10 bar with the results in Table 2.1, find the analytical equation that expresses these results (Fig. 2.9).

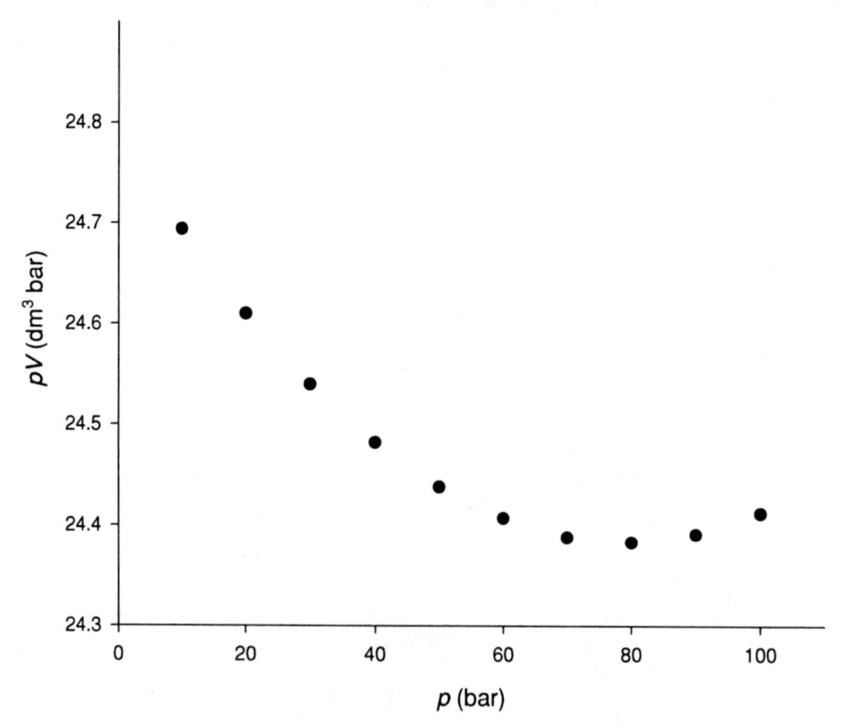

FIGURE 2.9 Experimental values of pV_m vs. p for one mole of a real gas.

TABLE 2.2 Observed Real Gas Behavior Expressed as (p, pV_m).

p (bar)	pV_m (dm^3 bar)
10.0000	24.6940
20.0000	24.6100
30.0000	24.5400
40.0000	24.4820
50.0000	24.4380
60.0000	24.4070
70.0000	24.3980
80.0000	24.3990
90.0000	24.4180
100.0000	24.4600

Solution 2.2 This problem and its solution are expressed using SigmaPlot 11.0© plotting software. Load the data set in the form of Table 2.2. Click on statistics → nonlinear regression → regression wizard → quadratic, → next, specify columns as x and y variables, and click finish.

The graph in Fig. 2.10 is shown with its fitted quadratic curve. The curve parameters are, as they should be, the parameters we started with in the previous problem:

y0	24.7906
a	−0.0103
b	6.5341E-005

The parameters are expressed in the form $y = y_0 + ax + bx^2 + cx^3 + \cdots$. Please notice the slight change in notation: The y intercept is called y_0, the slope is a, and the quadratic coefficient is b. The general independent variable is x, which is the pressure p in our case. Translated to the terms of the problem, we have

$$\begin{aligned} pV &= RT + B[T]\,p + C[T]\,p^2 + D[T]\,p^3 + \cdots \\ &= 24.7906 - 0.0103p + 6.5341 \times 10^{-5} p^2 \end{aligned}$$

where we have truncated the series at the quadratic term.

These two exercises make the important point that any data set can be expressed as a collection of numbers (table of observations), a graph, or an analytical equation. Tables, graphs, and empirical equations are merely different ways of saying the same thing. Difficulties may be encountered (imaginaries, singular points, multiple real roots, discontinuities, etc.), but they often point to interesting phenomena and their explanation may lead to new concepts in science. Larger data sets and more complicated functional behavior can be treated in the same way as this quadratic, except that the curve fit may be cubic, quartic, and so on.

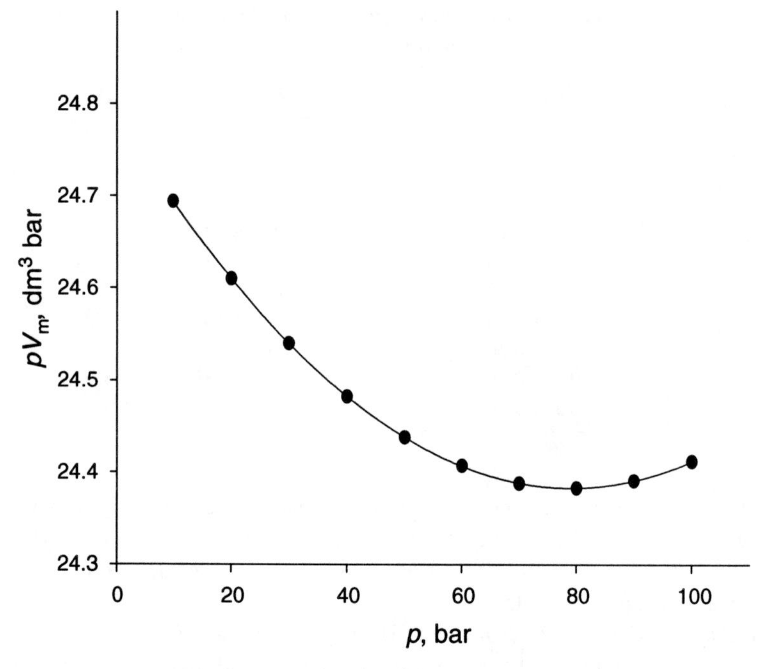

FIGURE 2.10 Quadratic real gas behavior.

Problem 2.1

The van der Waals constants for n-octane (a component of gasoline) are $a = 37.81$ and $b = 0.2368$. Find V for 1.00 mol of n-octane confined at 2.0 bar pressure and 450 K.

Problem 2.3

Find all three roots in the previous problem.

Problem 2.4

What are the units of van der Waals constants a and b?

Problem 2.5

Using commercial graphing software, produce a 3-D plot of p–V–T for the van der Waals gas N_2, where $a = 1.39$ and $b = 0.039$. What are the units of a and b? What happens to p as V becomes very small?

Problem 2.6

Find the van der Waals constant b in terms of the critical constants p_c, V_c, and T_c.

Problem 2.7

Show that the van der Waals parameter a is $a = \frac{9}{8} R T_c V_c$ where the subscript c denotes the critical point.

Problem 2.8

Find the molar volume of ethane at 50 bar pressure and 400 K, from the van der Waals parameters $a = 5.562$ and $b = 0.0638$.

Problem 2.9

Draw the pV isotherms for an ideal gas at 300, 400, 600, and 800 K. The results should resemble the corresponding isotherms in the text.

Problem 2.10

The second virial coefficient $B\,[T]$ of toluene (methylbenzene) is –1641 $cm^3\,mol^{-1}$ at 350 K. First, convert this unit to the more modern unit of $dm^3\,mol^{-1}$, and then find the compressibility factor Z at this temperature and 1.00 bar.

Problem 2.11

Continuing with the data above, what is the molar volume of toluene vapor at 1.00 bar at 350 K?

Problem 2.12

A real gas follows the equation $pV = 24.79 - 0.0103p + 6.52 \times 10^{-5} p^2$ for one mol. Plot the curve of pV vs. p from 0 to 100 bar and locate the minimum pV product.

Problem 2.14

Suppose the data set in the previous problem were as shown in Table 2.2. The last four entries in Table 2.2 differ from those of the previous problem in the second and third digits beyond the decimal point. The graph in Fig. 2.11 turns up slightly above $p = 70$ bar. Fit this data set with a cubic equation.

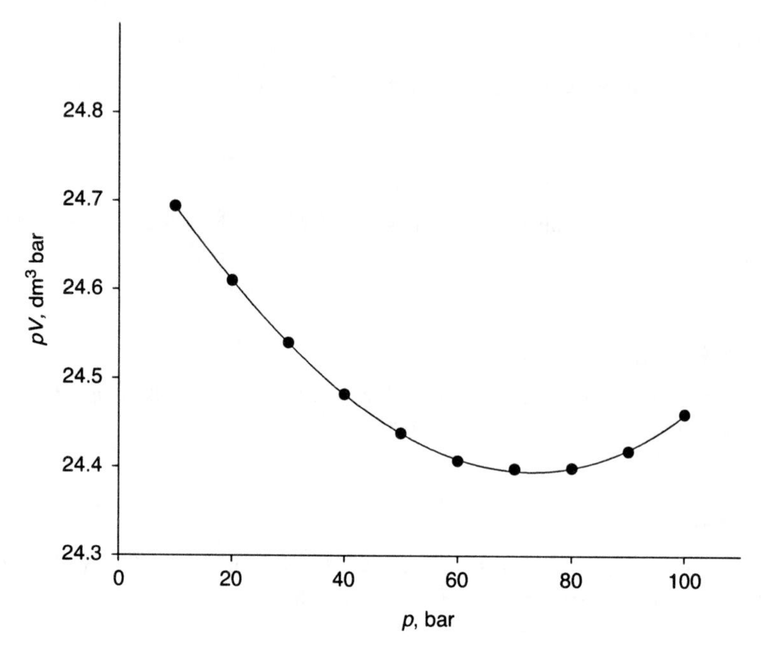

FIGURE 2.11 Cubic real gas behavior.

Problem 2.15

(a) What is the volume of one mole of CO_2 at 366 K and $p = 111$ bar according to the ideal gas law?

(b) The critical temperature of CO_2 is 305.1 K. The critical pressure and volume of CO_2 are 73.8 bar and 0.0956 dm^3(Laidler and Meiser, 1999). If a compressibility factor of 0.68 is read from a chart of isotherms, what is the reduced volume of one mole of CO_2 at 366 K and $p = 111$ bar according to the chart?

(c) The van der Waals parameters for CO_2 are $a = 3.66$ and $b = 0.0429$. What is the volume of one mole of CO_2 at 366 K and $p = 111$ bar according to the van der Waals equation?

(d) What is the volume of one mole of CO_2 at 366 K and $p = 111$ bar according to the corresponding states equation? What is the volume?

(e) The critical constants for helium are $T_c = 5.3$ K, $p_c = 2.29$ bar, and $V_c = 0.0577\ dm^3\ mol^{-1}$. What is the volume of one mole of He at 7.95 K and $p = 2.75$ bar according to the corresponding states equation? What is the volume?

3

THE THERMODYNAMICS OF SIMPLE SYSTEMS

Thermodynamics, literally heat motion, is one of the pillars of physical chemistry and one of the great achievements of modern science. Classical thermodynamics is the study of those quantities that are conserved, like energy, enthalpy, and free energy. Statistical thermodynamics is the link between this great theoretical edifice and that of quantum mechanics.

3.1 CONSERVATION LAWS AND EXACT DIFFERENTIALS

Much of nineteenth-century thermodynamics was devoted to the discovery, definition, and characterization of physical quantities that are conserved. The simple statement *energy is conserved* is one of the ways of stating the *first law of thermodynamics*. Conservation can be illustrated by carrying a rock up a hill and then bringing it back down again. The potential energy of the rock in the gravitational field of the earth increases on the way up and decreases on the way down, but it is the same at the end of the process as it was at the beginning; it is conserved. By contrast, work is not conserved. Carrying the rock around the circular path up and down the hill can be easy or hard, depending on the path. Also, the work done on the way up is not the same as the work that can be obtained by allowing the rock to drive some kind of primitive motor as it rolls down the hill.

Work is not the only factor that has to be taken into account. The difference between an easy path and a hard path up the hill is illustrated by the simplified model of a mass being pushed up over the same change in height (altitude) but over two different

Concise Physical Chemistry, by Donald W. Rogers

inclined planes, one rough and the other smooth. The potential energy change is the same, but the rough plane requires more work and it produces some amount of *heat* due to friction. Heat is not conserved over the cyclic path either. We symbolize the nonconservation of work or heat by the sum, that is, the integral, of infinitesimal heat or work increments dq or dw over the cyclic path $\oint dw \neq 0$ or $\oint dq \neq 0$. But we have already said that energy U is conserved over a cyclic path so $\oint dU = 0$.

One of the great discoveries of Western science is that the infinitesimal increment of the energy of a thermodynamic system is the *sum* of an infinitesimal increment in work done on the system and an infinitesimal increment of the heat put into a system[1]

$$dU = dw + dq$$

The statement $\oint dU = dq + dw$ is another of the many equivalent ways of stating the first law of thermodynamics. Although the sum of work w and heat q is conserved, we don't know the ratio of w to q or even if they have the same sign, except by experiment. The law of conservation of energy is the accumulated knowledge gained from very many controlled observations over three centuries. It cannot be derived from simpler principles.

3.1.1 The Reciprocity Relationship

For some functions $u = u(x, y)$, the differential

$$du = M(x, y)dx + N(x, y)dy$$

has the property that

$$M(x, y) = \left(\frac{\partial u}{\partial x}\right)_y \quad \text{and} \quad N(x, y) = \left(\frac{\partial u}{\partial y}\right)_x$$

that is,

$$du = \left(\frac{\partial u}{\partial x}\right)_y dx + \left(\frac{\partial u}{\partial y}\right)_x dy$$

If this is true, then du is called an *exact differential.* If it is not true, then du is an inexact differential. By the Euler reciprocity relationship, we have

$$\frac{\partial^2 u}{\partial x \partial y} = \frac{\partial^2 u}{\partial y \partial x}$$

$$\left[\frac{\partial}{\partial x}\left(\frac{\partial u}{\partial y}\right)_x\right]_y = \frac{\partial}{\partial x}N(x, y) = \left[\frac{\partial}{\partial y}\left(\frac{\partial u}{\partial x}\right)_y\right]_x = \frac{\partial}{\partial y}M(x, y)$$

[1]In the example of the mass being pushed up a rough plane, frictional heat is lost (goes out of the system); hence the sign on dq is reversed: $dU = dw_{\text{in}} - dq_{\text{out}}$.

so the condition

$$\frac{\partial}{\partial x}N(x, y) = \frac{\partial}{\partial y}M(x, y)$$

is a test for exactness.

3.2 THERMODYNAMIC CYCLES

The problem of inexact and exact differentials can be expressed in several ways. Taking work as an example, work w is *not a thermodynamic function*. The differential of the work dw is *not exact*. The work done in a thermodynamic process depends upon the path. The integral

$$w = \int_{V_1}^{V_2} f(T)p\, dv$$

is a *line integral*. The line integral cannot be evaluated until we know $f(T)$, an arbitrary function of the temperature. There is an infinite number of paths (ways of getting from V_1 to V_2), hence an infinite number of solutions to the work integral.

Consider two ways of taking a gas from thermodynamic state A with molar volume V_1 to state B with molar volume V_2. Let the substance be in the gaseous state. Let the first path be an isothermal compression from 1.00 bar to 10.0 bars at 290 K followed by an *isobaric* (constant pressure) temperature rise from 290 K to 310 K. The second path will start with the isobaric temperature rise from 290 K to 310 K, followed by an isothermal compression from 1.00 bar to 10.0 bars at 310 K. The beginning and end points of the process are the same for both paths. For simplicity, assume that the gas is nitrogen, which is nearly ideal over this temperature and pressure range. The two paths are shown in Fig. 3.1.

The volume $V = RT/p$ can be calculated after each step. The results are shown at each corner of the rectangular diagram below. The volumes at each corner of the diagram are larger on the left (before compression) than on the right and larger at the top (after heating) than on the bottom.

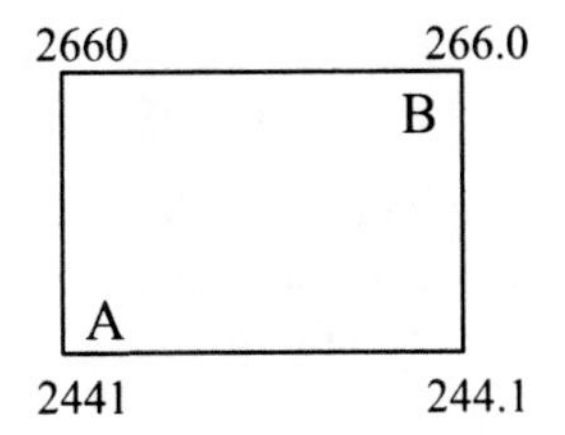

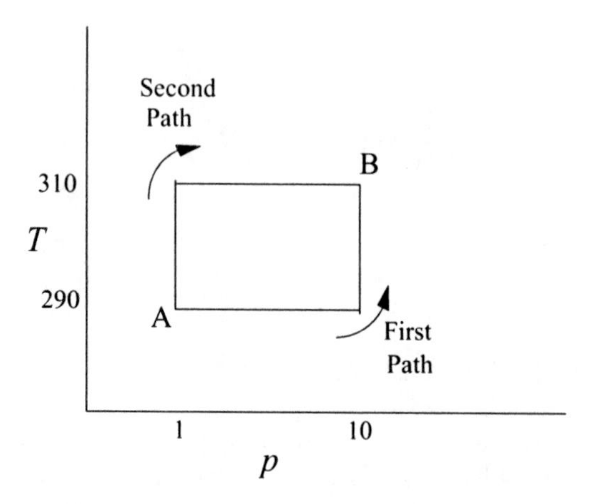

FIGURE 3.1 Different path transformations from A to B.

The work done on the system in the first compression (lower horizontal) is

$$w = -\int_{V_1}^{V_2} p\,dV = -\int_{V_1}^{V_2} \frac{RT}{V} dV = -\int_{V_1}^{V_2} RT\frac{dV}{V} = -RT\ln\frac{V_2}{V_1}$$
$$= -RT\ln 0.100 = -8.314(290)\ln 0.100 = 5552\,\mathrm{J}$$

The work done in the heating step (rightmost vertical) is isobaric with $p = 10.0$ bar:

$$w = p(V_2 - V_1) = 10.0\,(266.0 - 244.1) = 219\,\mathrm{J}$$

so the total over the first path is 5771 J.

Over the second path, the leftmost vertical gives $w = 1.00(2660 - 2441) = 219\,\mathrm{J}$; thus the two isobaric steps require the same work into the system. The isothermal compressions are not the same. The topmost horizontal is

$$-RT\ln 0.100 = -8.314(320)\ln 0.100 = 6126\,\mathrm{J}$$

so the total work done over the second path is obviously not the same as by the first path. The difference is 574 J.

3.2.1 Hey, Let's Make a Perpetual Motion Machine!

Let's run the thermodynamic system in Fig. 3.1 around a cycle such that the transformation over the second path runs backwards; that is, the top horizontal is an expansion instead of a compression and the leftmost step is a temperature decrease. The first path remains as before.

The work of the leftmost and rightmost steps cancel $-219 + 219$, but the topmost horizontal produces -6126 J of work and the bottom horizontal takes up only 5552 J of work to return the machine to its original state. The cycle produces $-6126 + 5552 = -574$ J of work (negative because work goes from the machine into the outside world) and has been returned to its original state, ready to produce work -574 J, -574 J, -574 J, ... over infinitely many work cycles, running forever.

Five or six generations of garage scientists have made the sad discovery that such machines ***never work***. However sad it may be for would-be millionaire inventors, the impossibility of a perpetual motion machine leads the physical chemist to a treasure trove of thermodynamic theory through the *second law of thermodynamics* to the inspired work of the American thermodynamicists J. W. Gibbs and G. N. Lewis.

3.3 LINE INTEGRALS IN GENERAL

If you have a straight brass rod—say, 70 cm in length and 1.0 cm^2 in cross section—it is a simple matter to determine its weight even though the rod may be inaccessible to you as part of a machine, so that you cannot simply weigh it. Multiply the density of the brass used in its manufacture, ρ in g/cm^3, by $70 \times 1.0 = 70\ cm^3 = 70\ \rho$ and you have the answer in grams. We have essentially set up an x axis and performed the integration

$$M = \rho \int_0^{70} dx = \rho(70 - 0)$$

Most machinists (and lots of other people) would laugh at how we have made a simple problem complicated.

But suppose the rod is bent (Fig. 3.2). Now setting up an x axis for the bar is not the best way of finding its mass. One x increment is not the same as another because the rod is not collinear with the axis.

The corresponding integral is

$$M = \rho \int_a^b ds = \rho(b - a)$$

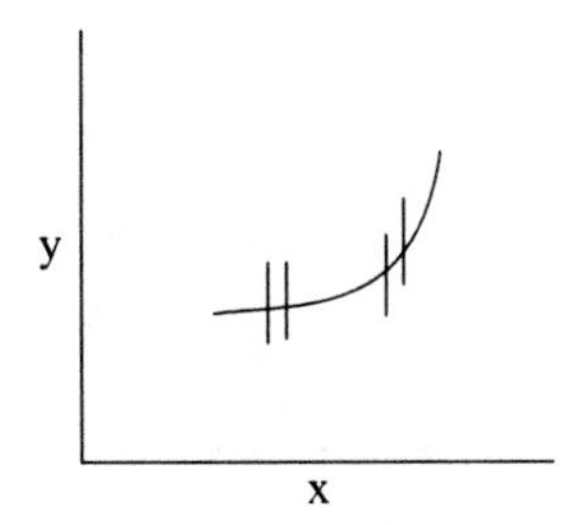

FIGURE 3.2 Different segments of a curved rod.

We still use the practical method of length times cross-sectional area to get the total volume which we then multiply by ρ, but the corresponding integral is no longer a conventional integral over dx; rather, it is the summed infinitesimal increments along the length of the rod ds, not coincident or even parallel to the x axis.

Now suppose that the density of brass is not constant in the rod but that it varies along the length of the rod according to some known function $\rho(s)$. The integral is

$$M = \int_a^b \rho(s)\, ds = \rho(s)\,(b - a)$$

Integrals taken along some curve C with arc length s, not one of the coordinate axes, are called *curvilinear* or *line integrals*. In general, the integral of some function of arc length $f(s)$ along a curve C is

$$I = \int_C f(s)\, ds$$

3.3.1 Mathematical Interlude: The Length of an Arc

By Pythagoras's theorem for the hypotenuse of a right triangle, the length Δs of a short part of a curve (arc) in x–y space is approximately (Barrante, 1998)

$$\Delta s \approx \left(\Delta x^2 + \Delta y^2\right)^{1/2}$$

Multiply and divide by Δx to find

$$\Delta s \approx \left(1^2 + \left(\frac{\Delta y}{\Delta x}\right)^2\right)^{1/2} \Delta x$$

or, in the infinitesimal limit,

$$ds = \left(1 + \left(\frac{dy}{dx}\right)^2\right)^{1/2} dx$$

For the length from a to b of a *finite* arc, find the integral $s = \int_a^b ds$.

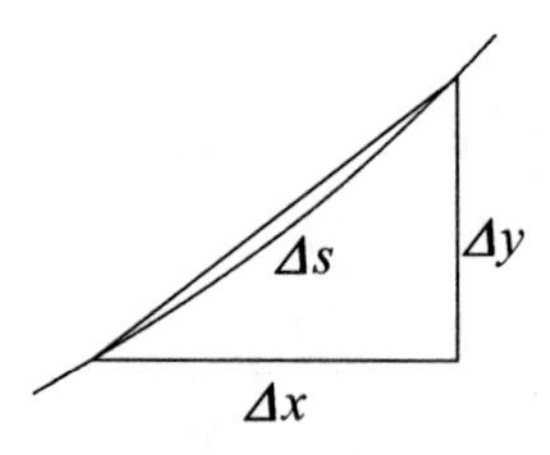

FIGURE 3.3 Pythagorean approximation to the short arc of a curve.

3.3.2 Back to Line Integrals

Suppose some property $f(x, y)$ is associated with a curve C in the x–y plane and we wish to find the line integral between certain limits along that curve. The appropriate integral is

$$I = \int_C f(x, y)\, ds$$

The integration is to be carried out, not along an axis as in the case of a simple integral, but along the curve in x–y space. If we have $y = f(x)$ that specifies the path in the x–y plane, we also have ds along the curve from Section 3.3.1 so the integral is

$$I = \int_C f(x, y)\left(1 + \left(\frac{dy}{dx}\right)^2\right)^{1/2} dx = \int_C f(x, f(x))\left(1 + \left(\frac{dy}{dx}\right)^2\right)^{1/2} dx$$

3.4 THERMODYNAMIC STATES AND SYSTEMS

A thermodynamic *system* is any part of the *universe* we want to look at. The rest of the universe is its *surroundings*.

$$\textit{system} + \textit{surroundings} = \textit{universe}$$

A system is usually defined in such a way as to be manageable; for example, a reaction flask containing chemicals is a system. The surroundings may be a constant temperature bath in which the system is immersed, along with the rest of the universe. An *isolated system* does not exchange energy or matter with its surroundings. A *closed system* exchanges energy, but not matter, with the surroundings. A piston fitted to a cylinder that can do work and exchange heat with its surroundings, but does not leak, is a closed system. An *open system* exchanges both energy and matter with its surroundings. An animal or plant is an open system.

3.5 STATE FUNCTIONS

A *thermodynamic state function* is one of several *conserved* mathematical functions describing a property of a system. Energy is a thermodynamic state function. The term *thermodynamic property* is used in the same way. To define the *state* of a system, we must describe it in *all* of its physical properties, but we already know that there are only three degrees of freedom for a pure substance, two if you specify *molar* properties as we normally do. In nonuniform systems, one of the variables may be different at different locations. For example, the temperature may not be the same everywhere in a closed room. Other variables will then be nonuniform as well. These systems are more difficult to treat mathematically, but they have been described in a series of brilliant twentieth-century researches by Prigogine (Nobel Prize, 1977) and coworkers. Here we shall restrict ourselves to uniform systems and processes that are carried out very slowly so that thermodynamic variables change smoothly and do not suffer the discontinuities and multiple-valued problems encountered in sudden processes like explosions.

Each point on a p–V graph represents a thermodynamic state. Moving from one thermodynamic state to another impelled by a change of one or more independent variables is described as a *thermodynamic transition*. Each transition from V_1 to V_2 brought about by a change from p_1 to p_2 is represented by a curve on the p–V graph. Real transitions can be carried out very slowly. When this is done, they can be made to approach a limiting process of an infinitely slow transition that moves from one state to another by infinitesimal steps, each of which is an *equilibrium state*. A finite transition carried out in such a way as to approach an infinite sequence of equilibrium states is called a *reversible transition*.

All this talk about transitions through equilibrium states is, of course, self-contradictory; if a state is at equilibrium, it isn't transiting anywhere.[2] The reversible process is an idealization which is taken seriously by otherwise skeptical scientists because it brings the tremendous power of calculus to bear on thermodynamics. With calculus and the concept of reversible transformations, we can build a majestic framework that encompasses all classical thermodynamic change, including the change in thermodynamic properties during chemical reactions. In addition to the utility of the idealized reversible process, it is astonishing how nearly some real chemical systems can be brought to true reversibility (see especially electrochemistry).

3.6 REVERSIBLE PROCESSES AND PATH INDEPENDENCE

One can determine the change of a thermodynamic state function if one knows the initial and final states of the system. Let U_i and U_f be the energy of a system in its initial and final states. The change in U is

$$\Delta U = U_f - U_i$$

Neither U_f nor U_i is known (classically) in an absolute sense, but this difficulty is easily circumvented by defining $U_i \equiv 0$ (or $U_f \equiv 0$) in some arbitrary state. Suppose we define a potential energy $U_i \equiv 0$ in a coordinate system with the z axis in the direction of a field. Then we increase z by 1000 meters (lift the mass). This changes the energy in the field. The unit acceleration in the *gravitational field* is very roughly 10 newtons (N); hence a mass of 1.0 kg increases in *potential* energy by about

$$\Delta U = mg\Delta z = (1.0)\,10(1000) = 10000 = 10^4\,\text{J} = 10\,\text{kJ}$$

[2]We ignore quantum fluctuations in classical thermodynamics.

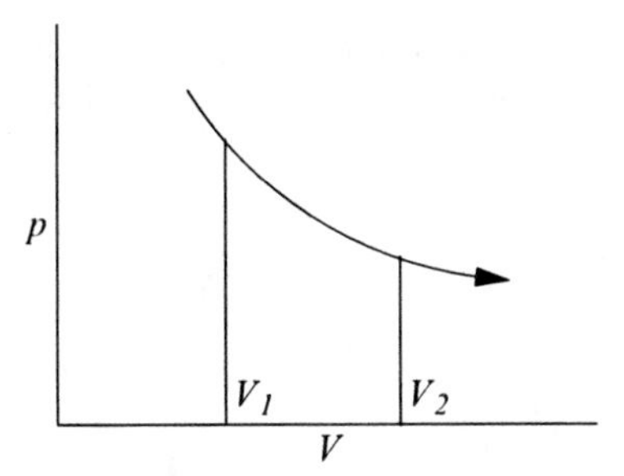

FIGURE 3.4 The energy change for reversible expansion of an ideal gas. The area under the p–V curve is the p–V work done by an expansion from V_1 to V_2.

where the unit of energy is the joule (J). One does not know how much work it will take to bring this process about because the amount of work lost to friction is not known. Work is not a thermodynamic function.

The ideal expansion of a gas driving a piston in the absence of frictional or other heat loss produces work equal to the energy change. The energy change can be found by integrating over the work which now follows a defined path and which in this respect behaves like a state function (Fig. 3.4). In the absence of heat loss, work is defined as the integral of force F over displacement ds from position s_1 to s_2:

$$w = \int_{s_1}^{s_2} F\, ds$$

For an idealized piston, this is the same as the integral of p over the change dV. The amount of work done by expansion against a piston is represented by the area under a p–V curve in Fig. 3.4. It can be written analytically as

$$w = \int_{V_1}^{V_2} p\, dV$$

The amount of work done by the system on the surroundings during an expansion at constant pressure is

$$w = p \int_{V_1}^{V_2} dV = p\Delta V$$

At constant temperature (isothermal conditions) we have

$$w = \int_{V_1}^{V_2} \frac{nRT}{V} dV = nRT \int_{V_1}^{V_2} \frac{dV}{V} = nRT \ln \frac{V_2}{V_1}$$

for an ideal gas. We can also define an *adiabatic* process, which is a perfectly insulated process. Before doing so, we shall need the concept of heat capacity.

3.7 HEAT CAPACITY

When heat q is put into a system its temperature rises. The change in temperature is proportional to the amount of heat

$$q = C\Delta T$$

where C is a proportionality constant. C depends on both the nature of, say, a chemical substance and the amount taken; it is an extensive property (Section 1.6). C is really a parameter because it is different for different substances. It is common experience that metals heat up faster over a flame than does water. They have different *capacities* to absorb *heat*; hence the parameter C is called the *heat capacity*.

To use calculus in working with the heat capacity, it is necessary to replace the approximate macroscopic observation $C = q/\Delta T$ with the infinitesimal $C = dq/dT$. Furthermore, we chemists carry our ordinary bench reactions under conditions of constant (atmospheric) pressure, and thermochemists carry out combustion reactions inside a closed bomb. The heat capacity under constant volume conditions is not exactly the same as the heat capacity under constant pressure conditions, so we distinguish between the two heat capacities C_V and C_p as

$$C_V = \left(\frac{\partial q}{\partial T}\right)_V \qquad \text{and} \qquad C_p = \left(\frac{\partial q}{\partial T}\right)_p$$

The infinitesimals in the heat capacity equations are partials because each parameter is defined holding either V or p constant.

3.8 ENERGY AND ENTHALPY

Having stipulated constant volume for the first of the heat capacity expressions, the work $p\,dV$ disappears for a system that can do only work of expansion dV against a pressure p

$$dU = dq + p\,dV$$

Consequently,

$$C_V \equiv \left(\frac{\partial q}{\partial T}\right)_V = \left(\frac{\partial U}{\partial T}\right)_V$$

It would be very convenient here to have a thermodynamic function that plays the same role for constant pressure transformations that energy plays for constant volume transformations. We can define such a property. It is called the *enthalpy*, and it is the heat produced or absorbed under the usual constant pressure conditions that characterize the reactions we carry out in the lab. The enthalpy has the analogous definition

$$C_p \equiv \left(\frac{\partial q}{\partial T}\right)_p = \left(\frac{\partial H}{\partial T}\right)_p$$

Now the volume is not constant but may vary in such a way as to do work on the surroundings or for the surroundings to do work on the system, so $dU = dq + p\,dV$and

$$C_p = \left(\frac{\partial q}{\partial T}\right)_p \equiv \left(\frac{\partial H}{\partial T}\right)_p = \left(\frac{\partial U - p\,dV}{\partial T}\right)_p$$

We have in effect *constructed* a new thermodynamic variable by specifying the path

$$H \equiv U - pV$$
$$dH = dU - p\,dV - \cancel{V\,dp}$$

where $V\,dp$ is zero over the stipulated path because $p =$ constant. The $p\,dV$ work over a stipulated path is conserved; that is, it sums to zero over a circular path. Enthalpy is thus the sum of a thermodynamic function and a conserved function, and therefore it must be conserved.

Enthalpy is a thermodynamic state variable.

Enthalpy is the constant pressure analog of energy in processes like chemical reactions. It can be handled mathematically in the same way that energy is handled and should be thought of almost as its twin. In many semiquantitative discussions the distinction between energy and enthalpy is ignored but it should be borne in mind, because it is important in rigorous treatments like the correction of quantum mechanical values for the energy of a molecule at 0 K to the enthalpy at 298 K.

The difference between enthalpy and energy is the pV work done on or by the system. For ordinary chemical reactions this is often negligible but if a gas is taken up or produced during a chemical reaction, U and H will be different because production of a gas represents an expansion against the atmosphere and consumption of a gaseous reactant is a contraction of the system driven by pressure of the surroundings.

3.9 THE JOULE AND JOULE–THOMSON EXPERIMENTS

The Joule experiment is important because it showed that there are *no* thermal effects arising from expansion of a gas. The Joule–Thomson experiment is important because it showed that there *are* thermal effects arising from expansion of a gas. What?

Wait! There's an explanation . . . .

The Joule series of experiments was the earlier of the two. It was carried out with simple apparatus and relatively insensitive thermometers. The Joule apparatus consisted of two chambers, one filled with a gas at pressure p and the other evacuated, with the chambers being connected by a short tube with a stopcock. The apparatus was placed in a bath and allowed to come to thermal equilibrium. The stopcock was opened, allowing *expansion of the gas* into the evacuated half of the apparatus. The temperature change was measured and a null result was recorded—hence the conclusion that expansion of a gas produces no thermal effect. The Joule experiment is analogous to Boyle's law in that it is almost correct for most gases under mild conditions (pressure change not too great). It is an ideal gas law. Joule happened to be rich (he never had to be distracted by earning a living) and he was also smart. He did not really believe the results of his experiment.

The Joule–Thomson experiment is a refinement of the Joule experiment intended to find the very thing that the Joule experiment failed to find: the thermal effect of expanding a gas. By means of a piston, a gas was driven through a porous plug from one chamber at high pressure into a second chamber at low pressure, thus expanding the gas (for more detail, see Klotz and Rosenberg, 2008). He then measured the temperature difference between the two chambers and found that it was not null. The result was more in accord with modern experience with highly pressurized gases. The expanding gas cools. The cooling factor is called the Joule–Thomson coefficient μ_{JT}

$$\mu_{JT} \equiv \left(\frac{\partial T}{\partial p}\right)_H$$

The partial on the right-hand side is subscripted H because the process, though it involves a change in T, involves no change in H . It is *isenthalpic*. The Joule–Thomson coefficient is usually positive (for an expansion, $dp < 0$ and $\mu_{JT} > 0$, so the gas cools). There are a few exceptions, which warm up on expansion starting from somewhere around room temperature. For these few gases, $\mu_{JT} < 0$ at room temperature; but if they are expanded at low temperatures, μ_{JT} changes to a positive value. As specific examples, nitrogen has $\mu_{JT} \cong 0.6$ at 200 K; hence N_2 can be cooled by expansion starting at 200 K and brought to a temperature so low that it is liquefied. Helium has $\mu_{JT} \cong -0.06$ and cannot be liquefied by expansion starting at 200 K. Helium must be precooled far below 200 K to where μ_{JT} becomes positive for it to be liquefied by expansion. If expansion is carried out at a sufficiently low temperature, all gases can be liquefied.[3] The point at which μ_{JT} changes from + to – is called the *inversion temperature*.

[3]Some special effects arise with helium isotopes.

Not surprisingly, μ_{JT} is related to the van der Waals a and b, most importantly the van der Waals parameter of attraction a. For gases having attractive interactions (most of them at room temperature) expansion against their attractive forces does internal work to separate the gaseous particles, which is why the gas cools and $\mu_{JT} > 0$. For helium, neon, hydrogen, and so on, the dominant forces are repulsive hence $\mu_{JT} < 0$ at room temperature. At low temperatures, attractive forces become dominant for all gases, so μ_{JT} changes sign.

The Joule–Thomson inversion temperature T_i can be related to van der Waals a and b by the equation

$$\frac{2a}{RT_i} - \frac{3abp}{R^2T_i^2} - b = 0$$

This equation is a quadratic in T_i; hence double roots are possible. Indeed, two inversion temperatures, upper and lower, are found at some pressures. At very high pressures, the two roots approach each other and become identical as shown in Fig. 3.5. On the high branch of Fig. 3.5, the upper inversion temperature, the second term in the inversion temperature equation becomes unimportant because it has T_i^2 in the denominator. Now

$$\frac{2a}{RT_i} \cong b$$

$$T_i \cong \frac{2a}{bR}$$

Hence the upper T_i can be calculated if a and b are known. The second equation above is used to estimate the upper inversion temperature of real gases from a and b (which are themselves estimates). Though approximate, this value is important in practical problems such as production of liquefied gas for cooling certain low-temperature experimental instruments.

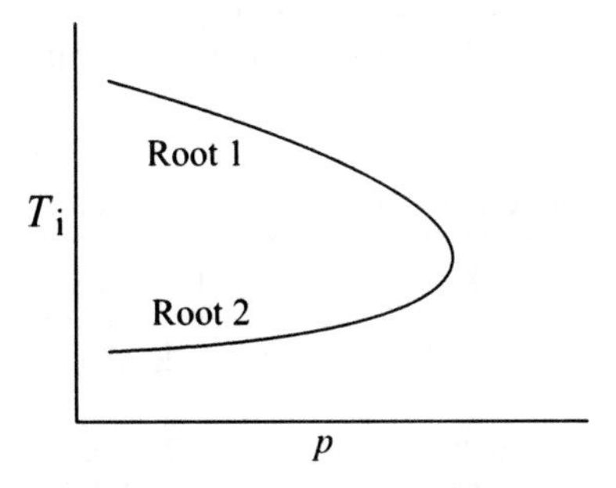

FIGURE 3.5 Inversion temperature T_i as a function of pressure. The temperature extremum dT_i/dp for nitrogen is at about 370 atm.

3.10 THE HEAT CAPACITY OF AN IDEAL GAS

An ideal gas consists of point particles that cannot vibrate or rotate. It can have only kinetic energy $U_{\text{kin}} = \frac{1}{2}m\mathbf{v}^2$. For a large collection of particles, the total kinetic energy is

$$U_{\text{kin}} = \tfrac{1}{2}Nm\bar{u}^2$$

where N is the number of atoms or molecules and $\bar{u}$ is their average speed. We also know from the kinetic theory of gases that

$$pV = RT = \tfrac{1}{3}N_A m\bar{u}^2 = \tfrac{2}{3}\left(\tfrac{1}{2}N_A m\bar{u}^2\right) = \tfrac{2}{3}U_{\text{kin}}$$

This (slightly circular) reasoning leads to the simple statement

$$U_{\text{kin}} = \tfrac{3}{2}RT$$

for one mole of an ideal gas. No direction is favored over any other in Cartesian 3-space, so we can split the kinetic energy components into three equal parts of $\frac{1}{2}RT$ per degree of freedom along any arbitrary x, y, z space coordinates. This division is general. On a molar basis, we expect to find $\frac{1}{2}RT$ of energy per mole per degree of freedom or $\frac{1}{2}k_BT$ per particle (molecule or atom) per translational, rotational, vibrational or, rarely, electronic degree of freedom. (Recall that k_B is the gas constant per particle.)

If the gas is ideal, we obtain a *molar heat capacity*

$$C_V = \left(\frac{\partial U}{\partial T}\right)_V = \frac{\partial\left(\frac{3}{2}RT\right)}{\partial T} = \tfrac{3}{2}R$$

Since $R \cong 2\ \text{cal K}^{-1}\text{mol}^{-1}$, the heat capacity of an ideal gas is about $3\ \text{cal K}^{-1}\text{mol}^{-1} = 12.5\ \text{J K}^{-1}\text{mol}^{-1}$. Table 3.1 shows that this is true for the monatomic gases helium He and mercury vapor Hg but that it is not true for more complicated molecular species.

TABLE 3.1 Heat Capacities and γ for Selected Gases.

Gas	C_V ($\text{J K}^{-1}\text{mol}^{-1}$)	C_p ($\text{J K}^{-1}\text{mol}^{-1}$)	γ (unitless)
He	12.5	20.8	1.67
Hg	12.5	20.8	1.67
H_2	20.5	28.9	1.41
NH_3	27.5	36.1	1.31
Diethyl ether	57.5	66.5	1.16

To find C_p we notice that, since $H = U + pV$, we have

$$C_p - C_V = \left(\frac{\partial H}{\partial T}\right)_p - \left(\frac{\partial U}{\partial T}\right)_V$$
$$= \left(\frac{\partial U}{\partial T}\right)_p + \left(\frac{\partial pV}{\partial T}\right)_p - \left(\frac{\partial U}{\partial T}\right)_V$$

For an ideal gas, we recall the Joule experiment which shows that the energy is a function of T only $U = f(T)$, hence

$$\left(\frac{\partial U}{\partial T}\right)_p = \left(\frac{\partial U}{\partial T}\right)_V$$

and

$$C_p - C_V = \left(\frac{\partial pV}{\partial T}\right)_p = \left(\frac{\partial RT}{\partial T}\right)_p = R\left(\frac{\partial T}{\partial T}\right)_p = R$$

Now we see that

$$C_p - C_V = R = 8.3\,\mathrm{J\,K^{-1}\,mol^{-1}}$$

and

$$C_p \cong 12.5 + 8.3 \cong 20.8\,\mathrm{J\,K^{-1}\,mol^{-1}}$$

A useful unitless parameter is $\gamma = C_p/C_V$. From the thermodynamics we have developed so far, $\gamma = C_p/C_V = 20.8/12.5 = 1.66$. Indeed, for He and Hg vapor, γ is close to the ideal value. For more complicated molecules, however, γ begins to fall off significantly. For ammonia vapor, γ has dropped to 1.3 and it appears to be approaching 1.0 for the relatively complicated molecule diethyl ether.

The problem is that more complicated molecules can rotate and vibrate. The NH_3 molecule resembles a pyramid with a triangular base. It can rotate with 3 degrees of rotational freedom in 3-space. The total number of degrees of freedom for translational plus rotational motion is $3 + 3 = 6$; hence the molar kinetic energy is $U_{kin} = = 6(1/2RT)$. From this, by the same reasoning as before, we get $C_V = 3R$ and $C_p = C_V + R = 3R + R = 4R$. Taking $R = 8.3\,\mathrm{J\,K^{-1}mol^{-1}}$, $C_V = 3R \cong 24.9$ $\mathrm{J\,K^{-1}mol^{-1}}$, and $C_p = 4R = 33.2$. This leads to $\gamma = 33/25 = 1.33$ (unitless) in good agreement with the experimental value for ammonia, which is 1.31 for γ, but the results are not so good for C_V and C_p. Hydrogen is an intermediate case because it normally has enough thermal energy to rotate end-over-end (in x, y 2-space) but not enough to spin on its axis.

Evidently, hydrogen and ammonia rotate but do not vibrate under the conditions of Table 3.1. Most chemical bonds also stretch and bend. These motions yield extra

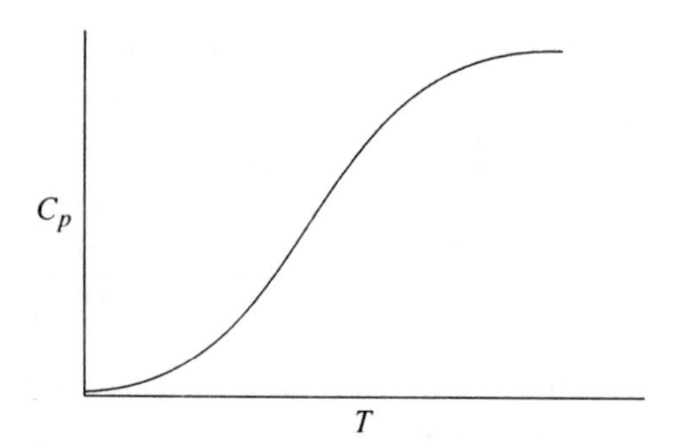

FIGURE 3.6 Typical heat capacity as a function of temperature for a simple organic molecule. Allowed modes of motion are gradually activated as the gas is warmed and more thermal energy becomes available. (See Klotz and Rosenberg, 2008 for more detail.)

degrees of freedom, one for each possible mode of motion. As in the hydrogen case, not all modes of motion are activated. At any given temperature, a molecule may have many degrees of freedom available to it but not enough thermal energy to fully activate all modes. For this reason, heat capacity curves are sigmoidal (S-shaped) starting from zero at 0 K, where there is no motion at all, and rising gradually, as modes of motion are activated, to a limiting value determined by how complicated the molecule is.

3.11 ADIABATIC WORK

Because partial derivatives like $(\partial U/\partial T)_p$, and so on, can be handled just as though they were algebraic variables, it is possible to develop quite an arsenal of equations relating the first law quantities described so far and to expand them to include other variables (Klotz and Rosenberg, 2008). An important concept is that of *adiabatic* (perfectly insulated) work done on or by a gas. The work dw behaves like a thermodynamic function because the path has been specified by setting $q = 0$. Now $dU = dq + dw = p\,dV$ for a system restricted to pressure–volume work. The energy U is a state variable $U = f(V, T)$ for one mole, so

$$dU = \left(\frac{\partial U}{\partial V}\right)_T dV + \left(\frac{\partial U}{\partial T}\right)_V dT$$

and

$$\left(\frac{\partial U}{\partial V}\right)_T dV + \left(\frac{\partial U}{\partial T}\right)_V dT + p\,dV = 0$$

The first term above drops out if we consider expansion of an ideal gas because the functional dependence on V disappears (Joule experiment). Also, $(\partial U/\partial T)_V = C_V$

and $p = {}^{RT}/_{V}$ so

$$C_V\, dT + \frac{RT}{V} dV = C_V \frac{dT}{T} + R\frac{dV}{V} = 0$$

Since $C_p - C_V = R$, we obtain

$$C_V \frac{dT}{T} + R\frac{dV}{V} = \frac{dT}{T} + \frac{C_p - C_V}{C_V}\frac{dV}{V} = \frac{dT}{T} + (\gamma - 1)\frac{dV}{V} = 0$$

Integrating between limits, with a little algebraic manipulation, yields

$$TV^{\gamma-1} = k \qquad \text{and} \qquad pV^{\gamma} = k'$$

where k and k' are constant.

The expression $pV^{\gamma} = k'$ looks like Boyle's law except for the parameter γ, which is always greater than 1. The presence of γ causes the pressure $p = k'/V^{\gamma}$ to be lower at any point during the expansion than it is during the isothermal (Boyle's law) expansion (upper curve in Fig. 3.7).

The difference between the two expansions is in the heat that flows or does not flow into the system to maintain $T = \text{const}$. In the isothermal case heat transfer is allowed. Heat is not allowed into the system in the adiabatic case where $dq = 0$ by definition. Without a compensating heat flow, the adiabatic system cools during expansion and the pressure is always lower at any specified volume than it is in the isothermal case ($p = k'/V^{\gamma},\ \gamma > 1.0$). The entire pV curve falls below the isothermal curve in Fig. 3.7.

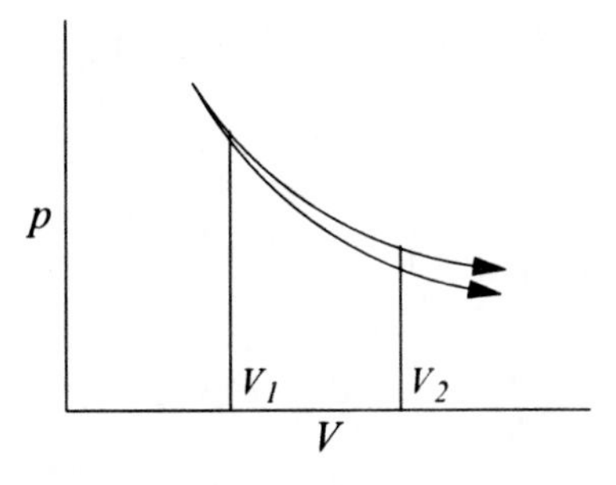

FIGURE 3.7 Two expansions of an ideal gas. The upper curve is isothermal and the lower curve is adiabatic. The adiabatic expansion does less work because there is no heat flow into the system.

PROBLEMS AND EXAMPLE

Example 3.1 Line Integrals

What is the line integral of the function $f(x, y) = xy$ over the parabolic curve $y = f(x) = x^2/2$ from $(x, y) = (0, 0)$ to $\left(1, \frac{1}{2}\right)$?

Solution 3.1 One way of writing a line integral of the function $I = \int_C f(x, y)\, ds$ over the curve C is to specify $y = f(x)$ and $ds = \left(1 + \left(\frac{dy}{dx}\right)^2\right)^{1/2} dx$. For example, integrating the function $f(x, y) = xy$ over the parabolic curve $y = f(x) = x^2/2$ from $(x, y) = (0, 0)$ to $\left(1, \frac{1}{2}\right)$. We have (Steiner, 1996)

$$f(x, y) = xy = x\left(\frac{x^2}{2}\right) = \frac{x^3}{2}$$

and

$$\frac{dy}{dx} = \frac{d\left(\frac{x^2}{2}\right)}{dx} = x$$

Thus,

$$I = \int_C f(x, f(x))\left(1 + \left(\frac{dy}{dx}\right)^2\right)^{1/2} dx = \int_0^1 \frac{x^3}{2}\left(1 + x^2\right)^{\frac{1}{2}} dx$$
$$= \frac{1}{2}\int_0^1 \left(x^6 + x^8\right)^{1/2} dx$$

where the limits of integration are the limits on x. Integration by Mathcad© gives

$$\frac{1}{2}\int_0^1 \left(x^6 + x^8\right)^{.5} dx = 0.161$$

Problem 3.1

One expression of a line integral is

$$\int_C F(x, y)\, dx + G(x, y)\, dy$$

where the subscript C indicates a line (or curve) integral. If $F(x, y) = -y$, $G(x, y) = xy$, and the line is the diagonal from $x = 1$ to $y = 1$ (Fig. 3.8). Carry out the integration.

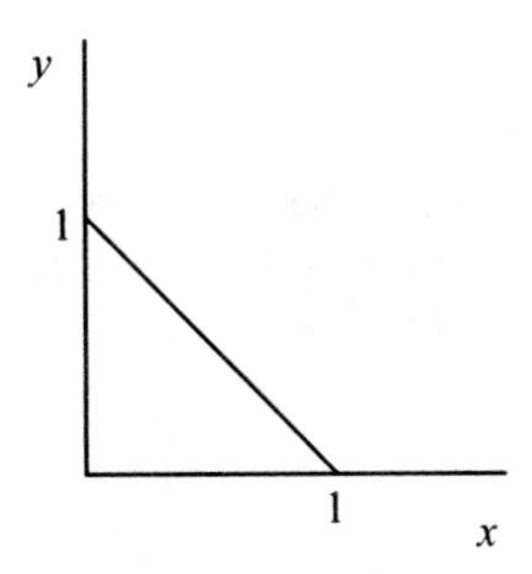

FIGURE 3.8 $C =$ Diagonal along $x = 1$ to $y = 1$.

Problem 3.2

If $F(x, y) = -y$ and $G(x, y) = xy$, evaluate the line integral over the quarter-circular arc from $x = 1$ to $y = 1$ (Fig. 3.9). Notice that the beginning and end points are the same and the functions are the same as in the previous problem but *the path is different*.

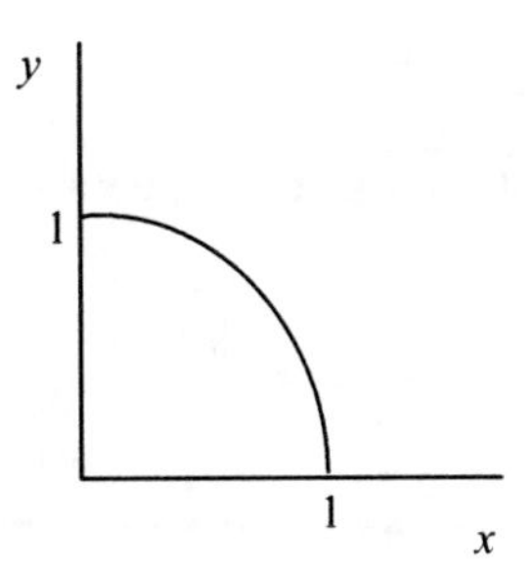

FIGURE 3.9 $C =$ Quarter-circular arc.

Problem 3.3

A mass m of 20.0 kg is raised to a height h of 20.0 m and allowed to drop. Ignoring air resistance, what is its speed when it hits the ground? What is its kinetic energy?

Problem 3.4

A 20-kg rock was carried up a hill that is 20 m high.

(a) What was the energy increase in the rock?
(b) What amount of work was done on the rock when carrying it up the hill?
(c) What is the kinetic energy gained by the rock when it rolls down the hill?

Problem 3.5

In a Joule experiment, two 20.3-kg weights fell 1.524 m to drive a paddle wheel immersed in 6.31 kg of water. The experiment was repeated 20 times, after which the temperature of the water bath was found to have risen by 0.352 K. What is the mechanical equivalent of heat in $J\,K^{-1}$ according to this experiment?

Problem 3.6

Show that dw is an inexact differential for one mole of an ideal gas undergoing a reversible expansion.

Problem 3.7

The definition of the calorie is that amount of heat that is necessary to raise 1 g of water 1 K. As it stands, this definition is approximately valid for temperature changes not too far from room temperature. Use this approximate definition to predict the final temperature of 200 g of water at 283 K mixed with 450 g of water at 350 K.

Problem 3.8

Show that, for an ideal gas with constant C_V, we have

$$\frac{T_2}{T_1} = \left(\frac{V_1}{V_2}\right)^{\frac{R}{C_V}}$$

Problem 3.9

From the kinetic theory of gases, we get the expression $U = \frac{3}{2}RT$ for the energy of an ideal monatomic gas. Show that dU is an exact differential.

Problem 3.10

(a) How much energy is required to heat 1.0 mol of an ideal monatomic gas at constant V from 25.0 to 75.0 K?

(b) What is the heat input if the process is carried out at constant pressure?

Problem 3.11

A 10.0-g piece of iron was heated to 100.0°C by immersing it in boiling water and then quickly transferring it to an insulated beaker containing 1000 g of water at 25°C. What was the final temperature of the water? The specific heat of iron is $0.449\ Jg^{-1}\ K^{-1}$. The specific heat of water is $4.184\ Jg^{-1}\ K^{-1}$.

Problem 3.12

The ratio $\gamma = C_p/C_V$ is 1.40 for nitrogen, N_2. The speed of sound in air (mostly nitrogen) is said to be

$$v_{\text{sound}} \cong \sqrt{\frac{\gamma RT}{\text{M}}}$$

where M is the molar mass of nitrogen. Find the value of v_{sound} at 273 K and compare it with the experimental value of 334 m s^{-1}.

4

THERMOCHEMISTRY

Einstein once said, "Some things are simple but not easy." Although he did not have thermochemistry in mind, his comment applies to this time-honored branch of physical chemistry. Thermochemistry is simple: Run a chemical reaction and measure the temperature change (if any). But it is not easy. Anyone trying to do this job at state-of-the-art precision will soon be enmeshed in technical problems that try the patience of Job. Entire institutes of experimental science exist just for precise measurement of the heat of chemical reactions. If governments are willing to spend millions of dollars to support acquisition of thermochemical data, there must be some significant advantage to be gained from them. That, in part, will be discussed in this chapter. Today, computers play a large role in this field. The most significant advance in thermochemistry in the last decade is the calculation of thermochemical quantities from quantum mechanical first principles. That also will be introduced in this chapter.

4.1 CALORIMETRY

A calorimeter is a device intended to measure heat given out or taken up when a chemical reaction or a physical change, such as a change of state, takes place. A Styrofoam coffee cup with a thermometer is a crude calorimeter. The word has historical significance; it was once supposed to measure the amount of "caloric" flowing into or out of a system. We have long since discarded the caloric theory but

Concise Physical Chemistry, by Donald W. Rogers

we still measure the amount of heat in calories or kilocalories, related through the conversion factor 4.184 to the number of joules or kilojoules.

4.2 ENERGIES AND ENTHALPIES OF FORMATION

Many compounds can be formed by direct combination of their elements. An example is carbon dioxide (CO_2). When a small measured amount of carbon in the common form of graphite, C(gr), is burned in O_2 in a closed steel container called a *bomb*, the process causes a small temperature rise ΔT in the immediate surroundings called the *bath*. The heat capacity of the bath having been previously determined by an electrical calibration, one can find q_V, the amount of heat given off by the combustion at constant volume:

$$\mathrm{C(gr)} + \mathrm{O_2(g)} \rightarrow \mathrm{CO_2(g)} + q_V$$

Straightforward proportional calculation gives the amount of heat that would have been given off if the measured amount of C(gr) had been one gram ($q_V = 32.76\,\mathrm{kJ\,g^{-1}}$) or if it had been one mole ($q_V = \Delta U^{298} = -393.51\,\mathrm{kJ\,mol^{-1}}$). The latter q_V gives the *molar energy change* ΔU^{298} of the system, which is negative because the system gives off heat to the surroundings. The heat is given off at constant volume of the closed bomb; hence it is an *energy* change. In this reaction, the number of moles of gas used up is the same as the number produced $\Delta n_{\mathrm{gas}} = 0$, so the energy change is the same as the *enthalpy* change:

$$\Delta H = \Delta U + \Delta(pV) = \Delta U + \cancel{\Delta n_{\mathrm{gas}}} RT$$

We can write $\Delta_f U^{298}(\mathrm{CO_2}) = \Delta_f H^{298}(\mathrm{CO_2}) = -393.5\,\mathrm{kJ\,mol^{-1}}$ to indicate the energy or the *enthalpy of formation* of CO_2 for an experiment carried out at 298 K.

The heat of combustion of a gas—for example, hydrogen—can be found using a flame calorimetric apparatus in which a known amount of gas is burned and the heat given off is measured by measuring the temperature rise of a suitably positioned bath. The apparatus is a fancy Bunsen burner heating up a container of water equipped with a thermometer. It functions at constant pressure, so the heat given out is the enthalpy decrease of the system:

$$\mathrm{H_2(g)} + \tfrac{1}{2}\mathrm{O_2(g)} \rightarrow \mathrm{H_2O(l)} + q_p$$

which leads to the molar enthalpy of formation of liquid water:

$$q_p = \Delta_f H^{298}(\mathrm{H_2O(l)}) = -285.6\,\mathrm{kJ\,mol^{-1}}$$

In the formation reaction, $\frac{3}{2}$ mol of gas are consumed to produce 1.0 mol of *liquid* water, which has a negligible volume compared to the gas burned. Energy and enthalpy are related by $H = U + pV$; hence $\Delta H = \Delta U + p\Delta V$ at constant pressure and

$\Delta H = \Delta U + \Delta n RT$ under the ideal gas assumption. For the formation of one mole of water, $n = -\frac{3}{2}$

$$\Delta_f U^{298} = \Delta_f H^{298} - \Delta n RT = \Delta_f H^{298} + \tfrac{3}{2} RT$$
$$= -285{,}600 \text{ J mol}^{-1} + \tfrac{3}{2}[8.31(298)] \cong -281.9 \text{ kJ mol}^{-1}$$

For comparison with theoretical calculations, one often needs to know the thermodynamic properties of molecules in the *gaseous* state—for example, $\Delta_f H^{298}(H_2O(g))$. This is handled by adding the heat of vaporization of water to $\Delta_f H^{298}(H_2O(l))$ to obtain

$$\Delta_f H^{298}(H_2O(g)) = -285.6 + 44.0 = -241.6 \text{ kJ mol}^{-1}$$

However, water vapor is not in its *standard state*.

4.3 STANDARD STATES

One can burn a diamond, C(dia), in an oxygen bomb calorimeter. When this is done, the measured enthalpy of formation of $CO_2(g)$ is about 2 kJ mol^{-1} more negative than the $\Delta_f H^{298}(CO_2(g))$ found when C(gr), carbon in the standard state, is burned. The difference is not in the $CO_2(g)$ produced, but in the crystalline form of diamond, which is not the standard state for carbon. Since the path from C to $CO_2(g)$ is about 2 kJ mol^{-1} longer in the diamond combustion, the starting point C(dia) must have been about 2 kJ mol^{-1} higher in enthalpy than C(gr). Differences like this lead us to define the standard state of all elements as the *stable* form at 1.000 atm pressure. By the nature of enthalpy and energy (thermodynamic properties), we can set any arbitrary point to zero as a reference point. Hence we define the enthalpy of formation of *any element in its standard state as zero at all temperatures*. This definition works because elements are not converted from one to another in ordinary chemical reactions.

Our previous observation that $\Delta_f H^{298}(\text{C(dia)}) \neq 0$ but is about 2 kJ mol^{-1} suggests that, given this small enthalpy change, it might be possible to convert common graphite into the nonstandard state of diamond. Indeed it is. Production of small industrial diamonds for cutting tools is commercially feasible.

4.4 MOLECULAR ENTHALPIES OF FORMATION

According to the first law of thermodynamics, the enthalpy of formation of a molecule can be determined even if the formation reaction from its elements cannot be carried out in an actual laboratory experiment. This is illustrated by the simple example of carbon monoxide CO(g). We know from painful examples that this poisonous gas is produced by incomplete combustion of hydrocarbon fuels, but the simple controlled

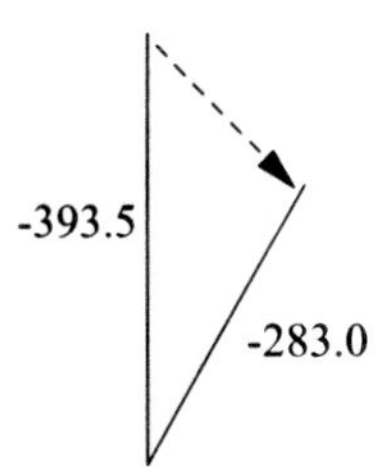

FIGURE 4.1 Combustion of C(gr) and CO(g).

experiment $C(gr) + \frac{1}{2}O_2(g)$ in a limited supply of $O_2(g)$ gives a mixture of products. It is not a "clean" reaction. The reaction

$$CO(g) + \tfrac{1}{2}O_2(g) \rightarrow CO_2(g)$$

is clean, however, and gives well-defined q_p and q_V. We find that the enthalpy change is $\Delta_r H^{298} = -283.0\,\mathrm{kJ\,mol^{-1}}$ for this reaction (Fig. 4.1). Thus we have two paths connecting the same initial and final thermodynamic states. By a thermochemical principle equivalent to the first law of thermodynamics known as *Hess's law*, the enthalpy change over both paths must be the same. The third leg of the triangle must be

$$\Delta_f H^{298}(CO(g)) = -393.5 - (-283.0) = -110.5\,\mathrm{kJ\,mol^{-1}}$$

This method of indirect determination of $\Delta_f H^{298}$ is capable of wide extension. For example, knowing $\Delta_f H^{298}(H_2O(l))$ and $\Delta_f H^{298}(CO_2(g))$, one can determine $\Delta_f H^{298}$ of a hydrocarbon like methane by burning it in a flame calorimeter and setting up a state diagram that is a little more complicated than the triangle just described but that works on the same principle. The combustion reaction

$$CH_4(g) + 2O_2(g) \rightarrow CO_2(g) + 2H_2O(l) \qquad \Delta_r H^{298} = -890\,\mathrm{kJ}$$

has the same final state as

$$2H_2(g) + C(gr) + 2O_2(g) \rightarrow CO_2(g) + 2H_2O(l) \qquad \Delta_r H^{298} = -966\,\mathrm{kJ}$$

In Fig. 4.2, the thermodynamic *state* of the products of combustion of methane, $CO_2(g) + 2H_2O(l)$ is reproduced by burning 2 mol of hydrogen and 1 mol of C(gr). The heats of combustion are arranged in an enthalpy diagram so as to make everything come out even except for one missing leg of the quadrangle, that of methane. The thermodynamic *state* of methane is found by difference; it is 76 $\mathrm{kJ\,mol^{-1}}$ below that of the elements. Formation of methane from its elements would give off 76 kJ per mole of methane produced; therefore $\Delta_f H^{298}(\text{methane}) = -76\,\mathrm{kJ\,mol^{-1}}$. (The tabulated NIST value is $-74.9\,\mathrm{kJ\,mol^{-1}}$.)

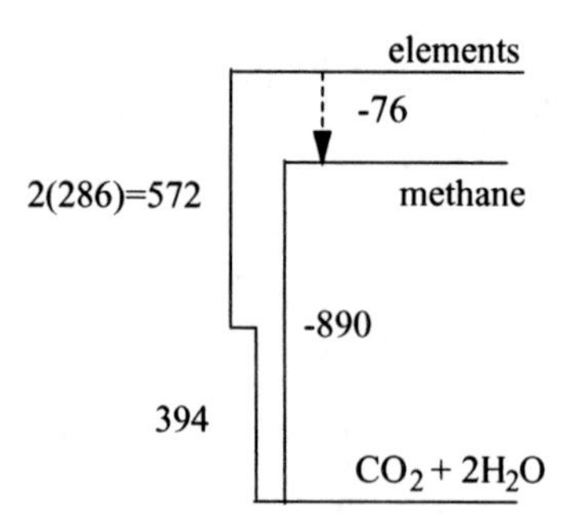

FIGURE 4.2 A Thermochemical cycle for determining $\Delta_f H^{298}$(methane). Not to scale.

The indirect method shown for methane has been extended to very many hydrocarbons and other organic compounds. Although many inorganic substances do not burn, they do react. Inorganic reaction cycles similar to Fig. 4.2 can often be set up to obtain thermochemical data. A free thermochemical database is maintained by the National Institutes of Standards and Technology (NIST) of the US government (.gov). Go to webbook.nist.gov.

4.5 ENTHALPIES OF REACTION

Suppose we know the enthalpies of formation $\Delta_f H^{298}$ of acetylene $CH{\equiv}CH$, ethene $CH_2{=}CH_2$, and ethane CH_3CH_3 (226.7, 52.5, and −84.7 kJ mol^{-1}) by the combustion method shown in Fig. 4.1. The enthalpy change $\Delta_r H^{298}$ of the reaction

$$CH{\equiv}CH(g) + 2H_2(g) \rightarrow CH_3CH_3(g)$$

can be found by comparing the level given by $\Delta_f H^{298}(CH{\equiv}CH(g))$ with that of $\Delta_f H^{298}(CH_3CH_3(g))$. (Remember that $\Delta_f H^{298}$ is zero for elemental hydrogen.)

$$\begin{aligned}\Delta_r H^{298} &= \Delta_f H^{298}(CH_3CH_3(g)) - \Delta_f H^{298}(CH{\equiv}CH(g)) \\ &= -84.7 - 226.7 = -311.4\,\text{kJ mol}^{-1}\end{aligned}$$

This reaction has been carried out with the result $\Delta_r H^{298} = -312.1 \pm 0.6$ kJ mol^{-1}.

A similar reaction, the partial hydrogenation,

$$CH{\equiv}CH(g) + H_2(g) \rightarrow CH_2{=}CH_2(g)$$

cannot be carried out in the laboratory. Hydrogenation doesn't stop at $CH_2{=}CH_2(g)$, but goes on to the fully hydrogenated product $CH_3CH_3(g)$ or gives a mixed product under all known experimental conditions. That doesn't matter because of Hess's law. The enthalpies of formation yield

$$\begin{aligned}\Delta_r H^{298} &= \Delta_f H^{298}(\mathrm{CH_2{=}CH_2(g)}) - \Delta_f H^{298}(\mathrm{CH{\equiv}CH(g)}) \\ &= 52.5 - 226.7 = -174.4\,\mathrm{kJ\,mol^{-1}}\end{aligned}$$

An equivalent way of finding $\Delta_r H^{298}$ of partial hydrogenation is by measuring the enthalpy change $-136.3\ \mathrm{kJ\,mol^{-1}}$ for the reaction

$$\mathrm{CH_2{=}CH_2(g) + H_2 \rightarrow CH_3CH_3(g)}$$

When this is done, one has a first law triangle like that of Fig. 4.1, with only $\Delta_r H^{298}$ of partial hydrogenation yet to be determined. It is

$$\Delta_r H^{298}(\text{partial}) = -312.1 - (-136.3) = -175.8\,\mathrm{kJ\,mol^{-1}}$$

as compared to the previous value of $-174.4\ \mathrm{kJ\,mol^{-1}}$.

One can also find the enthalpy of hydrogenation of ethene:

$$\begin{aligned}\Delta_r H^{298} &= \Delta_f H^{298}(\mathrm{CH_3CH_3(g)}) - \Delta_f H^{298}(\mathrm{CH_2{=}CH_2(g)}) \\ &= -84.7 - 52.5 = -137.2\,\mathrm{kJ\,mol^{-1}}\end{aligned}$$

This is a complete reaction to a well-defined final state. It has been carried out in the laboratory. The experimental value is $-136.3 \pm 0.2\,\mathrm{kJ\,mol^{-1}}$.

Although, for simplicity, we shall use hydrocarbons to illustrate the principles of thermochemistry, there is no such restriction in practice. For example, Pitzer et al. (1961) give the standard enthalpies of formation $\Delta_f H^\circ(\mathrm{KCl}) = -435.9$ and $\Delta_f H^\circ(\mathrm{KClO_3}) = -391.2\,\mathrm{kJ\,mol^{-1}}$. (The superscript $^\circ$ indicates the standard state in this notation; please do not confuse it with a superscripted zero of temperature 0.) These values lead to the enthalpy of reaction $\Delta_r H^\circ$ necessary to convert solid KCl(s) to solid $\mathrm{KClO_3(s)}$:

$$\mathrm{KCl(s)} + \tfrac{3}{2}\mathrm{O_2(g)} \rightarrow \mathrm{KClO_3(s)}$$

$$\Delta_r H^\circ = -391.2 - (-435.9) = 44.7\,\mathrm{kJ\,mol^{-1}}$$

Including the notation (s) to indicate the solid state is a precaution rather than a necessity because in the standard states, both KCl and $\mathrm{KClO_3}$ *are* solids.

Even from these few simple examples, it should be clear that with a sufficient database of enthalpies of formation, it is possible to calculate the enthalpies of an almost limitless array of chemical reactions of industrial, medical, and pharmaceutical importance. This is the reason why so much effort has gone into populating the database with a variety of entries (webbook.nist.gov) and why the most meticulous care is exercised to be sure that the data entered are accurate.

All of the previous examples are consistent with the equations

$$\Delta_r U^{298} = \sum \Delta_f U^{298}(\text{products}) - \sum \Delta_f U^{298}(\text{reactants})$$

and

$$\Delta_r H^{298} = \sum \Delta_f H^{298}(\text{products}) - \sum \Delta_f H^{298}(\text{reactants})$$

where it is to be understood that products and reactants are multiplied by their stoichiometric coefficients and that they are in their standard states. We shall soon see that all thermodynamic properties follow analogous equations, which is why thermodynamics is so useful.

Although combustion thermochemistry on C(gr), CH_4 (methane), and so on, is the prime source of $\Delta_f H^{298}$ data, other reactions can contribute as well. For example, suppose we know that $\Delta_f H^{298}(\text{ethane}) = -83.8 \pm 0.4\,\text{kJ mol}^{-1}$ but we don't know $\Delta_f H^{298}(\text{ethene})$. Measurement of the enthalpy of the hydrogenation reaction ethene $\rightarrow$ ethane gives $-136.3 \pm 0.2\,\text{kJ mol}^{-1}$; hence $\Delta_f H^{298}(\text{ethene})$ must be higher in enthalpy than $\Delta_f H^{298}(\text{ethane})$ by just that amount. By this reasoning,

$$\Delta_f H^{298}(\text{ethene}) = -83.8 \pm 0.4 + 136.3 \pm 0.2 = 52.5 \pm 0.4\,\text{kJ mol}^{-1}$$

The tabulated value is $52.5 \pm 0.4\,\text{kJ mol}^{-1}$.

Molar enthalpies of physical processes like phase changes (vaporization and melting), solution of a solute in a solvent, mixing of miscible solvents, and dilution are treated by slight modifications of the methods shown, always in accord with the first law of thermodynamics, which has never been violated in a controlled, reproducible experiment.

4.6 GROUP ADDITIVITY

Upon scanning a series of heats of combustion $\Delta_c H^{298}$, one notices a regular increase with molecular weight. For example, for the alkanes in their standard states (g), we have the following:

	Methane		Ethane		Propane		*n*-Butane
$\Delta_c H^{298}$ (g)	−890.8		−1560.7		−2219.2		−2877.6
Difference		−669.9		−658.5		−658.4	

If we were asked to predict $\Delta_c H^{298}$(n-pentane(g)), a reasonable answer would be to add another negative change just like the last two:

$$\Delta_c H^{298}(n\text{-pentane(g)}) = -2877.6 - 658.4 = -3536\,\text{kj mol}^{-1}.$$

The experimental value is 3535.4 ± 1.0.

Even simpler, we can note that, given the reference point $\Delta_c H^{298}$(ethane(g)) = −1560.7 kJ mol^{-1}, the heats of combustion of n-alkanes (except for methane) obey a linear function of the *number* of additional hydrogen atoms over those in the reference compound with slope equal to −658.5/2 = −329.2 kJ mol^{-1}. These CH_2 hydrogen atoms are called *secondary* hydrogens as distinct from CH_3, hydrogens, which are *primary*, and isolated C—H atoms, which are *tertiary*. Now any $\Delta_c H^{298}$(n-alkane(g)) can be found by counting secondary hydrogen atoms and adding the count times −329.2 to the base value of −1560.7.

Because the enthalpies of formation $\Delta_f H^{298}$ of n-alkanes(g) are proportional to $\Delta_c H^{298}$, one can estimate $\Delta_f H^{298}$ values in the same way that we used in the combustion case. Simply count hydrogen atoms and add to a base value for ethane:

	Ethane		Propane		n-Butane
$\Delta_f H^{298}$ (g)	−84.0		−104.7		−125.6
Difference		−20.7		−20.9	

Counting secondary hydrogens for n-octane and multiplying by −20.8/2 = −10.4 gives 12(−10.4) = −124.8. Adding the base value, −84.0 gives −208.8 kJ mol^{-1}. The experimental value is −208.4 ± 0.8 kJ mol^{-1}.

Zavitsas et al. (2008) have extended this method to cover primary, secondary, and tertiary hydrogens by the equation

$$\Delta_f H^\circ = -14.0n_p + (-10.4n_s) + (-6.65n_t) = \sum c_i n_i$$

where n_p, n_s, and n_t are the numbers of primary, secondary, and tertiary hydrogen atoms in any alkane or cycloalkane. By this system, the enthalpy of formation of 4-methylheptane, an isomer of n-octane, is

```
               H
               |
CH3CH2CH2CCH2CH2CH3
               |
              CH3
```

$$\Delta_f H^\circ = -14.0n_p + (-10.4n_s) + (-6.65n_t)$$

$$\Delta_f H^{298}(4\text{–methylheptane(g)}) = -14.0(9) + (-10.4(8)) + -6.65(1)$$

$$= -215.8\,\text{kJ mol}^{-1}$$

The experimental value is −212.1.

4.7 $\Delta_f H^{298}$(g) FROM CLASSICAL MECHANICS

One of the drawbacks of the hydrogen-atom counting method is that account is not taken of classical mechanical properties of molecules such as mechanical strain energy induced in distorted or crowded molecules. Other mechanical features not accounted for are those due to the quantum mechanical influences on electron probability densities which, in turn, influence both molecular energy and molecular structure.

The influence of molecular strain can be seen in the discrepancy between the hydrogen-atom counting estimate of $\Delta_f H^{298}$(cyclopentane(g)) $= -24.9\,\text{kJ mol}^{-1}$ as contrasted to the experimental value of -16.3 kJ mol^{-1}. The experimental value is higher (less negative) than the estimate. Evidently, crowding 10 hydrogen atoms into the small space afforded by a 5-membered ring increases the interatomic interference energy that contributes to $\Delta_f H^{298}$. N. L. Allinger at the University of Georgia has developed a series of molecular mechanics programs called MM1 to MM4 that incorporate energies due to bond bending, bond stretching, ring distortion, and so on, into the calculation of molecular structure and $\Delta_f H^{298}$. The method as implemented on a computer is extremely fast and is therefore well-suited to large molecules.

4.8 THE SCHRÖDINGER EQUATION

In 1926 Erwin Schrödinger published an equation that gives correct solutions for the energy levels of the hydrogen atom. Shortly afterward, Heitler and London showed that the Schrödinger equation, as it has come to be called, predicts the existence of a chemical bond between H and H and that it gives an approximate strength of the H—H bond. A powerful refinement and extension of this *molecular orbital* calculation called GAMESS is available as *a site license at no cost* for your microcomputer (academic or similar affiliation must be specified):

http://www.msg.ameslab.gov/GAMESS/GAMESS.html

We shall use the GAMESS program to determine the bond energies of a number of molecules starting with the simplest case, that of the hydrogen molecule when it is formed from two hydrogen atoms:

$$2\text{H}\cdot \rightarrow \text{H—H}$$

The GAMESS output for calculation of the total energy of H· is $E = -0.4998\ E_h$, where E_h is the *hartree*, a unit of energy. A GAMESS output for the molecule H—H is

$$E = -1.1630349978 E_h$$

Hence the bond energy is the energy of the final state (the molecule) minus the energy of the initial state (the two atoms):

$$E(\text{H–H}) - 2E(\text{H}\cdot)$$
$$= -1.1630 - (2 \times -0.4998) = -0.1630 E_h = -102.3\,\text{kcal mol}^{-1} = -428\,\text{kJ mol}^{-1}$$

as compared to the experimental value of $-431\ \text{kJ mol}^{-1}$. The bond energy is negative because the molecule is more stable than the isolated atoms (energy goes downhill). Thus our first molecular orbital calculation comes to within about 1% of the experimental value. The conversions $1.0\ E_h = 627.51\ \text{kcal mol}^{-1}$ and $1.0\ \text{kcal mol}^{-1} \equiv 4.184\,\text{kJ mol}^{-1}$ have been used. This calculation can be carried out on much more complicated molecules, ions, and free radicals, some of which will be described in later chapters The advent of these powerful computer programs and the hardware to run them has made it possible to study reactions that are not accessible by experimental means. Computational thermochemistry is an active research area at present.

4.9 VARIATION OF ΔH WITH T

The definition of the heat capacity at constant pressure $C_p = (\partial H/\partial T)_P$ leads to the infinitesimal enthalpy change with temperature of a pure substance $dH = C_P\,dT$. Over a reasonably short temperature interval ΔT, the equation $\Delta H = C_P \Delta T$ is approximately true. The heat capacity of a mixture is the sum of the molar heat capacities of its components multiplied by the number of moles of each component present. When a chemical reaction takes place, the number of moles of the reactants decreases and the number of moles of products appears with different heat capacities. The difference in heat capacities between the reactant state and the product state is

$$\Delta C_P = \sum C_P(\text{products}) - \sum C_P(\text{reactants})$$

Applying the definition of C_p to all of the component species of a chemical reaction, we get

$$\Delta C_p = \left(\frac{\partial \Delta H}{\partial T}\right)_P$$

Selecting the hydrogenation of ethene to ethane for simplicity, experimental values (webbook.nist.gov) are $C_p = 42.90\,\text{J K}^{-1}\text{mol}^{-1}$ for ethene and $C_p = 52.49\,\text{J K}^{-1}\text{mol}^{-1}$ for ethane; and taking an experimental value (Atkins, 1994) of $C_p = 28.87\,\text{J K}^{-1}\text{mol}^{-1}$ for hydrogen, we obtain

$$\Delta C_p = 52.49 - 42.90 - 28.87 = -19.28\,\text{J K}^{-1}\,\text{mol}^{-1}$$

The change in $\Delta_r H$ for relatively small changes in T is $\Delta C_P \Delta T$. We can test this result against experimental values for the simple hydrogenation of ethene (Kistiakowsky et al., 1935). Using the computed result for $\Delta C_P \Delta T$ over the temperature range 355–298, the range between the temperature at which his experiments were carried out (355 K) and standard temperature, we get

$$\Delta\Delta_{\mathrm{hyd}} H^{355} = -19.28\,[-(355 - 298)] = 1.099\,\mathrm{kJ\,mol^{-1}}$$

for the change in $\Delta_{\mathrm{hyd}} H^{355}$, from 355 K, the temperature at which the hydrogenation was actually carried out to room temperature 298 K. This small temperature correction gives

$$\Delta_{\mathrm{hyd}} H^{298} = -137.44 + 1.10 = -136.34\,\mathrm{kJ\,mol^{-1}}$$

for the enthalpy of hydrogenation of ethene at 298 K. This value is to be compared to the result of a series of experiments carried out at 298 K by a different experimental method which led to $\Delta_{\mathrm{hyd}} H^{298} = -136.29 \pm 0.21$ kJ mol^{-1}. The computed heat capacities are probably reliable over a temperature range of ± 50 K or so. Furthermore, the enthalpy of hydrogenation itself is insensitive to temperature, so we may take experimental determinations of $\Delta_{\mathrm{hyd}} H$ carried out under normal laboratory conditions as essentially the same as the standard state $\Delta_{\mathrm{hyd}} H^{\circ}_{298}$.

When calculations are carried out over larger temperature ranges, as they often are in industrial applications, a polynomial approximation to the heat capacity is used.

$$C_p = \alpha + \beta T + \gamma T^2 + \cdots$$

This enables one to determine C_p for the reactants and products of a chemical reaction at some new temperature other than 298 K, thereby enabling one to determine the new change in heat capacity ΔC_p for the reaction:

$$\Delta C_P = \sum C_P(\text{products}) - \sum C_P(\text{reactants})$$

Variation in theΔH of physical and chemical processes with variation in pressure can be calculated from equations of state or by acquisition of experimental data and curve fitting. Many reactions are less sensitive to pressure change than to temperature change over comparable ranges. Metiu (2006) has treated both temperature and pressure variation in the industrial production of ammonia from its elements.

4.10 DIFFERENTIAL SCANNING CALORIMETRY

When moderate amounts of heat are supplied to a solution of simple salts in water, we expect a smooth heating curve between, say 290 and 320 K, like the lower curve in Fig. 4.3. The heat capacity of water is nearly constant over this temperature range, and it will be little affected by small amounts of dissolved salts. Electrical circuitry exists that permits one to supply heat to a dilute solution in an adiabatic (insulated) calorimeter in very small pulses which may be regarded as infinitesimals dq. We normally carry out the experiment at constant pressure, so the definition of heat capacity at constant pressure $C_p = dq_p/dT$ is satisfied. The gradual temperature rise over many small pulses can be followed by means of a thermistor circuit or its equivalent.

If, instead of a dilute solution of simple salts, the calorimeter contains a solute that is capable of undergoing a *thermal reaction*, which is a reaction brought about by heat, the heating curve is more complicated. Thermal reactions are important in many areas, especially in biochemistry. Proteins undergo heat denaturation. Heat denaturation involves unfolding of the native protein and requires breaking of some or many of the bonds holding it in its native structure. Heat denaturation may be quite specific as to the temperature at which it occurs, and it may bring about subtle changes in the protein, like changes in physiological activity, or it may bring about gross changes in the form of the protein as in the cooking of an egg.

Because heat denaturation involves breaking of internal bonds in the protein, it requires an enthalpy input at constant pressure. The reaction is *endoenthalpic*. A dilute solution of salt and protein takes more heat to bring about a small temperature change than would the solution without the protein. The difference is observed only at or near the temperature of denaturation. Thus we have a normal temperature rise until denaturation begins, after which the heat capacity of the solution is abnormally large until we achieve complete thermal denaturation whereupon the temperature rise drops back to the normal baseline of a salt solution. Plotting C_p as a function of T, we see a peak at the denaturation temperature. This is the upper line in Fig. 4.3. It is a simple matter to interface a computer to the scanning calorimeter output and to integrate under the experimental curve:

$$\Delta_{\text{den}} H = \int_{T_i}^{T_f} C_p \, dT$$

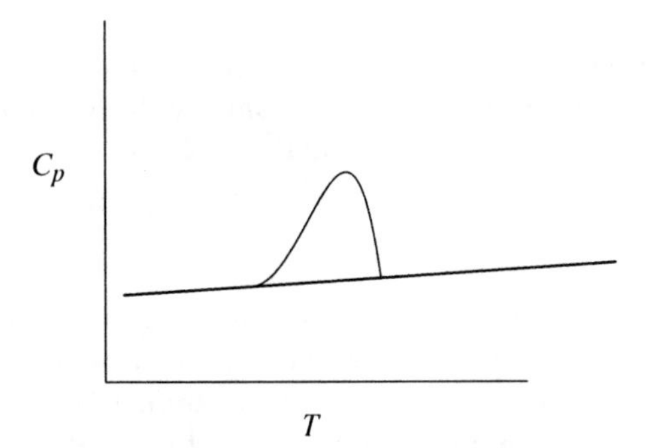

FIGURE 4.3 Schematic diagram of the thermal denaturation of a water-soluble protein. The straight line is the baseline of salt solution without protein. The peak is due to endoenthalpic denaturation of the protein.

The heat capacity curve of the simple salt solution (the baseline) is subtracted from the experimental result. There may be multiple peaks if there is more than one protein in the test solution or if the protein is capable of unfolding in sequential steps.

PROBLEMS AND EXAMPLE

Example 4.1 Oxygen Bomb Calorimetry

Exactly 0.5000 g of benzoic acid C_6H_5COOH were burned under oxygen. The combustion produced a temperature rise of 1.236 K. The same calorimetric setup was used to burn 0.3000 g of naphthalene ($C_{10}H_8$) and the resulting temperature rise was 1.128 K. The heat of combustion of benzoic acid is $q_V = \Delta_c U^{298} = -3227\ \mathrm{kJ\ mol^{-1}}$ (exothermic). What is the heat of combustion $q_V = \Delta_c U^{298}$ of naphthalene?

Solution 4.1 The corresponding molar masses are: benzoic acid, 122.12 g mol^{-1}; and naphthalene, 128.19 g mol^{-1}. The temperature rise for each combustion was: benzoic acid, 1.236/0.5000 = 2.472 K g^{-1}; and naphthalene, 1.128/0.3000 = 3.760 K g^{-1}. Multiplying by the molar masses in each case, one obtains 301.9 K mol^{-1} for benzoic acid and 482.0 K mol^{-1} for naphthalene. This gives us the ratio

$$\frac{301.9}{482.0} = \frac{-3227}{x}$$

$$x = -3227\frac{482.0}{301.9} = -5152\ \mathrm{kJ\ mol^{-1}}$$

$$q_V = \Delta_c U^{298} = -5152\ \mathrm{kJ\ mol^{-1}}$$

Notice that the units cancel on the left; thus x has the units of kJ mol^{-1}, not K mol^{-1}. The handbook value is $\Delta_c U = -5156\ \mathrm{kJ\ mol^{-1}}$.

Problem 4.1

A resistor of precisely 1 ohm is immersed in a liter (1 dm^3) of water in a perfectly insulated container. Suppose that precisely 1 ampere flows through the resistor for precisely 1 second. What is the temperature rise of the water?

Problem 4.2

Exactly one gram of a solid substance is burned in a bomb calorimeter. The bomb absorbs as much heat as 300 g of water would absorb. (Its water equivalent is 300 g.) The bomb was immersed in 1700 g of water in an insulated can. During combustion of the sample, the temperature went from 24.0°C to 26.35°C. What is the heat of combustion per gram of the sample? What is the molar enthalpy of combustion if the molar mass of the substance is 60.0 g mol^{-1} and 2 mol of gas are formed in excess of the O_2 burned?

Problem 4.3

The enthalpy of formation of liquid acetic acid $CH_3COOH(l)$ is $\Delta_f H^\circ = -484.5\ \text{kJ mol}^{-1}$. What is $\Delta_c H$?

Problem 4.4

The enthalpy of combustion of solid α-D-glucose, $C_6H_{12}O_6(s)$ is $-2808\ \text{kJ mol}^{-1}$. What is its enthalpy of formation?

Problem 4.5

Estimate $\Delta_c H^{298}(n\text{-octane(g)})$ of *n*-octane by the hydrogen-atom counting method for alkanes.

Problem 4.6

Find $\Delta_f H^{298}$(2,4-dimethylpentane(g)) by the hydrogen atom counting method. What is the enthalpy of isomerization of *n*-heptane(g) to 2,4-dimethylpentane(g) according to this method? Compare your answer with the experimental result of $-14.6 \pm 1.7\ \text{kJ mol}^{-1}$.

Problem 4.7

The input file for a Gaussian© quantum mechanical calculation will be discussed in later chapters. Briefly, it consists of a few lines of instructions to the computer concerning memory requirements, the number of processors to be used, and the Gaussian procedure to be used, followed by an approximate geometry of the molecule. In simple cases, the input geometry can be merely a guess based on what we learned in general chemistry. The machine specifications will vary from one installation to another. Our input file for the water molecule is

```
%mem=1800Mw
%nproc=1

# g3

water

  0   1
H      -1.012237     0.210253     0.097259
O      -0.260862     0.786229     0.119544
H       0.489699     0.209212     0.142294
```

Adapt this input file for your system and run the water molecule. What is the optimized geometry in the form of Cartesian coordinates (like the input file)? What is the O–H bond length? What is the H–O–H bond angle? What energy is given for water?

Problem 4.8

Plot the heat capacity of ethylene from the following data set:

Temperature (K)	Heat capacity, C_p (J K^{-1} mol^{-1})
300.0000	43.1000
400.0000	53.0000
500.0000	62.5000
600.0000	70.7000
700.0000	77.7000
800.0000	83.9000
900.0000	89.2000
1000.0000	93.9000

Problem 4.9

Suppose that the heat capacities C_p for N_2, H_2, and NH_3 in the standard state are constant with temperature change (they aren't) at 29.1, 28.8, and 35.6 J K^{-1} mol^{-1}. Suppose further that $\Delta_r H^\circ$ for the reaction

$$N_2 + 3H_2 \rightarrow 2NH_3$$

is -92.2 kJ mol^{-1} (of N_2 consumed) at 298 K. What is ΔC_p for the reaction? What is $\Delta_r H^\circ$ at 398 K? What is the (hypothetical) $\Delta_r H^\circ$ at 0 K?

5

ENTROPY AND THE SECOND LAW

The second law of thermodynamics and the concept of entropy are firmly based on Sadi Carnot's somewhat abstract demonstration that the work done by drawing heat from a hot reservoir and expelling it to a cold reservoir is independent of the nature of the engine that carries out the work (Fig. 5.1). An especially readable description of the progression of Carnot's abstract reflections on the efficiency of steam engines to the form of the entropy function we use today was given by Kondipudi and Prigogine (1998). We shall not follow the historical development of this idea, interesting though it is; rather we shall jump right to Clausius's definition of the entropy, Ludwig Boltzmann's statistical interpretation, and the influence of the second law on physical phenomena and chemical reactions.

5.1 ENTROPY

In 1865, Rudolf Clausius showed that the cyclic integral $\oint dq/T$ is zero for an abstract Carnot engine operating *reversibly* around a cyclic path:

$$\oint \frac{dq}{T} = 0$$

Thus the integrand, which Clausius named *entropy* and gave the symbol dS, is a *thermodynamic function*. This definition is one statement of the second law of

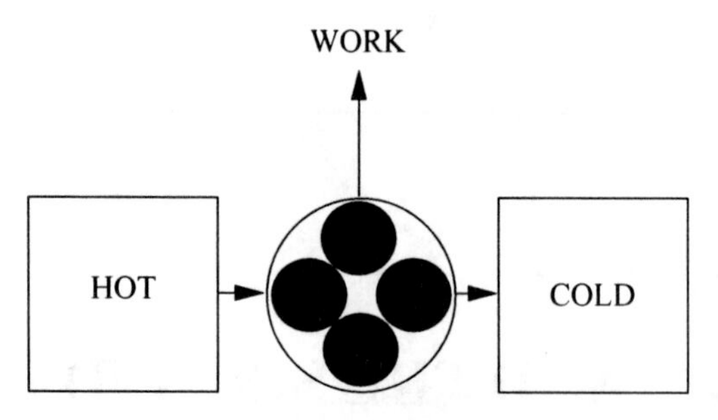

FIGURE 5.1 An Engine.

thermodynamics, which, like the first law, can be stated in many ways. It follows that $\Delta S = \int_a^b dS = \int_a^b dq/T$ is the entropy change of a system carried reversibly over an arbitrary path from a to b and is *independent of the path*. This powerful definition constitutes the second of the two great pillars of thermodynamics. If we can devise a way of calculating ΔS for a reversible chemical reaction, we shall know it for all chemical reactions having the same initial and final states (reactants and products) because of path independence.

Clausius expanded upon the concept of entropy by writing the complete statement as

$$dS \geq \frac{dq}{T}$$

which takes both reversible and *irreversible* changes into account. The irreversible change $dS_{irr} > dq_{irr}/T$ is the real case, a change that takes place in finite time.

If we attempt to take an engine around an irreversible cycle to reproduce its initial state, we shall fall short. We have received a certain amount of work from the engine, but when it comes to the payback (in heat) we see the following with regard to the second law:

$$dS \geq \frac{dq}{T} \quad \text{implies that} \quad dq_{irr} < TdS_{irr}$$

The system will not be returned to its original state, violating the principal stipulation that the system operate around a cyclic path. We shall have to take some heat from the hot reservoir in Fig. 5.1 to complete the cycle and bring the entropy back to its initial value. Where does the extra heat over and above the reversible heat eventually end up? It can go only one place. Since it hasn't done any work, it must have passed through the engine and gone directly to the low-temperature reservoir. The efficiency, *work out relative to heat in*, of a real engine operating irreversibly is less than 1.0 because some heat is doing work and some is not. The important concept is that, of the heat taken from the hot reservoir, not all of it can do work. Some heat *must* pass through the engine from the hot reservoir directly to the cold reservoir doing nothing but restoring the system entropy to its original state. Because of this necessity, some of the heat drawn from the hot reservoir in an irreversible cycle is *unavailable* to do work.

5.1.1 Heat Death and Time's Arrow

Because there is always heat transferred to the surroundings in an irreversible cycle, the entropy of the system plus the surroundings always increases. If we take the universe as the surroundings, then since all real processes are irreversible we have a consequence of the second law: *The entropy of the universe tends to a maximum.* When the universe has reached its maximum entropy, no more irreversible change will be possible. The driving force of change will be gone. This is called the *heat death* of the universe. In case you are worried, it is calculated to be in the far distant future.

The entropy of the universe must be greater after an irreversible change has occurred than it was before the change, so we have a thermodynamic definition of the direction of time (which the first law doesn't give). Time must go from before the change to after the change; it cannot go in the reverse direction.[1] Entropy is sometimes called "time's arrow."

5.1.2 The Reaction Coordinate

Prigogine has defined a *reaction coordinate* ξ which progresses as a chemical reaction takes place. Starting with pure reactant A, the reaction coordinate increases as product B is produced:

$$A \rightarrow B$$

At some point, the time derivative of ξ becomes zero and the reaction stops insofar as macroscopic concentration measurements are concerned.[2] When the time derivative of the reaction coordinate is zero, the system consisting of $n_A + n_B$ is at equilibrium. Because there are no macroscopic concentration changes, the ratio of the mole numbers of reactant and product n_B/n_A is constant. It is called the *equilibrium constant* K_{eq}:

$$\left(\frac{\partial \xi}{\partial t}\right)_{T,p} = 0$$

$$K_{eq} = n_B/n_A$$

It would be appealing to think that the reaction coordinate is determined solely by the energy U or enthalpy H of the system flowing from a high level to a low level,

[1] Classical thermodynamics does not include QED.

[2] There are microscopic exchanges between species A and B, but they are equal and opposite on average so they do not bring about measurable concentration changes.

but this is not the case. There is more to think about in a chemical or physical change than just minimization of U or H. There is the question of order and disorder of the reactant state and the product state.

5.1.3 Disorder

When we look at a chemical reaction

$$A \rightarrow B$$

or an analogous physical change

$$A(l) \rightarrow A(g)$$

we must look at the driving force that produces the change. Part of that force comes from the tendency to seek a minimum (water flows downhill), but part of it comes from the universal tendency of thermodynamic systems to seek maximum disorder. A familiar example is vaporization of a liquid such as water.

The liquid state, though not perfectly ordered, is held together by strong intermolecular forces. These are the very forces that we say are nonexistent or negligible in the vapor state. The entropy change for many liquids is about 88 J K^{-1} mol^{-1}, a rule known as Trouton's rule, that has been verified many times over for liquids as diverse as liquid Cl_2, HCl, chloroform, and the n-alkanes. Liquids that deviate from this rule do so, not because of any failure of the entropy concept for vaporization, but because of abnormal forces in the liquid state. An example is water, which deviates a little due to hydrogen bonding, and hydrogen fluoride HF, which deviates a lot.

5.2 ENTROPY CHANGES

In general, the entropy change for any change of state, including melting and change in crystalline form in the solid state, is given by the enthalpy change *and* the temperature appropriate to the change or *transition* considered:

$$\Delta S = \frac{\Delta H_{\text{trans}}}{T_{\text{trans}}}$$

5.2.1 Heating

The entropy of heating of an ideal gas is positive because thermal agitation makes the high-temperature state more disordered than the low-temperature state. From the expression for molar heat capacity, one has

$$C_p = \left(\frac{\partial H}{\partial T}\right)_p$$

Hence

$$dS_{\text{rev}} \equiv \frac{dq_{\text{rev}}}{T} = \frac{dH}{T} = C_p \frac{dT}{T}$$

Taking the integrals over the interval T_1 to T_2, we obtain

$$\Delta S = \int_{T_1}^{T_2} C_p \frac{dT}{T} = C_p \int_{T_1}^{T_2} \frac{dT}{T} = C_p \ln \frac{T_2}{T_1}$$

for the molar entropy change of heating at constant pressure. There is an analogous equation for the energy change of heating at constant volume.

5.2.2 Expansion

For expansion at constant temperature, U is constant. The first law $dU = dq + dw = 0$ gives us $dq = dw$ and the second law gives us

$$dS_{\text{rev}} = \frac{dq_{\text{rev}}}{T} = \frac{dw}{T} = \frac{pdV}{T}$$

for reversible pressure–volume work. Taking the gas to be ideal for simplicity, we obtain

$$dS_{\text{rev}} = \frac{p\,dV}{T} = \frac{\frac{RT}{V}\,dV}{T} = R\frac{dV}{V}$$

Integrating between limits as before, we obtain

$$\Delta S_{\text{rev}} = R \int_{V_1}^{V_2} \frac{dV}{V} = R \ln \frac{V_2}{V_1}$$

for the reversible expansion of one mole of an ideal gas from V_1 to V_2. The right-hand side of the equation should be multiplied by n for expansion of n moles of gas. If the gas is not ideal, a real gas equation can be substituted for the ideal gas law in these equations. Thus, the mathematical complexity will be increased, but the principle is the same.

5.2.3 Heating and Expansion

Because dS is an exact differential and $S = f(T, p)$, we have

$$dS = \left(\frac{\partial S}{\partial T}\right)_p dT + \left(\frac{\partial S}{\partial p}\right)_T dp$$

We know from the second law that at constant pressure, we obtain

$$dS = \frac{dq_p}{T} = \frac{C_p}{T}$$

so

$$dS = \frac{C_p}{T} dT + \left(\frac{\partial S}{\partial p}\right)_T dp$$

By the Euler reciprocity relation, for exact differentials du written in differential form

$$du = M(x, y) + N(x, y)$$

we have the equality

$$\frac{\partial M(x, y)}{dy} = \frac{N(x, y)}{dx}$$

In the case of the Gibbs thermodynamic function (next chapter) $\mu = f(S, p)$, we have

$$d\mu = -S\,dT + V\,dp$$

so

$$-\left(\frac{\partial S}{\partial p}\right)_T = \left(\frac{\partial V}{\partial T}\right)_p$$

which leads to

$$dS = \frac{C_p}{T}\,dT + \left(\frac{\partial S}{\partial p}\right)_T dp = \frac{C_p}{T} dT - \left(\frac{\partial V}{\partial T}\right)_p dp$$

Both of these coefficients C_p/T and $(\partial V/\partial T)_p$ can be measured, so the infinitesimal dS can be found at any T and V. The finite change ΔS is

$$\Delta S = \int_{T_1}^{T_2} \frac{C_p}{T}\,dT - \int_{p_1}^{p_2} \left(\frac{\partial V}{\partial T}\right)_p dp$$

Starting with the Helmholtz free energy in place of the Gibs function, a comparable derivation yields

$$\Delta S = \int_{T_1}^{T_2} \frac{C_p}{T}\,dT - \int_{V_1}^{V_2} \left(\frac{\partial p}{\partial T}\right)_V dV$$

These derivations are given in more detail in Metiu (2006) along with results calculated from accurate equations of state for real gases and comparisons to National Institutes of Standards and Technology tabulations (webbook.nist.gov).

5.3 SPONTANEOUS PROCESSES

5.3.1 Mixing

Consider two chambers of the same volume, one containing 0.500 mol of ideal gas A and the other containing 0.500 mol of ideal gas B. The gases are at the same temperature and 1.0 bar pressure, and the chambers are connected by a valve. When the valve is opened, the gases mix spontaneously just as the smell of perfume gradually permeates all regions of a closed room. The diffusion process is like an expansion for the gases considered individually because the volume in which the gas molecules move is doubled relative to what it was before the valve was opened. For gas A we can say

$$\Delta S_{\mathrm{A}} = 0.500\, R \ln \frac{V_2}{V_1} = 0.500\, R \ln \frac{2}{1} = 2.88\ \mathrm{J}$$

The equivalent calculation for gas B gives the same result, so the total entropy change is the sum

$$\Delta S_{\mathrm{A}} + \Delta S_{\mathrm{A}} = 2(2.88) = 5.76\ \mathrm{J}$$

If the volumes of the chambers are changed arbitrarily so that they are not equal and if quantities of gas are taken that are not 0.500 mol, nor are they equal, the entropy increase for gases A and B is different from what we have just calculated, but it is always an *increase* because each gas sees a larger volume after the mixing process than it did before. Mixing is always spontaneous and the entropy of mixing is *always positive* because, at constant p, V_2 is always larger than V_1 from the point of view of each gas.

5.3.2 Heat Transfer

Consider now two bricks in contact with one another. One is a hot brick and the other is a cold brick. We know, from millennia of experience, that after sufficient time we will have two warm bricks and that we never observe the reverse process, two warm bricks spontaneously undergoing a transformation that produces a hot brick and a cold brick. This is a crude statement of the second law. In fact an engine is merely a device placed between a hot reservoir and a cold reservoir to siphon off some of the heat flow and use it to do work.

Spontaneous flow of heat from hot to cold cannot be explained by the first law because for each joule of heat lost by the hot brick there is exactly one joule gained by the cold one. The process involves no change in energy or enthalpy; hence the first law tells us nothing about it. To look at the situation more quantitatively, suppose that the bricks are of equal size and have the same heat capacity C. Suppose further that one brick is at 400 K and the other is at 200 K. After sufficient time, both will be at 300 K, assuming no heat is lost to the surroundings.

$$\Delta S = C \ln \frac{T_2}{T_1} = C \ln \frac{300}{400} = -0.288C$$

for the hot brick, but

$$\Delta S = C \ln \frac{T_2}{T_1} = C \ln \frac{300}{200} = 0.405C$$

for the cold brick. The result that the positive entropy change is greater than the negative change is independent of the initial temperatures, heat capacities, sizes of the bricks, and so on. Entropy for the spontaneous process *always increases* for spontaneous heat transfer. The only way to get away from the inequalities of ΔS for hot and cold bricks would be to make $T_1 = T_2$ but then no heat would flow.

5.3.3 Chemical Reactions

The order or disorder within a system undergoing chemical reaction changes, sometimes dramatically. For example, combination of equal volumes of $H_2(g)$ and $O_2(g)$ produces a negligible volume of $H_2O(l)$ plus $^1/_2$ volume of $O_2(g)$ left over after the hydrogen is all used up. The volume of the system goes from 2 to $^1/_2$, so we expect an entropy change on that basis alone. The initial volume is now larger than the final volume at constant pressure, so, in contrast to the expansion case, the entropy change of this reaction is negative; the final state is more *ordered* than the initial state. The negative $\Delta_r S$ is in opposition to the direction of the chemical reaction. The first and second laws operating simultaneously on the system through $\Delta_r H - T\Delta_r S$ give a term $-T\Delta_r S$[3] which is positive but which is smaller than $\Delta_r H$. Hence $\Delta_r S$ is negative for this spontaneous (sometimes explosive) reaction. This is an example of the very common tendency of reactions to go in the direction of a *spontaneous creation of order* from a disordered system, provided that there is enough enthalpy decrease to drive the entropy change "backwards." Some creationists believe that a spontaneous ordering process is impossible, but they are wrong.

5.4 THE THIRD LAW

The third law of thermodynamics states that the entropy, unlike the energy and enthalpy, has a natural zero point. The entropy of a perfect crystal is zero at 0 K.

[3]The product of two negative numbers.

Because of the third law, it is possible to obtain a standard molar entropy (often called the "absolute" entropy) of any pure substance at any temperature. The task is simple but not easy. One must determine the molar heat capacity at constant pressure C_p for the crystal at many temperatures until it undergoes the first phase transition. By the methods shown in Section 5.2.1, the integral taken down to low T

$$S = \int_0^{T_1} \frac{C_p}{T}\, dT$$

gives the standard entropy at the transition temperature T_1. Since C_p is a molar heat capacity, S is an standard molar entropy. Normally one wants the entropy at some higher temperature, say 298 K, and often the phase transition takes place at a temperature T_1 lower than 298 K. Therefore we must add the entropy contribution from the phase transition to the value of S that we already have to obtain the standard molar entropy after the phase change at T_1. The new phase then is heated over a temperature interval $T_1 \rightarrow T_2$ where the new temperature may be 298 K or may be the temperature of a new phase change. Melting and vaporization are handled in the same way as crystalline phase changes. Eventually, over a few or many phase changes, one arrives at the desired temperature. The standard entropy is the summation of all the contributions along the way:

$$S = \int_0^{T_1} \frac{C_p}{T}\, dT + \sum \frac{\Delta H_{\text{trans}}}{T_{\text{trans}}} + \sum \int_{T_{\text{lower}}}^{T_{\text{higher}}} \frac{C_p}{T}\, dT$$

5.4.1 Chemical Reactions (Again)

The change in entropy of a chemical reaction can be determined by carrying out a determination of the standard molar entropies of all of the reactants and all of the products as just described and taking the sum

$$\Delta S_r = \sum S(\text{products}) - \sum S(\text{reactants})$$

The entropy of ordering or disordering that occurs when, for example, the product state is in the gaseous phase and the reactants are in a condensed phase (liquid or solid) is included in this sum because terms like $\Delta H_{\text{vap}}/T_b$ are included in $\sum S$(products) but not in $\sum S$(reactants). It would be attractive to adopt the simplistic attitude that all spontaneous chemical and physical reactions produce an entropy increase for the reacting system, but, once again, things are more complicated than that. A spontaneous change is driven both by the tendency of a system to reduce its energy and enthalpy and by the tendency of a system to increase its disorder. A composite function is needed that includes both the enthalpy and entropy, and this is the function found and described in mathematical detail by the great American thermodynamicist J. Willard Gibbs. The composite function the Gibbs free energy or, more simply, the Gibbs function $G = H - TS$ now bears his name. There is a comparable function involving the energy and the entropy used more by engineers than by chemists; this is called the Helmholtz free energy, $A = E - TS$.

PROBLEMS AND EXAMPLE

Example 5.1 The Standard Entropy of Silver

The C_p/T vs. T data set for silver from 15 to 300 K is

T	C_p	C_p/T	$\ln T$
15.0000	0.6700	2.7100	0.0447
30.0000	4.7700	3.4000	0.1590
50.0000	11.6500	3.9100	0.2330
70.0000	16.3300	4.2500	0.2333
90.0000	19.1300	4.5000	0.2126
110.0000	20.9600	4.7000	0.1905
130.0000	22.1300	4.8700	0.1702
150.0000	22.9700	5.0100	0.1531
170.0000	23.6100	5.1400	0.1389
190.0000	24.0900	5.2500	0.1268
210.0000	24.4200	5.3500	0.1163
230.0000	24.7300	5.4400	0.1075
250.0000	25.0300	5.5200	0.1001
270.0000	25.3100	5.6000	0.0937
290.0000	25.4400	5.6700	0.0877
300.0000	25.5000	5.7000	0.0850

The C_p/T vs. T curve for silver is given as Fig. 5.2. Select from packaged software (for example, SigmaPlot©) or write a short program of your own that will enable you to integrate the data set for silver to find the standard entropy S at 300 K. The problem is simplified by the lack of phase transitions in solid Ag over the temperature range, including the melting and boiling points, which are well above 298 K. The *CRC Handbook of Chemistry and Physics*, 2008–2009 (89th ed.) value for the standard S^{298}_{Ag} at 298 is 42.67 J K^{-1} mol^{-1}.

Solution 5.1 SigmaPlot© contains a macro that carries out integration under curves that are displayed as a smooth function. First plot your function, then execute `Tools → macro → run → compute`. Be sure to designate your plot below the macro. The SigmaPlot output for this integration is 42.2076 over the interval from 15 K to 298 K.

The result is pretty close to the standard entropy in the handbook, but it lacks a contribution below 15 K. This problem is usually handled by the Debye method (Problem 5.7), which assumes a third-power equation leading to

$$C_p = AT^3 = 0.67\,\mathrm{J\,K^{-1}\,mol^{-1}}$$

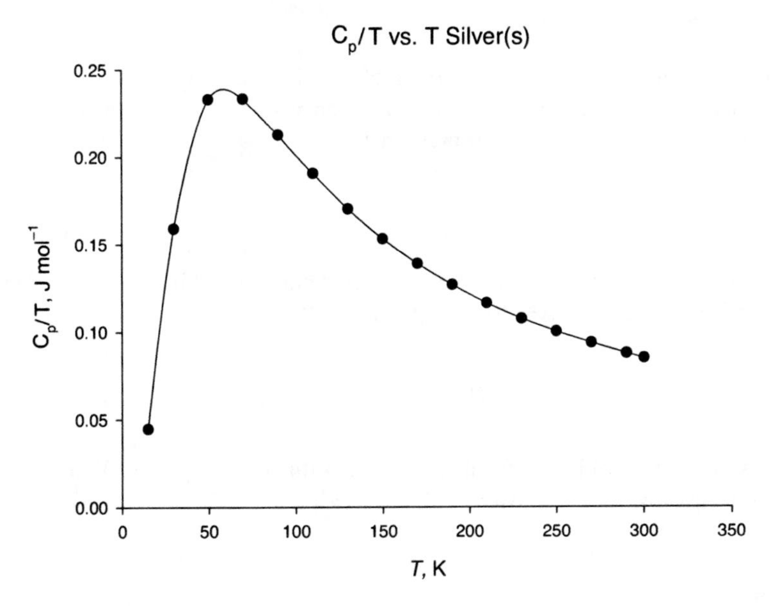

FIGURE 5.2 C_p/T vs. T for metallic silver Ag(s). There are no phase transitions for solid Ag over this temperature range.

at 15 K and leads to

$$S_0^{15} = \frac{C_p}{3} = \frac{0.67}{3} = 0.22\,\mathrm{J\,K^{-1}\,mol^{-1}}$$

This small addition yields $S_0^{298} = 42.43\,\mathrm{J\,K^{-1}\,mol^{-1}}$, which is within 0.6% of the handbook value.

Problem 5.1

Hexa-1,3,5-triene has a boiling point of 355 K under atmospheric pressure. Estimate the enthalpy of vaporization of hexa-1,3,5-triene.

Problem 5.2

(a) What is the entropy change brought about by heating 2.5 mol of helium from 300 to 400 K at constant volume?

(b) What is the entropy change brought about if the same heating process takes place at constant pressure?

Problem 5.3

What is the entropy of isothermal mixing of 1 mol of helium with 1 mol of argon if the two gases start out in separate chambers, each at 1 bar pressure, and they produce 2 mol of mixed gases also at 1 bar pressure?

Problem 5.4

Industrial production of ammonia NH_3 is carried out by combination of the elements at about $T = 650$ K and $p = 400$ bar (Metiu, 2006):

$$3H_2 + N_2 \rightarrow 2NH_3$$

What is the enthalpy change for this reaction at this temperature and pressure? For simplicity, assume ideal behavior of all three gases.

Problem 5.5

What is the enthalpy change for the pressure change from 1.0 bar to 40 MPa?

Problem 5.6

Sulfur dioxide has a heat of fusion of 7.41 kJ mol^{-1} at its melting point of 200 K. Find the entropy change for the melting process

$$SO_2(s) \rightarrow SO_2(l)$$

How does this compare with the entropy change for melting ice which has ΔH_{fusion}= 333.6 J g^{-1} (*CRC Handbook of Chemistry and Physics*, 2008–2009, 89th ed.)

Problem 5.7

On theoretical grounds, Peter Debye proposed what is known as the Debye third-power law for the entropy of perfectly crystalline solids near absolute zero K.

$$C_p = AT^3, \qquad T < 15\,\text{K}$$

Solid chlorine $Cl_2(s)$ has a heat capacity of $C_p = 3.72\,\text{J K}^{-1}\,\text{mol}^{-1}$. What is the entropy of $Cl_2(s)$ at $T = 15$ K?

Problem 5.8

The C_p/T vs. T data set for solid lead Pb(s) is written in BASIC as follows:

```
DATA 0,0,5,.061,10,.28,15,.4666,20,.54,25,.564,30,.55,50,.428,
70,.333,100,.245,150,.169 200,.129,250,.105,298,.089
N = 14
```

The data are in 14 pairs. The first number of the pair is T and the second is C_p. Devise a program of your own or use a canned program to estimate the standard entropy of Pb(s).

Problem 5.9

Phase changes occur reversibly. For example, the transition from solid ice to liquid water, which occurs at a temperature that is infinitesimally above the melting point of 273.15 K, can be reversed by lowering the ambient temperature to slightly less than 273.15 K.

(a) The standard enthalpy of fusion $\Delta_{\text{fus}} H^\circ$ is 6.01 kJ mol^{-1}. What is the entropy of fusion, $\Delta_{\text{fus}} S^\circ$?

(b) The standard enthalpy of vaporization of water is $\Delta_{\text{vap}} H^\circ = 40.7$ kJ mol^{-1}. What is the entropy of vaporization of water, $l \rightarrow v$?

(c) Why are the results so different?

6

THE GIBBS FREE ENERGY

Our brief consideration of engines and heat transfer showed that only part of the heat transferred from a hot reservoir to a cold one by a real process is available to do work. This is the available or "free" energy that we have been seeking. In the theory of steam engines, the Helmholtz free energy $A = U - TS$ is central. In chemical reactions the closely analogous Gibbs free energy $G = H - TS$ is central. We shall be mainly concerned with the Gibbs free energy.

6.1 COMBINING ENTHALPY AND ENTROPY

We seek a combined function that expresses the spontaneous tendency of a chemical system to undergo a change in enthalpy simultaneously with a change in its entropy. In other words, we seek a factor that governs chemical reactions. Clearly, the signs of enthalpy and entropy must be opposite because one function tends to a maximum and the other tends to a minimum. We might write $X = U - S$ for our unknown energy function such that matter flows from a point of high potential to one of low potential in the way that water flows downhill, but we chemists are usually interested in reactions that are carried out at constant pressure, so we substitute H for U in the equation. Also we notice that the units are wrong. Enthalpy has the unit J and entropy has the unit $\mathrm{J\,K^{-1}}$. To bring everything into consistent units of energy, we multiply the entropy by T to get the central equation of chemical thermodynamics:

$$G \equiv H - TS$$

Concise Physical Chemistry, by Donald W. Rogers

This equation implies constant p by the use of H and constant T at thermal equilibrium. Multiplication of S by T is also consistent with the definition of entropy as $dS = dq/T$, where q is a reversible thermal energy or enthalpy. The infinitesimal and small finite expressions dG and ΔG are also implied:

$$dG = dH - T\,dS$$

and

$$\Delta G = \Delta H - T\Delta S$$

The Gibbs function is the algebraic sum of a thermodynamic state function H and a constant times a thermodynamic function $-TS$, hence it is a thermodynamic state function as well.

The terms *Gibbs state function, Gibbs free energy*, and *chemical potential* are all used for the thermodynamic property G. Usually the first term is used to stress the mathematical properties of the function, the second is used in general descriptions, and the third is used to stress the intensive nature of the *molar* free energy $\mu = G/n$. We shall use these terms more or less synonymously, relying on context to make the meaning clear or reminding the reader, from time to time, of the distinction between the molar Gibbs free energy G and the extensive Gibbs function G. The term chemical potential, for the molar (or partial molar) quantity μ, depicts well the property of chemical systems to flow down a gradient toward a minimum that we refer to as an equilibrium point.

6.2 FREE ENERGIES OF FORMATION

At this point, we are able to determine the enthalpy change of a chemical reaction by direct or indirect calorimetric measurement of $\Delta_f H$ of the participants in the reaction and we can find the corresponding entropy change by integration of heat capacity data for each of the participants. These results permit us to calculate *Gibbs free energies of formation*. Suppose that we select combustion of C(graphite) in $O_2(g)$ to form $CO_2(g)$ as the illustrative case and that we have determined the standard entropies of these three species to be 5.74, 205.138, and 213.74 J K^{-1} mol^{-1}, respectively. We apply the general formula

$$\Delta S_r^{298} = \sum S(\text{products}) - \sum S(\text{reactants})$$

to the reaction

$$C(gr) + O_2(g) \rightarrow CO_2(g)$$

to get

$$\Delta S_r^{298} = 213.74 - 5.74 - 205.138 = 2.862\,\mathrm{J\,K^{-1}mol^{-1}}$$

as the change in standard entropy of the reacting system.

The *enthalpy* of formation of CO_2(g) has been measured and found to be −393.51 kJ mol^{-1} (Section 4.2), so from the fundamental equation for a finite change (ΔG)

$$\Delta G = \Delta H - T\Delta S$$

we get

$$\Delta G_f = \Delta H_f - T\Delta S_f = -393.51 - 298(2.862 \times 10^{-3})$$
$$= -393.51 - 0.852 = -394.36\,\mathrm{kJ\,mol^{-1}}$$

which is the Gibbs function of formation at 298 K of CO_2(g).

Comparable calculations yield the Gibbs functions of many direct formation reactions of elements to their compounds. Armed with these Gibbs functions of formation, we can manipulate them to find the Gibbs functions of compounds not cleanly formed from their elements in the same way as we did for the enthalpies of formation.

6.3 SOME FUNDAMENTAL THERMODYNAMIC IDENTITIES

For a reversible change doing only pV work, the first law gives

$$dU = dq - dw = TdS - p\,dV$$

whence, knowing that $U = f(S,V)$, $dU = (\partial U/\partial S)_V\,dS + (\partial U/\partial V)_S\,dV$, we find that

$$\left(\frac{\partial U}{\partial S}\right)_V = T \qquad \text{and} \qquad \left(\frac{\partial U}{\partial V}\right)_S = -p$$

Also, $G = f(p,T)$, $dG = \left(\frac{\partial G}{\partial p}\right)_T dp + \left(\frac{\partial G}{\partial T}\right)_p dT$, and

$$dG = V\,dp - S\,dT$$

with the results that

$$\left(\frac{\partial G}{\partial p}\right)_T = V \qquad \text{and} \qquad \left(\frac{\partial G}{\partial T}\right)_p = -S$$

6.4 THE FREE ENERGY OF REACTION

We have procedures that give us the enthalpies of formation of the participants in any reaction at 1.0 atm pressure, and we can calculate the absolute entropies of the participants. Therefore we can construct a table of Gibbs free energies of formation for many compounds. The free energy change of reaction is found in the usual way:

$$\Delta G_r = \sum \Delta_f G(\text{products}) - \sum \Delta_f G(\text{reactants})$$

The Gibbs free energies of formation of all elements are defined as zero. This definition is possible because no element can be formed from any other element by an ordinary chemical reaction.

An insight into the difference between chemical potential and enthalpy can be found in the stepwise hydrogenation of acetylene, first to ethene and then to ethane:

$$HC{\equiv}CH(g) + H_2(g) \rightarrow H_2C{=}CH_2(g)$$

$$HC{=}CH_2(g) + H_2(g) \rightarrow H_3C{-}CH_3(g)$$

The enthalpies of these two reactions are, respectively, -174 and -137 kJ mol^{-1} and the Gibbs free energies of reaction are -141 and -101 kJ mol^{-1}, respectively. The numbers themselves are quite different, but the *difference* between them is comparable: 33 kJ mol^{-1} in the first case and 36 kJ mol^{-1} in the second. This is because each reaction involves "tying up" two moles of gas and releasing only one. Each reaction involves more or less the same reduction of disorder, hence there is less energy free to seek a minimum than one might expect considering the enthalpy change alone. The chemical potential well is less deep than it would be without the opposing entropy factor.

6.5 PRESSURE DEPENDENCE OF THE CHEMICAL POTENTIAL

The previous calculations were carried out for reactions at 1 atm pressure. All reactions are not carried out at 1 atm pressure, so we need a method of finding the change in chemical potential at any pressure. One can construct a reaction diagram (Fig. 6.1). If we can find ΔG_2 and ΔG_3 for a change in pressure from 1 bar (or 1 atm) to a new pressure p_2 for all the reaction components, the change in Gibbs chemical potential for the reaction ΔG_4 can be found at any selected pressure p_2. The problem is already solved, however, because we have the identity $(\partial G/\partial p)_T = V$ from Section 6.3. Explicitly stipulating constant temperature, we can go from partials to total derivatives for an ideal gas:

$$dG = V\,dp = \frac{RT}{p}\,dp$$

$$
\begin{array}{ccc}
A(g, p_2) & \xrightarrow{\Delta G_4} & B(g, p_2) \\
\uparrow \Delta G_2 & & \uparrow \Delta G_3 \\
A(g, p_{1\text{bar}}) & \xrightarrow{\Delta G_1} & B(g, p_{1\text{bar}})
\end{array}
$$

FIGURE 6.1 A reaction diagram for ΔG_4.

Integrating to find ΔG_2, we obtain

$$\Delta G_2 = \int_{G_{1\text{bar}}}^{G_{p_2}} dG = RT \int_{G_{1\text{bar}}}^{G_{p_2}} \frac{dp}{p} = RT \ln \frac{p_2}{1}$$

where the pressure at the lower limit of integration is 1 bar. A similar equation describes ΔG_3. Combination of ΔG_1, ΔG_2, and ΔG_3 with attention to the sign differences gives ΔG_4, the desired change in the Gibbs function for reaction at the new pressure p_2. One way of getting the signs straight is to set up the diagram as a cyclic process, and remember that the sum of changes in any thermodynamic function must be zero around a cycle.

6.5.1 The Equilibrium Constant as a Quotient of Quotients

In writing equilibrium constant expressions such as $K_{\text{eq}} = p_B/p_A$ for a reaction A $\rightarrow$ B in the gas phase, it should be remembered that the pressures are quotients relative to a pressure in the standard state p/p_0, hence they are unitless and so is K_{eq}. If the gas is not ideal, the fugacity f is used in place of the pressure or in the case of liquids and solutions, the activity a may be used. Fugacities or activities are also ratios f/f_0 or a/a_0 relative to a standard state so K_{eq} is still unitless, as it must be for some of the mathematical manipulations to come. Changes to K_{eq} expressed in terms of fugacities or activities bring about complication in the mathematical expression and in the experimental determination of the quantities involved; but once again, the principles are the same as in the ideal case.

6.6 THE TEMPERATURE DEPENDENCE OF THE FREE ENERGY

From the usual expression for the derivative of a quotient, $\dfrac{d\left(\frac{u}{v}\right)}{dx} = \dfrac{v\frac{du}{dx} - u\frac{dv}{dx}}{v^2}$, and stipulating constant pressure, we get

$$\frac{d\left(\frac{G}{T}\right)}{dT} = \frac{T\left(\frac{dG}{dT}\right) - G\left(\frac{dT}{dT}\right)}{T^2} = \frac{-TS - G}{T^2}$$

where we recall that $(dG/dT)_p = -S$. From $G = H - TS$, $H = G + TS$, and $-G - TS = -H$, so

$$\frac{d\left(\frac{G}{T}\right)}{dT} = \frac{-TS - G}{T^2} = \frac{-H}{T^2}$$

and

$$\frac{d\left(\frac{\Delta G}{T}\right)}{dT} = \frac{-\Delta H}{T^2}$$

It is convenient to remember that $d(1/T)/dT = -1/T^2$ or $dT/d(1/T) = -T^2$, so

$$\frac{dT}{d\left(\frac{1}{T}\right)} \frac{d\left(\frac{\Delta G}{T}\right)}{dT} = -T^2\left(-\frac{\Delta H}{T^2}\right) = \Delta H$$

or

$$\frac{d\left(\frac{\Delta G}{T}\right)}{d\left(\frac{1}{T}\right)} = \Delta H$$

This is one form of the *Gibbs–Helmholtz equation*. It is general and applies to chemical and physical changes. For relatively short temperature intervals, ΔH may be regarded as a constant. A plot of $\Delta G/T$ vs. $1/T$ gives ΔH. Conversely, knowing ΔH enables one to determine $\Delta G/T$, hence ΔG at temperatures other than 298 K. Generally speaking, reactions are more sensitive to temperature changes than to pressure changes of comparable magnitude, hence the Gibbs–Helmholtz equation is of overarching importance in practical and industrial chemistry.

We now have methods to determine ΔG at any temperature and pressure from the tabulated ΔG^{298} values under standard conditions. The usual technical complications arise for real systems. For example, ΔH may not be constant, but it may be followed closely by a power series of the form $\Delta H = a + bT + cT^2 + \cdots$. These extra terms make the equations look messy, but the principles remain the same.

PROBLEMS AND EXAMPLE

Example 6.1

Find the Gibbs free energy of formation for methane. The enthalpy of formation of methane is $\Delta H^{298}_{f,\text{methane}} = -74.81\ \text{J mol}^{-1}$. Standard entropies of H_2(g), C(graphite), and methane CH_4(g) are also known. They are 130.684, 5.74, and 186.26 J K^{-1} mol^{-1}, respectively.

Solution 6.1 For the reaction

$$\text{C(graphite)} + 2\text{H}_2\text{(g)} = \text{CH}_4\text{(g)}$$

we have

$$\Delta S^{298}_{r,\text{CH}_4} = 186.26 - 2(130.684) - 5.74 = -80.848\ \text{J K}^{-1}\text{mol}^{-1}$$

At 298.15 K, we obtain

$$T\Delta S = -0.080848(298.15) = -24.09\ \text{kJ mol}^{-1}$$

and

$$\Delta G^{298}_{f,\text{CH}_4} = -74.81 - (-24.093) = -50.72\ \text{kJ mol}^{-1}$$

(Notice the conversion from J to kJ in the penultimate step.) In this way, we can build up a table of Gibbs functions for as many compounds as time and money permit.

Problem 6.1

(a) Given that

$$G = H - TS$$

show that

$$dG = dH - T\,dS - S\,dT$$

(b) Given that

$$H = U + pV$$

show that

$$dH = dU + p\,dV + V dp$$

for $p \neq$ const.

(c) Show that $dU = T\,dS - p\,dV$.
(d) Combine (a), (b), and (c) to show that

$$dG = V\,dp - S\,dT$$

Problem 6.2

(a) Calculate the change in Gibbs free energy that takes place with the isothermal compression of water treated as an incompressible liquid at 298 K from 1.00 to 2.00 bars pressure.
(b) Calculate the change in Gibbs free energy that takes place with the isothermal compression of water treated as an ideal gas at 298 K from 1.00 to 2.00 bars pressure.
(c) Comment upon the difference.

Problem 6.3

The statement has been made: "Algebraic manipulation gives

$$dG = V\,dp - S\,dT\text{"}$$

Prove this statement.

Problem 6.4

A chemical reaction at 1.0 bar pressure has $\Delta G^{298}_{r,p=1} = -335\ \mathrm{kJ\ mol^{-1}}$. The Gibbs function of the reactant system changes by 7.5 kJ mol^{-1} when the pressure is changed from 1.0 bar to p_2. The product system changes by 8.4 kJ mol^{-1} over the same pressure change. What is $\Delta G^{298}_{r,p_2}$?

Problem 6.5

What is the entropy change of a Trouton's rule liquid at its boiling point?

Problem 6.6

The heat capacity of carbon disulfide CS_2(s) is 6.9 J K^{-1}mol^{-1} at 15.0 K. What is its standard entropy at 15.0 K, assuming that the solid is a perfect crystal.

Problem 6.7

Solid carbon disulfide shows the following experimental values of heat capacity as a function of temperature. Taking the Debye contribution into account (Problem 6.4), find the standard entropy of CS_2(s) at its normal melting point of 161 K.

T	C_p
15.0000	6.9000
20.0000	12.0000
29.8000	20.8000
42.2000	29.1000
57.5000	35.6000
75.5000	40.0000
89.4000	43.1000
99.0000	45.9000
108.9000	48.5000
119.9000	50.5000
131.5000	52.6000
145.0000	54.3000
157.0000	56.6000
161.0000	57.4000

Problem 6.8

The enthalpy of fusion of CS_2(s) to liquid carbon disulfide CS_2(l) is 4.38 kJ mol^{-1} Pitzer et al. 1961. What is the entropy of fusion of CS_2(s) at its normal melting point of 161 K?

Problem 6.9

The heat capacity at constant pressure of liquid CS_2(l) is nearly constant at 75.5 J K^{-1}mol^{-1}. What is the molar entropy increase of CS_2(l) from 161 K to 298 K?

Problem 6.10

Find the standard molar entropy of carbon disulfide at 298 K.

Problem 6.11

The Gibbs free energies of combustion of methane at 300K and 350 K are −815 and −802 kJ mol^{-1}. Find the enthalpy of combustion of methane at 325 K $\Delta_c H^{325}$(methane).

Problem 6.12

Solve the previous problem by setting up a pair of simultaneous equations and solving them.

7

EQUILIBRIUM

Chemists are often depicted in the popular media as sinister fellows who pour a solution from beaker A into beaker B with catastrophic results. In fact, we are neither more nor less sinister than the next person and we take elaborate precautions to avoid catastrophies. Nevertheless, the process of adding one or many components A of one kind to a system of another kind B to establish an equilibrium mixture *is* central to our art or science, and the theoretical examination of chemical equilibrium is the high point and culmination of classical chemical thermodynamics.

7.1 THE EQUILIBRIUM CONSTANT

Suppose that gases A and B are capable of equilibration

$$\mathrm{A(g)} \rightleftarrows \mathrm{B(g)}$$

and arbitrary amounts of A(g) and B(g) are introduced into a closed container at 298 K. Because the amounts are arbitrary, the quotient of the partial pressures $\mathbb{Q} = p_{\mathrm{B}}/p_{\mathrm{A}}$ in the container will not be the equilibrium constant K_{eq}. The chemical potentials G_{A} and G_{B} will probably not be in the standard state. Instead they will differ from G_{A}° and G_{B}°, by

$$G_{\mathrm{A}} = G_{\mathrm{A}}^{\circ} + RT \ln \frac{p_{\mathrm{A}}}{1.0}$$

Concise Physical Chemistry, by Donald W. Rogers

and

$$G_{\mathrm{B}} = G_{\mathrm{B}}^{\circ} + RT \ln \frac{p_{\mathrm{B}}}{1.0}$$

where the 1.0 in the denominators signify that the partial pressures are relative to the standard state of 1.0 bar. The difference between the chemical potentials of the reactant state and product state is

$$\Delta_r G = \sum G(\text{prod}) - \sum G(\text{react})$$

which in this simple case is

$$\Delta_r G = G_{\mathrm{B}} - G_{\mathrm{A}} = G_{\mathrm{B}}^{\circ} - G_{\mathrm{A}}^{\circ} + RT \ln \frac{p_{\mathrm{B}}}{1.0} - RT \ln \frac{p_{\mathrm{A}}}{1.0}$$

or

$$\Delta_r G = \Delta G^{\circ} + RT \ln \frac{p_{\mathrm{B}}}{p_{\mathrm{A}}}$$

As the reaction progresses, $\Delta_r G$ is not zero and G of the reacting system is not constant with time t, $(\partial G/\partial t)_{T,p} \neq 0$. When the chemical reaction has come to completion, the pressure quotient $\mathbb{Q} = p_{\mathrm{B}}/p_{\mathrm{A}}$ has arrived at a value such that $(\partial G/\partial t)_{T,p} = 0$, hence

$$\Delta_r G = \Delta G^{\circ} + RT \ln \frac{p_{\mathrm{B}}}{p_{\mathrm{A}}} = 0$$

The free energy change of the system has arrived at a Gibbs potential energy minimum. Under these and *only under these conditions*, we have $\Delta_r G = 0$, so that

$$\Delta G^{\circ} = -RT \ln \frac{p_{\mathrm{B}}}{p_{\mathrm{A}}} = -RT \ln K_{\mathrm{eq}}$$

The expression is frequently written in the equivalent form:

$$K_{\mathrm{eq}} = e^{-\Delta G^{\circ}/RT}$$

7.2 GENERAL FORMULATION

A more general formulation of the equilibrium expressions above is given by the reaction

$$a\mathrm{A} + b\mathrm{B} + \ldots = c\mathrm{C} + d\mathrm{D} + \ldots$$

where a, b, ... c, d, ... are the stoichiometric coefficients of the reaction. It is still true that

$$\Delta_r G = \sum G(\text{prod}) - \sum G(\text{react})$$

When equilibrium has been reached, we have

$$\Delta G^\circ = -RT \ln K_{\text{eq}}$$

Now the concentration quotient $\mathbb{Q}$ takes the form

$$\mathbb{Q} = \frac{[C]^c\,[D]^d \ldots}{[A]^a\,[B]^b \ldots}$$

leading to the familiar expression of the equilibrium constant as

$$K_{\text{eq}} = \frac{[C]^c\,[D]^d \ldots}{[A]^a\,[B]^b \ldots}$$

which is true only after the Gibbs free energy has come to a minimum and $\Delta_r G = 0$. The stoichiometric coefficients become exponents, and the square brackets [ ] indicate some kind of unitless concentration variable relative to a standard state. This notation is often used in solution chemistry to denote a concentration in moles/liter, where the standard state of the solute in the solvent is taken for granted.

As an example of a reaction in the gas phase, the expression

$$\Delta G^\circ = -RT \ln \frac{p^2_{NO_2}}{p_{N_2O_4}}$$

can be used to find the equilibrium constant for the reaction

$$N_2O_4(g) \rightleftarrows 2NO_2(g)$$

The standard state Gibbs chemical potential difference for this reaction is

$$\Delta_r G^\circ = \Delta G^\circ_{2(NO_2)} - \Delta G^\circ_{N_2O_4} = 2(51.31) - 97.89 = 4.73\ \text{kJ mol}^{-1}$$

The equilibrium constant is

$$K_{\text{eq}} = e^{-\Delta G^\circ/RT} = e^{-4730/8.314\times 298.15} = 0.148$$

which is in good agreement with the experimental value of 0.13.

It is difficult to obtain accurate K_{eq} values from calorimetric determination of $\Delta_r G^\circ$ (from $\Delta_r H^\circ$ and $\Delta_r S^\circ$) because of the exponential relationship $K_{\text{eq}} = e^{-\Delta G^\circ/RT}$. This mathematical form brings about a large error in K_{eq} when $\Delta_r G^\circ$ is in error by a small amount. To a certain degree, a "small" error or a "large" error is in the eye of the beholder; the terms are used in the literature as influenced by the "difficulty" of the experiment and how "good" previous measurements of the same kind have been. Clearly, these are subjective value judgments.

7.3 THE EXTENT OF REACTION

In a chemical reaction, the participant concentrations change, but they do not change independently, rather they are related by the stoichiometric coefficients of the reaction. For example, if I know that reactant A decreases by 0.1 mol in the reaction system

$$\mathrm{A + B = 2C}$$

I know that reactant B also decreases by 0.1 mol and product C increases by 0.2 mol. The infinitesimal changes in mole numbers dn_{A}, dn_{B}, and dn_{C} are linearly dependent. One frequently expresses the change in a reaction using the single variable ξ which, given the linear dependence of the reaction components, suffices to express all three in the specific case cited. In general, infinitesimal variations in the amount of any number of components dn_i can be expressed in terms of the *extent of reaction* ξ_i where

$$\xi_i = \frac{dn_i}{\nu_i}$$

and ν_i is the *stoichiometric coefficient*: -1, -1, and 2 in the specific example cited. Note that the stoichiometric coefficients of the two reactants are negative because their concentrations decrease as the reaction proceeds. This is a convention of course, because the reaction could equally well have been written

$$\mathrm{2C = A + B}$$

whereupon the signs of ξ_i would change.

Up to this point, we have stressed thermodynamic state functions for 1 mol of a pure substance, for example, the molar energy $U = f(V, T)$. When we express the energy of any mixture, specifically a mixture of reactants and products in a reacting system, the composition of the system in terms of mole numbers $n_i \neq 1.0$ influences the energy and other state functions. We can express this dependence in terms of the extent of reaction:

$$U = f(V, T, \xi_i)$$

This dependence means that along with V and T, ξ_i is a full-fledged degree of freedom and we can write the exact differential

$$dU = \left(\frac{\partial U}{\partial V}\right)_{T,\xi_i} dV + \left(\frac{\partial U}{\partial T}\right)_{V,\xi_i} dT + \left(\frac{\partial U}{\partial \xi_i}\right)_{T,V} d\xi_i + \ldots$$

where i extends over all degrees of freedom.

Notice that this implies a hyperspace in U, V, T, and ξ_i. Energy has been selected here to illustrate the principle that analogous equations exist for the other state variables—for example, G, H, and S.

7.4 FUGACITY AND ACTIVITY

For nonideal systems the concentration variable is replaced by a new variable that expresses the *effective* concentration of the species in a mixture. For example, a solute may be more chemically active in methanol solution than it is in water. Or it may be more active in water solution than in methanol. A pure gas may behave in a nonideal way, and its degree of nonideality may be influenced by other gases in a mixture. These deviations from ideal behavior are expressed by a *coefficient* γ which yields the *activity* of a solute or *fugacity* of a gas when multiplied into the concentration variable, for example,

$$a_A = \gamma_A [A] \qquad \text{or} \qquad f_A = \gamma_A p_A$$

The activity and fugacity coefficients are simply numbers telling us whether the behavior of the species is greater or less than it would be in the standard state. They are concentration- or pressure-dependent and are usually determined for real systems by rather painstaking empirical methods.

7.5 VARIATION OF THE EQUILIBRIUM CONSTANT WITH TEMPERATURE

Combining the Gibbs–Helmholtz equation for the temperature variation of free energy with the equation connecting the free energy in the standard state to the equilibrium constant gives

$$\left[\frac{\partial\left(\frac{\Delta G^\circ}{T}\right)}{\partial\left(\frac{1}{T}\right)}\right]_p = \left[\frac{\partial\left(\frac{-RT\ln K_{eq}}{T}\right)}{\partial\left(\frac{1}{T}\right)}\right]_p = -\left[\frac{\partial\left(R\ln K_{eq}\right)}{\partial\left(\frac{1}{T}\right)}\right]_p$$

but

$$\left[\frac{\partial\left(\frac{\Delta G^\circ}{T}\right)}{\partial\left(\frac{1}{T}\right)}\right]_p = \Delta H^\circ$$

so

$$\left[\frac{\partial\left(R \ln K_{\mathrm{eq}}\right)}{\partial\left(\frac{1}{T}\right)}\right]_p = -\Delta_r H^\circ$$

where $\Delta_r H^\circ$ is the standard enthalpy of reaction. Now, $dT^{-1}/dT = -T^{-2}$ and $d(1/T) = -T^{-2}dT$, so

$$d\left(R \ln K_{\mathrm{eq}}\right) = -\Delta H^\circ d\left(\frac{1}{T}\right) = \frac{\Delta H^\circ}{T^2}\, dT$$

This equation can be integrated as

$$\int d \ln K_{\mathrm{eq}} = -\frac{\Delta_r H^\circ}{R} \int \frac{1}{T^2}\, dT$$

to give the *van't Hoff equation*

$$\ln K_{\mathrm{eq}} = -\frac{\Delta_r H^\circ}{R}\left(-\frac{1}{T}\right) + \text{const.} = \frac{\Delta_r H^\circ}{R}\left(\frac{1}{T}\right) + \text{const.}$$

This is the equation of a straight line of $\ln K_{\mathrm{eq}}$ plotted against $(1/T)$ with $\Delta_r H^\circ/R$ as the slope. It applies so long as $\Delta_r H^\circ$ remains constant. The equation as written implies that the slope will be positive, but this does not follow because $\Delta_r H^\circ$ may be positive, negative, or zero for endothermic, exothermic, or null-thermal reactions. Common examples are melting, freezing, or mixing of two isomers of a liquid alkane which are endothermic, exothermic, and null-thermal.

The van't Hoff equation can be integrated between limits to give an expression for a new equilibrium constant at a new temperature from known values of $\Delta_r H^\circ$, K_{eq}, and the initial and revised temperatures, T and T':

$$\int_{K_{\mathrm{eq}}}^{K'_{\mathrm{eq}}} d \ln K_{\mathrm{eq}} = -\frac{\Delta_r H^\circ}{R} \int_T^{T'} \frac{1}{T^2} dT$$

$$\ln \frac{K'_{\mathrm{eq}}}{K_{\mathrm{eq}}} = -\frac{\Delta_r H^\circ}{R}\left(\frac{1}{T'} - \frac{1}{T}\right) = \frac{\Delta_r H^\circ}{R}\left(\frac{1}{T} - \frac{1}{T'}\right)$$

Conversely, the integrated form is a way of determining $\Delta_r H^\circ$ from two experimental determinations of K_{eq} at different temperatures.

7.5.1 Le Chatelier's Principle

Le Chatelier's principle states that, in a stressed chemical reaction, the equilibrium will be displaced in such a way as to relieve the stress. First, we need to define "stress." An exothermic reaction

$$A \rightleftarrows B + q$$

where q is heat given off, will be stressed backward by application of heat q—that is, by a temperature rise. Application of heat will drive the reaction backward to give more A and less B. The opposite will be true for an endothermic reaction. Looking at the integrated van't Hoff equation

$$\ln \frac{K'_{eq}}{K_{eq}} = \frac{\Delta_r H^\circ}{R} \left(\frac{1}{T} - \frac{1}{T'} \right)$$

we see that for a temperature rise $T' > T$ we have $(1/T) > (1/T')$. For $\Delta_r H^\circ < 0$, $\ln (K'_{eq}/K_{eq}) < 0$ and the new equilibrium constant K'_{eq} is smaller than the original one, K_{eq}. The thermal condition $\Delta_r H^\circ < 0$ is the characteristic of an exothermic reaction, so Le Chatelier's principle applied to the heat of reaction agrees with the van't Hoff equation. Le Chatelier's principle is a qualitative statement giving the sign but not the magnitude of the effect of a temperature change on the equilibrium constant.

7.5.2 Entropy from the van't Hoff Equation

If $\Delta_r H^\circ$ is independent of the temperature, then from $d\Delta_r H^\circ = \Delta_r C_p \, dT = 0$. Since dT is not zero, $\Delta_r C_p$ must be zero. From

$$\Delta_r C_p = \sum C_p(\text{prod}) - \sum C_p(\text{react})$$

the heat capacity of the reactant system must be equal to that of the product system. In practical terms, we take this equality to be approximately true over short temperature intervals. We have already found the equations

$$\Delta_r G^\circ = -RT \ln K_{eq}$$

and

$$\Delta_r G^\circ = \Delta_r H^\circ - T \Delta_r S^\circ$$

so

$$-RT \ln K_{eq} = \Delta_r H^\circ - T \Delta_r S^\circ$$

and

$$\ln K_{\text{eq}} = \frac{-\Delta_r H^\circ}{R}\left(\frac{1}{T}\right) + \frac{\Delta_r S^\circ}{R}$$

This is a "slope–intercept" problem for the linear function $\ln K_{\text{eq}}$ vs. $1/T$, where the slope of the plot is $-\Delta_r H^\circ/R$ and the intercept is $\Delta_r S^\circ/R$. Two or more equilibrium measurements at different temperatures yield both the enthalpy change of the reaction and the entropy change. The resulting thermodynamic functions are only as good as the input data, and this is a very sensitive experimental problem—especially in the $\Delta_r S^\circ$ determination, which depends on what may be a long and mathematically questionable extrapolation.

7.6 COMPUTATIONAL THERMOCHEMISTRY

Within a year of the discovery of quantum mechanics by Schrödinger and Heisenberg, Heitler and London had demonstrated that the energy of the H—H chemical bond can be approximated by a quantum mechanical calculation. The reason this calculation is *always approximate* is a reflection of one of the most profound laws of nature, Heisenberg's uncertainty principle. We are unable to know exactly where the moving electrons are. Hence their Coulombic potential energy, total energy, entropy, and Gibbs free energy are inaccessible to us in even the simplest chemical bond H—H.

Nevertheless, a diverse array of approximate computational methods exists and the results can be ranked in order of the quality of their approximations. Ranking enables us to decide which methods are best for a given job, and it points the way toward improved methods. Usually the quality of an approximate calculation is measured by how well the calculated value matches well-established experimental results. Recently, however, computed values have become so reliable that they can be used to pick out and discard errors in the tabulated experimental results. The present trend is from using computed results to approximate experimental results to using them to *supplant* or *replace* experimental results, especially for reactions that occur within hostile experimental environments (flames or explosions) or reactions that have fleeting intermediates (free radicals). Quantum mechanical methods produce results for molecular energy, enthalpy, entropy, Gibbs free energy, and equilibrium constants, but they are circumscribed by limits on molecular size and they are of varying reliability.

7.7 CHEMICAL POTENTIAL: NONIDEAL SYSTEMS

Given that the effective pressure fraction of a gas in a nonideal mixture of gases is its fugacity f and that the effective concentration fraction of a solute in a mixture is given by its activity a, equilibrium constants in simple A, B nonideal systems can be written

$$K_{eq} = \frac{f_B}{f_A} \qquad \text{or} \qquad K_{eq} = \frac{a_B}{a_A}$$

The Gibbs free energy of a *pure* substance is a function of the temperature and pressure only, $G = g(T, p)$, but the free energy of a many-component system is a function of T, p, and the amounts of each component n_i, $G = g(T, p, n_i)$. The total change in free energy for a mixture is

$$dG = \left(\frac{dG}{dT}\right)_{p,n_i} dT + \left(\frac{dG}{dp}\right)_{T,n_i} dp + \sum_i \left(\frac{dG}{dn_i}\right)_{p,T} dn_i$$

It is the last term that we are interested in. At constant T and p, we have

$$dG = \sum_i \left(\frac{dG}{dn_i}\right)_{p,T} dn_i$$

where the n_j are constant ($j \neq i$). In the simple illustrative reaction mixture,

$$\mathrm{A(g)} \rightleftarrows \mathrm{B(g)}$$

in the ideal case,

$$\Delta G = \Delta G^\circ + RT \ln \mathbb{Q} = \Delta G^\circ + RT \ln \frac{p_B}{p_A}$$

where the p values are partial pressures. In the nonideal case the form of the pressure quotient $\mathbb{Q}$ still holds, but the partial pressures are replaced by the fugacities f. The change in the Gibbs free energy is also given a different symbol $\Delta\mu$ to denote the nonideal case:

$$\Delta\mu = \Delta\mu^\circ + RT \ln \mathbb{Q} = \Delta\mu^\circ + RT \ln \frac{f_B}{f_A}$$

Evidently,

$$\Delta\mu = \mu_B - \mu_A$$

and

$$\Delta\mu^\circ = \mu_B^\circ - \mu_A^\circ$$

Although the Gibbs chemical potential applies to the many-component nonideal case as well as the ideal case, it is mathematically rigorous and it is a thermodynamic function.

From these definitions of the Gibbs chemical potentials, it follows that

$$\mu_{\mathrm{A}} = \mu_{\mathrm{A}}^{\circ} + RT \ln f_{\mathrm{A}}$$

and

$$\mu_{\mathrm{B}} = \mu_{\mathrm{B}}^{\circ} + RT \ln f_{\mathrm{B}}$$

In the general case, Gibbs chemical potentials are

$$\mu_{\mathrm{i}} = \mu_{\mathrm{i}}^{\circ} + RT \ln f_{\mathrm{i}}$$

for an indefinite number of fugacities.

Analogous definitions for the general case lead to

$$\mu_{\mathrm{i}} = \mu_{\mathrm{i}}^{\circ} + RT \ln a_{\mathrm{i}}$$

for activities a_{i}, with

$$\mathbb{Q} = \frac{[C]^c\,[D]^d \ldots}{[A]^a\,[B]^b \ldots} = \frac{\prod_i [X]^i}{\prod_j [X]^j}$$

where the $[X]^i$ and $[X]^j$ are relative concentration variables, pressure fractions, mole fractions, mass fractions, and so on, symbolized by either f or a and taken to the appropriate stoichiometric coefficient.

7.8 FREE ENERGY AND EQUILIBRIA IN BIOCHEMICAL SYSTEMS

Reactions of biochemical interest do not normally occur in the gas phase. Rather they occur in solution, usually saline solution. Therefore the correct expression of the free energy changes of reaction and the equilibrium constant is in terms of the corresponding activities and changes in chemical potential. Determination of activities and activity coefficients over a concentration range on nonideal solutions is not a simple matter, nor is it necessary for *in vitro* studies of biochemical reactions. Instead, concentrations are used with the stipulation that the "background" conditions must be constant over the course of the study and must be reproduced from one study to the next. For example, energy studies on dephosphorylation of adenosine

5′-triphosphate would be carried out at a specified and constant temperature, pressure, pH, pMg, and ionic strength where pH and pMg refer to the ion concentrations $pH = -\log\left[H^+\right]$ and $pMg = -\log\left[Mg^{2+}\right]$, and the ionic strength is constant for all dissolved salts in the reaction solution. Ionic strength is essentially the ionic charge concentration in solution calculated as the sum $\frac{1}{2}\sum c_i z_i^2$ over all salt concentrations c_i that dissolve to give ions of charge z_i.

These conditions specify a unique standard state which is not the thermodynamic standard state but which is adhered to throughout the experiment and experiments with which results will be combined or compared. Under these controls, it is proper to write

$$G_A = G_A^\circ + RT \ln [A]$$

for G_A° in the specified standard state with a concentration of reactant A, with similar expressions for B, C, ... The change in free energy for a reaction quotient $\mathbb{Q}$, now in terms of initial concentrations, is

$$\Delta G = \Delta G^\circ + RT \ln \mathbb{Q} = \Delta G^\circ + RT \ln \frac{[C]^\xi [D]^\xi \ldots}{[A]^\xi [B]^\xi \ldots}$$

Different stoichiometric coefficients are all expressed using the same symbol ξ for simplicity. It is understood that $\mathbb{Q}$ is a concentration quotient, which might fortuitously be equal to the equilibrium constant but in general will not. In general, ΔG will be different from zero but when the reaction has arrived at equilibrium, the concentrations will have arrived at K_{eq} and ΔG will have arrived at zero so that

$$\Delta G^\circ = -RT \ln K_{eq}$$

Thus the *form* of the free energy relation to the equilibrium constant is reproduced but only under rigorously controlled background conditions.[1] Change the ionic strength or the pH, for example, and you can expect to find a different ΔG° and K_{eq}.

How can ΔG° change when we usually think of it as a sum of rock-firm G° values? By changing our background conditions, we have changed the standard state, the benchmark to which we refer all G° values.

7.8.1 Making ATP, the Cell's Power Supply

The difference between $\Delta_r G$ and ΔG° in the metabolic degradation of glucose, a physiological energy source, to lactate and ATP can be broken down into two individual steps (Hammes, 2007), one taking up 2 mol of ATP and producing 2 mol of the diphosphate ADP

$$\text{Glucose} + 2\ \text{ATP} \rightarrow 2\ \text{ADP} + 2\ \text{glyceraldehyde-3-phosphate}$$

[1]See Treptow (1996).

and the other producing 4 mol of ATP

2 Glyceraldehyde-3-phosphate + 2 phosphate + 4 ADP → 2 lactate + 4 ATP + 2 H_2O

The summed reaction is

Glucose + 2 phosphate + 2 ADP → 2 lactate + 2 ATP + 2 H_2O

for a net gain of 2 mol of ATP. The first of these two reactions is unfavorable from the point of view of a positive $\Delta G^\circ = 2.2\,\text{kJ mol}^{-1}$ but at the initial concentrations chosen to be at or near physiological concentrations,

$$\mathbb{Q} = \frac{(0.14)^2\,(0.019)^2}{5.0\,(1.85)^2} = 0.0196\left(3.61 \times 10^{-4}\right) \approx 4 \times 10^{-7} \ll 1$$

This very small reaction quotient drives the metabolic conversion of ADP to ATP which then powers other reactions within the biological system as a whole.

PROBLEMS AND EXAMPLES

Example 7.1 Solution Calorimetry

A half-liter solution calorimeter system consisted of the calorimeter itself, a temperature measuring circuit, an electrical heating circuit, and calorimeter fluid. The system was prepared with a solution of magnesium ion at 0.001 M. The ionic strength was brought to 0.25 M with KCl, a neutral electrolyte, and the pH was adjusted to 7.0. When the heating circuit was activated at a current flow of I amperes for a time t, an amount of heat energy $q_p = 96.5$ J was delivered through a heating resistor R ($q = I^2Rt$) producing a temperature rise of $\Delta T = 0.166$ K.[2] What was the water equivalent (calorimeter constant) of the calorimeter system?

Solution to Example 7.1 The water equivalent of a calorimeter is the heat capacity of the entire system as if it were all water even though it consists of various parts made of various materials and contains a solution of reactants and products different from pure water. The water equivalent can be found by a straightforward heat capacity calculation even though we know that the heat capacity of the system is the sum of many parts. We calculate

$$C_p = \frac{dq_p}{dT} \simeq \frac{q_p}{\Delta T} = \frac{95.6}{0.166} = 576\ \text{J K}^{-1} = 0.576\ \text{kJ K}^{-1}$$

[2] 1 amp volt sec = 1 J and E in volts = IR.

Example 7.2 Adenosine 5′-triphosphate ATP

The hydrolysis of adenosine 5′-triphosphate ATP to adenosine 5′-diphosphate ADP plus an inorganic phosphate ion can be written

$$\mathrm{ATP} + \mathrm{H_2O} \rightarrow \mathrm{ADP} + \text{phosphate}$$

If 10 mL of ATP with concentration $[\mathrm{ATP}] = 0.200\,\mathrm{mol\,L^{-1}}$ are pipetted into a calorimeter with a heat capacity of $0.576\,\mathrm{kJ\,K^{-1}}$ (Example 7.1) and a temperature rise of 0.107 K is found, what is the enthalpy of hydrolysis of adenosine 5′-triphosphate to adenosine 5′-diphosphate plus an inorganic phosphate ion under these conditions? What is the sign of $\Delta_r H$?

Solution to Example 7.2

$$0.107\,\mathrm{K}\left(0.576\,\mathrm{kJ\,K^{-1}}\right) = 0.0616\,\mathrm{kJ}$$

$$10.0\,\mathrm{mL} = 0.0100\,\mathrm{L}$$

$$0.0100\,\mathrm{L}(0.200\,\mathrm{mol\,L^{-1}}) = 0.00200\,\mathrm{mol\ ATP}$$

$$\frac{0.0616\,\mathrm{kJ}}{0.00200\,\mathrm{mol}} = -30.8\,\mathrm{kJ\,mol^{-1}}$$

The sign of $\Delta_r H$ is negative because the reaction is exothermic. Heat flowing out of the reacting system at constant external pressure means that its enthalpy balance must decrease.

Problem 7.1

What is the change in the entropy S for one mole of a pure substance for an infinitesimal change in T and p? Given: Entropy is an exact differential

$$dS(T, p) = \left(\frac{\partial S}{\partial T}\right)_p dT + \left(\frac{\partial S}{\partial p}\right)_T dp$$

Problem 7.2

In an experiment on the nitrogen tetroxide reaction at 298 K

$$\mathrm{N_2O_4(g)} \rightleftarrows 2\mathrm{NO_2(g)}$$

pure N_2O_4(g) was introduced into a reaction vessel maintained at constant temperature and a total pressure of 2.500 bar. When equilibrium had been reached, the partial pressure of N_2O_4(g) had dropped to 1.975 bar. What is K_{eq} for this reaction? Compare your answer to the calculated value (Section 7.2). What are the units of K_{eq}?

Problem 7.3

Supposing that a simple A(g) $\rightleftarrows$ B(g) reaction has an equilibrium constant $K_{eq} = 0.100$ at $T = 200$ K and $K'_{eq} = 0.200$ at $T = 300$ K. What is $\Delta_r H^\circ$ for the reaction?

Problem 7.4

For the reaction

$$Br_2(g) \rightarrow 2Br(g)$$

the equilibrium constant has been measured and found to be 0.410 at 1125 K and 1.40 at 1175 K. What is K_{eq} at 1225 K?

Problem 7.5

Data quoted in Lewis and Randall (1961) for the formation reaction of H_2S

$$H_2(g) + \tfrac{1}{2}S_2(g) = H_2S(g)$$

include the values given in Table 7.1.

TABLE 7.1 *T*, 1/*T*, *K*, and ln *K* for formation of H_2S from its elements.

T	1023	1218	1405	1667
$1/T$	9.775×10^{-4}	8.210×10^{-4}	7.117×10^{-4}	5.999×10^{-4}
K	105.9	20.18	6.209	1.807
$\ln K$	4.662	3.005	1.826	0.5917

What is $\Delta_r H^\circ$ for this reaction? What is $\Delta_r S^\circ$?

Problem 7.6

A reaction in the gas phase has $K_{eq} = 10$ at 298 K and 10^{-1} at 500 K. Is the reaction endothermic or exothermic? Explain this in terms of Le Chatelier's principle.

Problem 7.7

An experiment was set up so that conditions (pH, pMg$^+$, ionic strength, etc.) were identical to those described in Exercises 7.1 and 7.2, except that, in this experiment, adenosine 5′-diphosphate ADP was substituted for ATP. The hydrolysis reaction producing adenosine 5′-monophosphate AMP

$$\text{ADP} + \text{H}_2\text{O} \rightarrow \text{ADP} + \text{phosphate}$$

resulted in a temperature change of $\Delta T = 0.100$ K. Find ΔH° for the reaction

$$2\text{ADP} \rightarrow \text{AMP} + \text{ATP}$$

(The symbol ΔH° is used to indicate a standard enthalpy change relative to a defined concentration under specific solution conditions.)

8

A STATISTICAL APPROACH TO THERMODYNAMICS

In the late nineteenth century, Ludwig Boltzmann made the connection between Maxwell's statistical-atomic equations and the deterministic equations of chemical thermodynamics, which were only emerging at the time of his work. (Gibbs was not widely read in Europe at that time.) A central concept in *statistical thermodynamics*, as we now call the new science, is the *partition function*. We shall see the relation between the partition function and the thermodynamic properties including the Gibbs free energy and the equilibrium constant. Actual calculation of partition functions falls anywhere within the range of easy to impossible. We shall calculate some of the easy ones and approximate some of the others.

8.1 EQUILIBRIUM

If two very simple gaseous systems, A and B, are in equilibrium and each system has only one energy level as shown in Fig. 8.1, the equilibrium constant is $K_{eq} = n_B/n_A = 3/5 = 0.600$. Knowing K_{eq}, we can calculate the energy separation between levels A and B from the *Boltzmann equation*:

$$K_{eq} = e^{-(\varepsilon_B - \varepsilon_A)/k_B T}$$

For example, at 298 K, in Fig. 8.1, $K_{eq} = 0.600$ leads to $(\varepsilon_B - \varepsilon_A) = 2.10 \times 10^{-21}$ J.

Concise Physical Chemistry, by Donald W. Rogers

$$A(g) \rightleftarrows B(g)$$

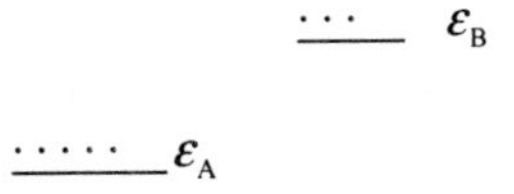

... ε_B

..... ε_A

FIGURE 8.1 A two-level equilibrium. There are 5 molecules in A and 3 molecules in B.

Conversely, knowing the energy separation, one can calculate the equilibrium constant. If the energy of B is *lower* than the energy of A by 2.10×10^{-21} J, level B will contain more molecules at equilibrium than will level A and K_{eq} will be larger than 1.0 (Fig. 8.2).

$$K_{eq} = \frac{n_B}{n_A} = e^{-(\varepsilon_B - \varepsilon_A)/k_B T} = \frac{5}{3} = 1.67$$

In these simple models, the equilibrium constants are inverses and $(\varepsilon_B - \varepsilon_A)$ of the first equilibrium is equal but opposite in sign to $(\varepsilon_B - \varepsilon_A)$ of the second reaction. This analysis gives a description of the relationship between energy and the equilibrium constant that is useful as far as it goes, but it has left out an important consideration: the influence of entropy.

8.2 DEGENERACY AND EQUILIBRIUM

If the upper level is split into two levels, each of which has the same energy, its capacity to accommodate molecules of kind B is doubled (Fig. 8.3). The equilibrium constant $K_{eq} = \frac{6}{5}$ has been made greater than 1.0. The reaction now favors products rather than reactants even though the energy change is uphill. The second law has come into play; the product state is more disordered. There are more places for the molecule to go. It is like having two stacks of paper on your desk rather than one. There are now two places where that important piece of paper you are looking for may be, whereas before there was only one. Increased disorder means increased entropy.

...

.....

A B

FIGURE 8.2 A two-level equilibrium.

FIGURE 8.3 A degenerate two-level equilibrium.

Even if the two levels of state B are not exactly at the same energy, their capacity to accommodate B molecules is greater than it would be if there were only one state as in the original equilibrium. Suppose now that both state A and state B consist of many levels. The distribution of molecules in state A will be controlled by one Boltzmann factor and the distribution of molecules in state B will be controlled by another Boltzmann factor, but the distribution between A and B will also be controlled by a Boltzmann factor. We now have three factors controlling the equilibrium: the distribution within A, the distribution within B, and the distribution between A and B. The distribution within the levels of state A is

$$\frac{n_{A_i}}{n_{A_0}} = e^{-(\varepsilon_{A_i} - \varepsilon_{A_0})/k_B T}$$

The distribution within state B is

$$\frac{n_{B_i}}{n_{B_0}} = e^{-(\varepsilon_{B_i} - \varepsilon_{B_0})/k_B T}$$

The distribution between the lowest levels in states A and B is

$$\frac{n_{B_0}}{n_{A_0}} = e^{-(\varepsilon_{B_0} - \varepsilon_{A_0})/k_B T}$$

The total number of molecules in state A is the summation over all the states in A

$$n_A = \sum_i n_{A_i} = \sum_i n_{A_0} e^{-(\varepsilon_{A_i} - \varepsilon_{A_0})/k_B T}$$

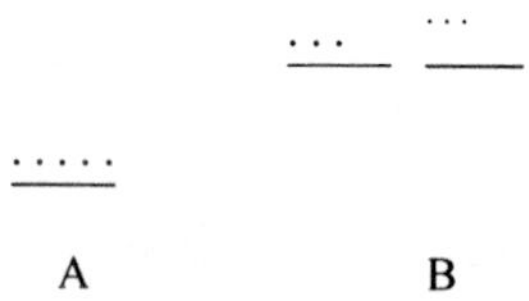

FIGURE 8.4 A degenerate two-level equilibrium. The two energies at the B level are not exactly the same.

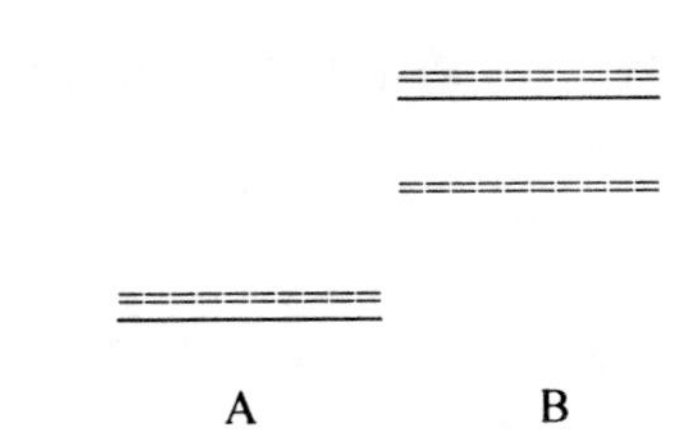

FIGURE 8.5 A two-level equilibrium with many A and many B levels.

and the total number in state B is given by a similar summation

$$n_{\mathrm{B}} = \sum_i n_{\mathrm{B}_i} = \sum_i n_{\mathrm{B}_0} e^{-(\varepsilon_{\mathrm{B}_i} - \varepsilon_{\mathrm{B}_0})/k_{\mathrm{B}}T}$$

The equilibrium constant can be written as the quotient of the number of molecules in B over the number of molecules in A:

$$K_{\mathrm{eq}} = \frac{n_{\mathrm{B}}}{n_{\mathrm{A}}} = \frac{\sum_i n_{\mathrm{B}_0} e^{-(\varepsilon_{\mathrm{B}_i} - \varepsilon_{\mathrm{B}_0})/k_{\mathrm{B}}T}}{\sum_i n_{\mathrm{A}_0} e^{-(\varepsilon_{\mathrm{A}_i} - \varepsilon_{\mathrm{A}_0})/kT}}$$

Factoring out $n_{\mathrm{B}_0}/n_{\mathrm{A}_0} = e^{-(\varepsilon_{\mathrm{B}_0} - \varepsilon_{\mathrm{A}_0})/k_{\mathrm{B}}T}$, which is common to each term in the sums, we obtain

$$K_{\mathrm{eq}} = \frac{n_{\mathrm{B}_0}}{n_{\mathrm{A}_0}} \frac{\sum_i e^{-(\varepsilon_{\mathrm{B}_i} - \varepsilon_{\mathrm{B}_0})/k_{\mathrm{B}}T}}{\sum_i e^{-(\varepsilon_{\mathrm{A}_i} - \varepsilon_{\mathrm{A}_0})/k_{\mathrm{B}}T}} = e^{-(\varepsilon_{\mathrm{B}_0} - \varepsilon_{\mathrm{A}_0})/k_{\mathrm{B}}T} \frac{\sum_i e^{-(\varepsilon_{\mathrm{B}_i} - \varepsilon_{\mathrm{B}_0})/k_{\mathrm{B}}T}}{\sum_i e^{-(\varepsilon_{\mathrm{A}_i} - \varepsilon_{\mathrm{A}_0})/k_{\mathrm{B}}T}}$$

which is to say

$$K_{\mathrm{eq}} = e^{-(\varepsilon_{\mathrm{B}_0} - \varepsilon_{\mathrm{A}_0})/k_{\mathrm{B}}T} \frac{Q_B}{Q_A}$$

where Q_A and Q_B are called the *partition functions* for the A and B states. Please do not confuse the partition function Q with the concentration quotients Q or $\mathbb{Q}$.

8.3 GIBBS FREE ENERGY AND THE PARTITION FUNCTION

Taking logarithms of the previous equation, we obtain

$$\ln K_{\text{eq}} = -\frac{\Delta\varepsilon_0}{RT} + \ln\frac{Q_B}{Q_A}$$

where $\Delta\varepsilon_0 = \varepsilon_{\text{B}_0} - \varepsilon_{\text{A}_0}$. Multiplying by $-RT$, we get

$$-RT\ln K_{\text{eq}} = \Delta\varepsilon_0 - RT\ln\frac{Q_B}{Q_A}$$

but $-RT\ln K_{\text{eq}} = \Delta G^\circ$, so

$$\Delta G^\circ = \Delta\varepsilon_0 - RT\ln\frac{Q_B}{Q_A}$$

Comparing this equation with the definition of the Gibbs chemical potential $\Delta G^\circ \equiv \Delta H^\circ - T\Delta S^\circ$, we see that the classical enthalpy change ΔH° for a reaction is largely controlled by the energy separating the ground states $\Delta\varepsilon_0$, and the entropy change $T\Delta S^\circ$ is largely controlled by the ratio of the multiplicity of levels available in each state $RT\ln(Q_B/Q_A)$. In very many cases, the enthalpy change is dominant in chemical reactions, which explains why some very good nineteenth-century scientists thought that ΔH° was the only factor controlling equilibrium and reaction spontaneity.

Given that reactions and physical processes exist that are not solely controlled or dominated by the enthalpy change, it is the ratio of partition functions that interests us. The partition function gives us the distribution of molecules over the energy levels within a state. We may think of these as *microstates* within a state. If the enthalpy balance is close, the reaction tends to go in the direction with the most microstates. Such a reaction or process is maximizing the number of choices a particle (molecule) has in which to reside within a state. A system providing a maximum number of choices provides the maximum disorder, just as the number of places you *may* find that piece of paper you are looking for is determined by the disorder of the desktop. A disordered desktop has more microstates than an orderly one.

Looking again at the equation $\Delta G^\circ = \Delta\varepsilon_0 - RT\ln(Q_B/Q_A)$, we can see another reason why thermodynamicists prior to Gibbs could have missed the importance of entropy in equilibrium. The second term includes T, so it may be small at ordinary temperatures. The ratio of microstates $Q_{\text{B}}/Q_{\text{A}}$ may be negligible at lower temperatures, but it becomes important at higher temperatures and it may favor or disfavor the product side of the equilibrium according to which is larger, Q_{A} or Q_{B}. As is true in classical thermodynamics, there are four possibilities for the two terms in $\Delta G^\circ = \Delta\varepsilon_0 - RT\ln(Q_B/Q_A)$: They may be FF, FD, DF, or DD, where F denotes a term that favors the product side of the equilibrium and D denotes a term that disfavors the product side.

8.4 ENTROPY AND PROBABILITY

If you drop marbles randomly into a large box with 75 compartments painted green and 25 compartments painted red, the ratio of occupation numbers of marbles in compartments tends toward 75/25 = 3/1 over very many trials. If you put all the marbles into the green compartments and shake the box, the distribution of marbles in compartments will approach 3/1, as before, even though the potential energy of the boxes in the gravitational field is the same (for a level box). The ratio of the number of red compartments to the number of green compartments is the ratio of microstates for the system. It is independent of the size of the system; 7500/2500 is the same as 75/25. Call this ratio W. Boltzmann made the connection between statistics (probability ratio) and thermodynamics by defining the proportionality

$$S \propto \ln W$$

but this is just the way we found the A, B equilibrium constant except that it doesn't contain the temperature T. *You* supply the "thermal energy" by shaking the box. If you don't shake the box, it does not approach equilibrium (no reaction at 0 K).

The universal constant of proportionality between S and $\ln W$ is now designated k_B and called Boltzmann's constant in honor of this great and tragic figure[1]:

$$S \equiv k_B \ln W$$
$$k_B = 1.38066 \times 10^{-23}\ \mathrm{J\,K^{-1}}$$

8.5 THE THERMODYNAMIC FUNCTIONS

If the number of particles at each energy level

$$n_i = \frac{n e^{-\varepsilon_i / k_B T}}{Q}$$

is multiplied by the energy at that level ε_i, then the total energy is the sum of the individual contributions

$$E = \sum_i \varepsilon_i n_i = \frac{n \sum_i e^{-\varepsilon_i / k_B T}}{Q}$$

Having the energy and entropy, we have all that is necessary to derive the rest of the thermodynamic functions in terms of the partition function Q. Irikura (1998) has given the necessary equations in compact form, which we present as Table 8.1.

[1] Suicide, 1906.

TABLE 8.1 Thermodynamic Functions (Irikura, 1998).

$S = Nk_B\left[\frac{\partial}{\partial T}(T\ln Q) - \ln N + 1\right]$
$C_V = Nk_B T\frac{\partial^2}{\partial T^2}(T\ln Q)$
$C_p = C_V + R$
$H(t) - H(0) = \int_0^T C_p dT = \frac{RT^2}{Q}\frac{\partial Q}{\partial T} + RT$
$\frac{\partial}{\partial T}(T\ln Q) = \ln Q + \frac{T}{Q}\frac{\partial Q}{\partial T}$
$\frac{\partial^2}{\partial T^2}(T\ln Q) = \frac{2}{Q}\frac{\partial Q}{\partial T} + \frac{T}{Q}\frac{\partial^2 Q}{\partial T^2} - \frac{T}{Q^2}\left(\frac{\partial Q}{\partial T}\right)^2$
$\frac{\partial Q}{\partial T} = \frac{1}{k_B T^2}\sum_i \varepsilon_i e^{-\varepsilon_i/k_B T}$
$\frac{\partial^2 Q}{\partial T^2} = \frac{-2}{T}\frac{\partial Q}{\partial T} + \frac{1}{k_B^2 T^4}\sum_i \varepsilon_i^2 e^{-\varepsilon_i/k_B T}$

The essential thermodynamic functions are expressed above the line in Table 8.1, and the derivatives are given below the line for somewhat simpler and more practical evaluation. Along with the equations in Table 8.1, "excess" free energy and enthalpy functions are defined as $(G^\circ - H_0)/T$ and $(H^\circ - H_0)/T$. The excess thermodynamic property is the amount of that quantity above the value per degree K the system would have if all its parts were in the lowest possible energy state. Excess free energy and enthalpy functions vary in a gradual way over a wide range of temperature T. This enables one to make accurate interpolations and to extrapolate from data one has to get information one does not have.

Now we need to find out how to calculate the partition function.

8.6 THE PARTITION FUNCTION OF A SIMPLE SYSTEM

Spectroscopy gives information as to the intervals between energy levels in molecules, hence it is a gateway to molecular partition functions. As a simple example, consider a noninteracting[2] system of chemical bonds treated as harmonic oscillators (Chapter 18). A harmonic oscillator has the peculiar property that its energy levels are

[2]Strictly, the system is *weakly interacting* because particle transfer from one level to another is a necessary condition of the statistical treatment.

equally spaced. Because of this, the sum, written without degeneracy,

$$Q_{\text{vib}} = \sum_i e^{-\varepsilon_i/k_B T}, \qquad i = 0, 1, 2, \ldots$$

can be represented by

$$Q_{\text{vib}} = 1 + e^{-\varepsilon/k_B T} + e^{-2\varepsilon/k_B T} + e^{-3\varepsilon/k_B T} + \cdots$$

where ε is the size of equal steps up an energy ladder. Multiplying both sides of this equation by $e^{-\varepsilon/k_B T}$, we get

$$e^{-\varepsilon/k_B T} Q_{\text{vib}} = e^{-\varepsilon/k_B T} + e^{-2\varepsilon/k_B T} + e^{-3\varepsilon/k_B T} + \cdots$$

but this sum is the same as the one above it except that it lacks the first term. The difference between the two sums is just unity:

$$Q_{\text{vib}} - e^{-\varepsilon/k_B T} Q_{\text{vib}} = 1$$

so

$$Q_{\text{vib}} = \frac{1}{1 - e^{-\varepsilon/k_B T}}$$

An ideal harmonic oscillator has only one vibrational spectral line at frequency ν, which gives its energy through the Planck equation $\varepsilon = h\nu$, where h is Planck's constant 6.626×10^{-34} J s. Knowing the energy ε for the oscillator, one knows Q_{vib}.

Spectroscopists give the frequency of vibration as $\tilde{\nu}$ in units of reciprocal cm, cm^{-1}. Converting this unit to ν, we have $\nu = c\,\tilde{\nu}$ where c is the speed of electromagnetic radiation, 3.00×10^8 m s^{-1}. The formula for Q_{vib} in spectrocopist's terms is

$$Q_{\text{vib}} = \frac{1}{1 - e^{-\varepsilon/k_B T}} = \frac{1}{1 - e^{-hc\tilde{\nu}\beta}} = \frac{1}{1 - e^{-a}}$$

where $\beta = 1/k_B T$ and constants have been gathered for convenience into

$$a = \frac{1.439\tilde{\nu}}{T}$$

The diatomic molecule Na_2(g) at 1000 K, for example, has a strong vibrational resonance at 159.2 cm^{-1}. This leads to $a = 1.439(159.2)/1000 = 0.229$ and

$$Q_{\text{vib}} = \frac{1}{1 - e^{-0.229}} = 4.88$$

Similarly, the *rotational partition function* Q_{rot}, in spectroscopist's terms is

$$Q_{\text{rot}} = \frac{1}{hcB\beta} = \frac{0.695T}{\sigma B}$$

where B is the rotational resonance frequency, again in units of cm^{-1}, and σ is a "symmetry number" included to take into account the fact that a symmetrical molecule has different configurations that are, from a geometric point of view, the same. For example, the linear molecule Na–Na rotated by *one half* a rotation is still just Na–Na, indistinguishable from the starting configuration. The difference between distinguishable and indistinguishable configurations changes the statistics of the problem. (The statistics of a poker game using a deck in which each ace has an identical twin is not the same as the statistics using an honest deck.)

The rotational resonance frequency of interest for $Na_2(g)$ is $B = 0.1547\ \text{cm}^{-1}$. Applying the spectroscopist's formula to the Na–Na molecule with its symmetry number of 2 gives

$$Q_{\text{rot}} = \frac{0.695T}{\sigma B} = \frac{0.695(1000)}{2(0.1547)} = 2246$$

This kind of calculation will be useful to us in Section 8.8. Further discussion of Q determination is given in more detailed works (Nash, 2006; Maczek, 1998).

8.7 THE PARTITION FUNCTION FOR DIFFERENT MODES OF MOTION

A molecule can absorb energy by increasing its translational motion (displacement of the entire molecule), increasing its rotational motion, and increasing its vibrational motion, as well as by electronic excitation. The total energy of a molecule is the sum of energy contributions from all of its modes of motion:

$$\varepsilon = \varepsilon_{tr} + \varepsilon_{rot} + \varepsilon_{vib} + \varepsilon_{el}$$

In terms of its different modes of motion, the partition function is

$$Q = \sum_i e^{-(\varepsilon_{tr}+\varepsilon_{rot}+\varepsilon_{vib}+\varepsilon_{el})/k_{\text{B}}T}$$

Exponential sums can be written as products $e^{a+b} = e^a e^b$, and so on, so Q can be broken up into contributing partition functions, one for each mode of motion:

$$Q = \sum_i e^{-(\varepsilon_{tr})/k_{\text{B}}T} \sum_i e^{-(\varepsilon_{rot})/k_{\text{B}}T} \sum_i e^{-(\varepsilon_{vib})/k_{\text{B}}T} \sum_i e^{-(\varepsilon_{el})/k_{\text{B}}T}$$

This leads to

$$Q = Q_{\text{tr}} Q_{\text{rot}} Q_{\text{vib}} Q_{\text{el}}$$

where

$$Q_{\text{tr}} = \sum_i e^{-(\varepsilon_{tr})/k_B T}, \qquad \text{etc.}$$

Thus, if one knows the energy absorption for each possible mode of motion or excitation, one can write down the partition function and obtain all the thermodynamic information for that system. In practice this may be a difficult task because systems may not be *weakly* interacting; instead, they may interact in such a way that the vibrational energy spacing for each bond in the molecule depends on its neighbors. Other interactions may occur. Also, real chemical bonds are not truly harmonic and their energy level spacings are equal only as an approximation.

8.8 THE EQUILIBRIUM CONSTANT: A STATISTICAL APPROACH

First taking into account the energy spacing between reactants and products and then considering their partition functions, we can write K_{eq} in terms of the product Π of molecular partition functions Q

$$\ln K_{\text{eq}} = -\frac{\Delta\varepsilon_0}{RT} + \ln \Pi \left(\frac{Q_i}{N_A}\right)^{\xi}$$

where ξ is the appropriate stoichiometric coefficient and N_A is Avogadro's number. Note the correspondence between the second term in this equation and the entropy change of reaction $\Delta_r S^\circ$ as it appears in the classical van't Hoff and related equations.

Sodium metal can be vaporized at moderately high temperatures. The vapor exists as diatomic sodium molecules Na_2(g) in equilibrium with atomic sodium vapor Na(g).

$$\text{Na}_2(\text{g}) \rightleftarrows 2\text{Na}(\text{g})$$

Maczek (1998) has applied the statistical thermodynamic equation for K_{eq} to several reactions, including the dissociation of Na_2 molecules in the gas phase at 1274 K. Not all modes of motion apply in this case. For the product state, there is no partition function Q_{vib}(Na) or Q_{rot}(Na) because there can be no internuclear vibration or rotation of atoms, which are essentially point masses. The reactant state consists of connected masses separated in space for which vibration and rotation *are* quantum mechanically allowed. Now, however, electronic excitation, while quantum mechanically permitted, does not occur because the thermal energy at this temperature is insufficient to drive the electrons out of their ground state. The equilibrium expression is somewhat simpler than one might expect:

$$K_{eq} = \frac{[Q_{tr}(\text{Na})Q_{el}(\text{Na})]^2}{Q_{tr}(\text{Na}_2)Q_{rot}(\text{Na}_2)Q_{vib}(\text{Na}_2)} e^{-\varepsilon_0/RT}$$

The first partition function we need is Q_{tr} for *translational* motion. That simply means the partition function for sodium molecules or atoms (or anything else) flying around and bouncing off the walls of some container. Imposition of the laws of quantum mechanics brings about an astonishing phenomenon: Simply by being confined to a container or "box," the particles suddenly have allowed and forbidden energy levels. Their energy is *quantized*. Energy levels of a particle in a box form exactly the same kind of energy-level manifold we have been talking about except that, unlike the harmonic oscillator example, they are not evenly spaced. Instead, they go up according to the equation (Chapter 19)

$$E = \frac{n^2h^2}{8ma^2}$$

The integers $n = 1, 2, 3, \ldots$ are called the *quantum numbers*, $h = 6.626 \times 10^{-34}$ J s is Planck's constant in joule seconds, m is the mass of the particle, and a is the length of one edge of the box (taken to be cubic for simplicity).

When this energy restriction is placed on translational motion, it influences the partition function for a cubic box of volume $V = a^3$ according to the equation

$$Q_{tr} = \left(\frac{2\pi m}{h^2\beta}\right)^{\frac{3}{2}} a^3 = \left(\frac{2\pi m}{h^2\beta}\right)^{\frac{3}{2}} V$$

where we have again used a convenience variable, ubiquitous in this field, $\beta = 1/k_B T$. Inverting terms in the parentheses, we obtain

$$Q_{tr} = \left(\frac{h^2\beta}{2\pi m}\right)^{-\frac{3}{2}} V = \frac{V}{\left(\frac{h^2\beta}{2\pi m}\right)^{\frac{3}{2}}} \equiv \frac{V}{\Lambda^3}$$

In the jargon of this field, the cube root of the denominator is called the *thermal wavelength* and given the special symbol Λ:

$$\Lambda = \left(\frac{h^2\beta}{2\pi m}\right)^{\frac{1}{2}}$$

The thermal wavelength has the units of length, m [or picometers (pm) in atomic–molecular problems]. If we let the volume chosen be the molar volume $a^3 = V_m$, then the volume per particle is V_m/N_A, where N_A is Avogadro's number. Now Λ is only a collection of constants times the inverse square root of the mass of the particle and the temperature:

$$\Lambda = \left(\frac{h^2\beta}{2\pi m}\right)^{\frac{1}{2}} = h\left(\frac{1}{2\pi m k_B T}\right)^{\frac{1}{2}} = h\left(\frac{1}{2\pi m}\right)^{\frac{1}{2}}\left(\frac{1}{k_B T}\right)^{\frac{1}{2}}$$
$$= 7.113 \times 10^{-23}\frac{1}{\sqrt{mT}}$$

The partition function is a measure of the accessibility of quantum states to particles distributed within the system. Because the cube of the thermal wavelength is inversely related to the partition function through the volume of the container in which the particles are trapped, Λ gives us an idea of how big a container has to be for its quantum states to be fully populated. For atoms and small molecules the volume Λ^3 is of the order of 10^{-30}. Any imaginable laboratory container is very much larger than this, so we conclude that all *translational levels are accessible*. Bear in mind that this volume restriction relates to translational motion only and not to any other mode of motion.

8.9 COMPUTATIONAL STATISTICAL THERMODYNAMICS

Further insight into this calculation can be gained from a computational solution for K_{eq}.

Several computer programs give calculated values for the thermodynamic functions and the related partition functions. The functions are usually broken up into their individual contributions as described above. An edited quantum mechanical output of this kind for the sodium atom is given in Table 8.2. These values were taken from a much larger output file from the program GAUSSIAN 03©. Computed values should be used with some caution because they often rely on approximations like the harmonic oscillator approximation.

One can take the total partition functions, plug them into the equilibrium constant quotient and multiply by the difference in ground state energies of the atoms relative to the molecules to obtain K_{eq}. The Q values chosen are from $V = 0$, the vibrational ground state.

$$K_{eq} = \frac{Q_{Na}^2}{Q_{Na_2}} e^{-\varepsilon_0/RT} = \frac{(3.44 \times 10^8)^2}{9.531 \times 10^{12}} e^{-6.51} = 18.5$$

TABLE 8.2 Some Computed Partition Functions for Molecular and Atomic Sodium.[a]

Na_2 Molecules	T = 1300 K	
	Q	
Total Bot	0.873633D+13	
Total V = 0	0.953147D+13	←
Vib (Bot)	0.573265D+01	
Vib (Bot) 1	0.573265D+01	
Vib (V = 0)	0.625441D+01	
Vib (V = 0) 1	0.625441D+01	
Electronic	0.100000D+01	
Translational	0.486489D+09	
Rotational	0.313256D+04	
Na atoms	**T = 1300**	
	Q	
Total Bot	0.344000D+09	
Total V = 0	0.344000D+09	←
Vib (Bot)	0.100000D+01	
Vib (V = 0)	0.100000D+01	
Electronic	0.200000D+01	
Translational	0.172000D+09	
Rotational	0.100000D+01	

[a]Note that Q is unitless.

PROBLEMS AND EXAMPLES

Example 8.1 The Thermal Wavelength

Calculate the thermal wavelength of sodium atoms at 1300 K.

Solution 8.1 The calculation for Na(g) is as follows:

$$\Lambda = \left(\frac{h^2\beta}{2\pi m}\right)^{1/2}$$

Evaluation of the β constant,

$$\beta = 1/k_B T = 1/1.38 \times 10^{-23} \cdot (1300) = 5.570 \times 10^{19}$$

leads to the thermal wavelength for Na(g) atoms at 1300 K.

$$\Lambda = \left(\frac{5.570 \times 10^{19} h^2}{2\pi m}\right)^{1/2} = \left(\frac{2.44 \times 10^{-47}}{2\pi m}\right)^{1/2} = \left(\frac{2.4 \times 10^{-47}}{2\pi \cdot 23.0 \cdot 1.66 \times 10^{-27}}\right)^{1/2}$$
$$= \left(1.02 \times 10^{-22}\right)^{1/2} = 1.01 \times 10^{-11} = 10.1 \times 10^{-12} = 10.1 \text{ pm}$$

For $Na_2(g)$ molecules, a similar calculation gives a result that is not very different from the result for atoms. At 1300 K, $\Lambda(Na_2(g)) = 7.14$ pm. The dimensions for both thermal wavelengths are picometers $(10^{-12}\ \text{m})$, so the volume Λ^3 is of the order of $10^{-33}\ \text{m}^3$. This dimension confirms our prior supposition that translational energy levels are very close together. They can be regarded as a continuum if mathematically convenient, and they are fully populated by a mole of particles.

Example 8.2 The Translational Equilibrium Constant

Given that the energy separation of ground states (Section 8.3) is $\varepsilon_0 = 70.5\ \text{kJ mol}^{-1}$ (from spectroscopic measurements), calculate what the equilibrium constant would be for translational motion only (ignoring all internal modes of motion).

Solution 8.2 The equilibrium expression is

$$K_{\text{eq}} = \frac{Q_{\text{Na}}/N_{\text{A}}^2}{Q_{\text{Na}_2}/N_{\text{A}}} e^{-\varepsilon_0/RT}$$

The exponents 2 and 1 arise from the stoichiometric coefficients ξ in the equilibrium expression for the reaction written as a dissociation

$$K_{\text{eq}} = \frac{(p_{\text{Na}})^2}{p_{\text{Na}_2}}$$

Expressing the partition functions in terms of the thermal wavelengths, we obtain

$$K_{\text{eq}} = \left(\frac{\Lambda^3_{\text{Na}_2}}{\Lambda^6_{\text{Na}}}\right) \frac{V_m}{N_{\text{A}}} e^{-\varepsilon_0/RT} = \frac{3.64 \times 10^{-34}}{1.06 \times 10^{-66}} \frac{0.082}{6.022 \times 10^{23}} e^{-6.51}$$
$$= 4.67 \times 10^7 e^{-6.51} = 6.95 \times 10^4$$

The exponential value $-\varepsilon_0/RT = -70.4/R\,(1300) = -6.51$ comes from spectroscopic data. The molar volume is $0.0820\ \text{m}^3\ \text{mol}^{-1}$and $e^{-70.4/RT} = e^{-6.51} = 1.488 \times 10^{-3}$.

But things are not quite that simple. The Na_2 molecule has internal modes of motion, one for vibration along the molecular axis and one for rotation about its center of mass. Both partition functions can be determined from spectroscopic data. We have already seen how the value of Q_{vib} arises from the resonance frequency for vibrational motion. The quantized levels for rotation of $Na_2(g)$ molecules are found in essentially the same way. The two calculations give

$$Q_{\text{vib}} = 6.254, \qquad Q_{\text{rot}} = 3.132 \times 10^3$$

The partition functions are unitless. One more factor comes into the calculation: integer 4, which is the multiplicity of the sodium atom electronic structure.

When these two degrees of freedom and the Na multiplicity $\left(2^2\right)$ are taken into account, the equilibrium constant calculation at 1300 K is complete

$$K_{eq} = 4\left(6.95 \times 10^4\right)\left[\frac{1}{3.132 \times 10^3 \cdot (6.254)}\right] = 14.2$$

which, considering the magnitude of the numbers revalued is in good agreement with the previous calculation.

Problem 8.1

Find the equilibrium constant K_{eq} at 300 K for a simple nondegenerate two-level system

$$A(g) \rightarrow B(g)$$

where the energy of level B, ε_B, is 1.25 kJmol^{-1} higher than ε_A. What are the percentage occupations of levels A and B? What is the probability of selecting a molecule from the lower level if the selection process is completely random and does not favor either level?

Problem 8.2

Calculate the thermal wavelength of Na_2(g) molecules at 1000 K. Compare your answer to the value 8.14 pm.

Problem 8.3

If we multiply the thermal wavelength of Na by $1/\sqrt{2}$, we get the thermal wavelength of Na_2. Why is that?

Problem 8.4

Why is $\nu = c\,\tilde{\nu}$, where ν is the frequency of electromagnetic radiation, c is its speed 3.0×10^8 m s^{-1}, and $\tilde{\nu}$ is the "wavenumber in spectroscopist's terminology with units of reciprocal seconds (s^{-1})?

Problem 8.5

No doubt you have noticed that solid iodine I(s) produces a purple vapor upon heating in a closed container. The vapor is diatomic I_2(g) which enters into an equilibrium with atoms I(g) in a way analogous to Na_2(g) and Na(g) already discussed. Calculate the thermal wavelength of molecular I_2(g) and I(g) atoms confined to a molar volume at 1000 K.

Problem 8.6

Given that

$$Q_{\text{tr}} = \left(\frac{2\pi m}{h^2\beta}\right)^{3/2} a^3 = \left(\frac{2\pi m}{h^2\beta}\right)^{3/2} V$$

arises from space quantization in a container of volume V,

$$E = \frac{n^2h^2}{8ma^2}$$

where does the 2π come from in the numerator?

Problem 8.7

Find the translational partition function for $I_2(g)$ and the translational partition function for $I(g)$.

Problem 8.8

Find the thermal wavelengths for $I_2(g)$ and $I(g)$. From these wavelengths and the dissociation energy of 152.3 kJ mol^{-1}, find the equilibrium constant at 1000 K ignoring internal modes of motion.

Problem 8.9

Refine the result of Problem 8.8 by taking into account the vibrational and rotational motions of the $I_2(g)$ molecule. The multiplicity factor is 4.

9

THE PHASE RULE

It is essential in science for experiments to be repeated, tested, and verified before being accepted into the body of theory.[1] To duplicate thermochemical experiments, one must be able to duplicate the thermochemical *system*, and to do that we must know how many of its infinite number of physical properties—mass m, energy, entropy, the heat capacities C_p and C_V, refractive indices, and so on—must be specified.

A fundamental truth that we usually take for granted is that there is a certain number of variables that completely describe any selected system. When we measure some of them, we know all we can ever know about the system and we can, in principle, calculate all the variables we haven't measured from those that we have. Which variables are they and how many of them are there? Must we know a substantial portion of the infinite number of possibilities in order to define a thermochemical system? If that were true, science as we know it would hardly be possible because we could never really verify (or contradict) an experimental result by replication. This is the problem that was solved by J. Willard Gibbs.

9.1 COMPONENTS, PHASES, AND DEGREES OF FREEDOM

Gibbs distinguished between components C of a system, which are *chemically* distinguishable, and the phases P of a system, which are *physically* distinct. Thus, the

[1]There are exceptions. The science of cosmology offers interesting contradictions because we would be hard pressed to duplicate the creation of the universe.

Concise Physical Chemistry, by Donald W. Rogers

system ice and water has one component, identifiable as H_2O, and two phases, solid and liquid. With his famous phase rule (see below), Gibbs specified the number of *degrees of freedom* of thermochemical systems. Degrees of freedom are like the independent variables of a system of algebraic equations or the basis vectors of a complete set. Once a limited number of them have been specified, the others are no longer free to take on arbitrary values. They are linear combinations of the set that is already known. For this reason it is the *number* of independent variables that we get from the phase rule, not their identity. We are free to choose which of the identifiable quantities shall be taken as independent variables, but only up to the number specified by the phase rule.

In the case of liquid water, one might argue that there are three components, H^+, OH^-, and H_2O, or even four, including H_3O^+. We ignore these and similar forms because they are obviously related to each other and are not independent. In pure water, we know that $[H^+] = [OH^-]$ because the only source of the ions is dissociation:

$$H_2O \rightarrow H^+ + OH^-$$

That makes it clear that if we measure either one but only one, we know the other. When we do this, we find $\left[H^+\right] = 10^{-7}$; therefore we know that $\left[OH^-\right] = 10^{-7}$and we can write a dissociation equilibrium constant for pure water[2]:

$$K_w = \frac{[H^+]\,[OH^-]}{[H_2O]} \approx \frac{[H^+]^2}{[H_2O]} = \frac{[10^{-7}]^2}{1} = 10^{-14}$$

where the concentrations are expressed as moles per mole of solvent (water).

Pursuing these ideas just a little further, what we call "concentrations" in solution chemistry are really activities that, as unitless numbers (ratios relative to a standard state), give a unitless K_w.

If we reformulate so as to include higher polymers such as $H_5O_2^+$, and so on, their concentrations are not independent either because dissociation constants exist by which we could calculate their concentration. These calculations are, in principle, possible, even though we may not have enough information to carry them out. However you look at it, specifying one molar concentration specifies them all. There is one component in a pure phase.

9.2 COEXISTENCE CURVES

The thermodynamic state of water vapor, $H_2O(g)$, like that of any pure substance, has two degrees of freedom. Any state function, including the Gibbs free energy, can be expressed as a mathematical function of two independent variables, for example, $\mu = f(p, T)$. We can also write $E = f(p, T)$ or $C_p = f(p, T)$. In principle, equations

[2]By our conventional definition, $\mathrm{pH} \equiv -\log\left[H^+\right]$; this leads us to say that an aqueous solution is neutral when its $\mathrm{pH} = 7$.

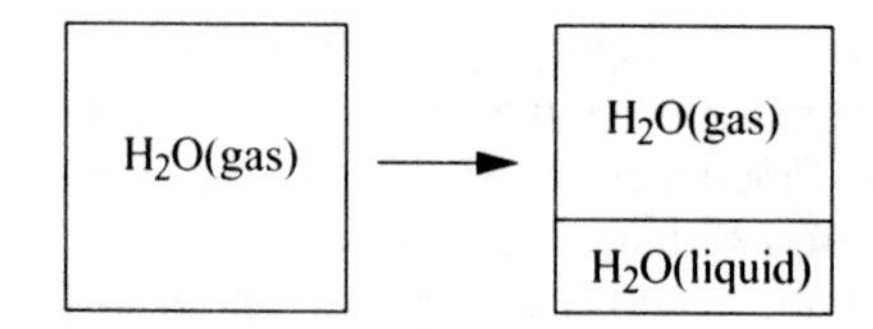

FIGURE 9.1 Pure water in one phase (*left*) and two phases (*right*). At some temperature, the phases coexist in equilibrium.

exist wherewith we can determine E and C_p from a knowledge of p and T even though we may not know exactly what these equations are. We could even write $p = f(C_p, E)$. We are not restricted as to what the independent variables are, so long as there are precisely two of them.

The number of degrees of freedom for a pure, one-phase system is easy. Identifying *some* phases is easy as well.[3] We can see the phase separation in mixtures of carbon tetrachloride and water because they have different refractive indices. The two phases, one as the top layer and one as the bottom layer, look different. The trick is to express the behavior of a completely general system, which need not be pure, and may consist of many chemical entities distributed over many phases.

If we choose water vapor as our example of a pure phase, examine the Gibbs free energy as our state function, and take pressure and temperature as our independent variables $\mu = f(p, T)$, the system as defined has a wide range of freedom. Within reasonable limits, we can change p and T arbitrarily. The state of the system can be represented as a point on a two-dimensional plot of p vs. T (or p vs. V, as in Chapter 1). Any point represents the system in some state.

Sooner or later though, we are bound to exceed reasonable limits; and the water, even in a closed container that previously contained only water vapor, is bound to condense to produce some liquid water in a state of equilibrium with gaseous water (Fig. 9.1)

We treat this physical equilibrium just as we do a chemical equilibrium. The vaporization equilibrium is

$$H_2O(l) \rightleftarrows H_2O(g)$$

The equilibrium constant is

$$K_{eq} = \frac{p_{H_2O}(g)}{a_{H_2O}(l)} = p_{H_2O}(g)$$

where pure water is in its standard state, resulting in a denominator of $a_{H_2O}(l) \equiv 1.0$ in the normal equilibrium expression. An *equilibrium* between pure liquid and pure

[3]Identification of some phases may not be easy. Solids can exist as phases that differ only in more subtle ways such as heat capacity or molar volume.

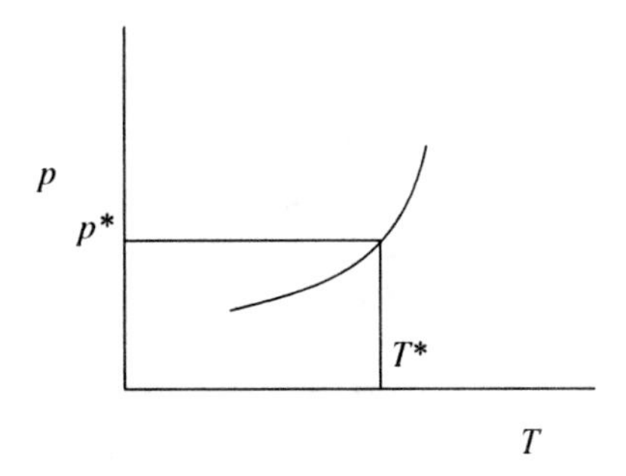

FIGURE 9.2 A liquid–vapor coexistence curve. Fixing the temperature at T^* automatically fixes the pressure at p^* for coexisting phases of a pure substance.

gas means that their free energies must be equal:

$$\mu_{H_2O}(l) = \mu_{H_2O}(g)$$

An equation restricting the variables to a fixed ratio reduces the number of *independent* variables by one. Now, by specifying either the temperature or the pressure, the other variable is no longer free. There is still an infinite number of possible free energies, but they are contiguous points on a line. Each specific T defines a coexisting state at pressure p. The locus of points at which liquid and vapor can coexist is called the *coexistence curve*. If T is changed by a small arbitrary amount, p automatically adjusts to an appropriate value to maintain the equilibrium and stay on the curve. The coexistence curve can now be completely described in two dimensions, p on the vertical axis as a function of an independent variable T on the horizontal axis (Fig. 9.2). The point representing the system is no longer free to move over a two-dimensional p–V plane; it is restricted to the curve. The system has lost one degree of freedom.

The liquid–vapor coexistence curve is not unique, nor is its exponential shape. Solids also exist in equilibrium with the vapor phase. That is why you can smell solids like naphthalene. Solids have an exponential (or approximately exponential) coexistence curve too, marking their equilibrium boundary with vapor, as seen at the left of Fig. 9.3. Of course, there is also a coexistence curve between solids and liquids (melting points), which normally has a positive slope as in Fig. 9.3. In the unusual case of water, the slope of the solid-liquid coexistence line is negative which is why you can ice skate.[4] The three curves taken together on a p–T surface constitute a *phase diagram*. As we shall see, phase diagrams are a very general way of representing all manner of coexistence curves.

At the *critical point* (Section 2.4), there is no longer a distinction between a liquid and a nonideal gas. Like the critical point, the *triple point* is unique to each pure substance. It cannot be changed by altering external conditions p and T. It has *no* degrees of freedom.

[4]How can there be a connection between a physical chemistry coexistence curve and ice skating? Think about it.

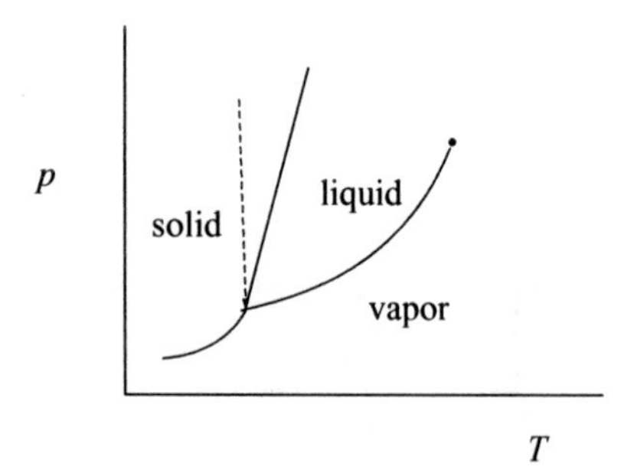

FIGURE 9.3 A single-component phase diagram. The unusual solid–liquid coexistence curve for water is shown as a dotted line. The terminus (•) of the curve on the right is the *critical point*. The intersection of the three curves is the *triple point*.

9.3 THE CLAUSIUS–CLAPEYRON EQUATION

The *Clapeyron equation* for a phase transition is

$$\frac{dp}{dT} = \frac{\Delta_{\text{trans}} H}{T\Delta V}$$

where $\Delta_{\text{trans}} H$ is the enthalpy change of the phase transition and ΔV is the corresponding volume change. It is customary to take both as molar quantities.

Clausius observed that the volume of a liquid in equilibrium with its vapor at the boiling point is normally very much smaller than the volume of the vapor itself: $V(\text{l}) << V(\text{g})$. Thus, taking the special case of vaporization of a liquid where $\Delta V = V(\text{g}) - V(\text{l})$, he ignored the smaller volume $V(\text{l})$ and substituted the larger volume $V(\text{g})$ for ΔV. He then assumed that the vapor over the liquid is an ideal gas, which permits the substitution

$$\frac{dp}{dT} = \frac{\Delta_{\text{vap}} H}{TV(g)} = \frac{\Delta_{\text{vap}} H}{T}\left(\frac{p}{RT}\right)$$

Rearranging, we obtain

$$\frac{dp}{p} = \frac{\Delta_{\text{vap}} H}{R}\left(\frac{1}{T^2}\right) dT$$

and taking the indefinite integral,

$$\int \frac{dp}{p} = \frac{\Delta_{\text{vap}} H}{R} \int \frac{1}{T^2}\, dT$$

we get

$$\ln p = \frac{\Delta_{\text{vap}} H}{R} \left(\frac{-1}{T} \right) + \text{const.}$$

This result can be written

$$p_{\text{vap}} = a e^{-\Delta_{\text{vap}} H / RT} = a e^{b(-1/T)}$$

where a arises from the constant of integration. This is a form of the Clausius–Clapeyron equation. If one can determine b in this exponential equation, then one has $\Delta_{\text{vap}} H$.

If an open container of liquid under an external pressure p_{ext} is heated to its boiling point, bubbles form at the bottom of the container, each of which forms a tiny closed "chamber" filled with pure vapor. That is why we observe that the normal boiling points of pure liquids observed in an open container follow the same equation (Clausius–Clapeyron) as that derived from the coexistence curve of the liquid in a closed container.

9.4 PARTIAL MOLAR VOLUME

Prior to Gibbs, thermodynamics was largely about the transfer of heat in the process of driving an engine. It was an age justly called the age of steam. Gibbs's departure was to focus on the transfer of *matter*. That is why Gibbs's work is so important to chemists. Our science is largely devoted to the transfer of matter from a reactant state to a product state. In classical physical chemistry, we ask whether the transfer occurs (thermodynamics) and, if it does, we want to know how long it takes (kinetics).

To answer the first of these questions, Gibbs used partial molar thermodynamic state functions. To start out, we shall consider the volume of a system and we shall restrict mixtures to two components. The volume of a system is easy to visualize, and restricting the system to two components simplifies the arithmetic. Later we shall release these restrictions.

The volume of an ideal two-component liquid solution at constant p and T depends on how much of each component is present:

$$V = f(n_1, n_2)$$

If the molar volume of pure component 1 is V_{m1}° and the molar volume of pure component 2 is V_{m2}°, then an ideal solution of equal molar amounts of 1 and 2 will have a molar volume that is the average of the two $\left(V_{m1}^{\circ} + V_{m2}^{\circ}\right)/2$ (Fig. 9.4). Solutions of other ratios of components 1 and 2 would be arrayed along a straight line connecting V_{m1}° to V_{m2}°, provided that no shrinkage or swelling takes place when 1 and 2 are mixed.

In general, for different ratios of the components, we obtain

$$V = V_{m1}^{\circ} n_1 + V_{m2}^{\circ} n_2$$

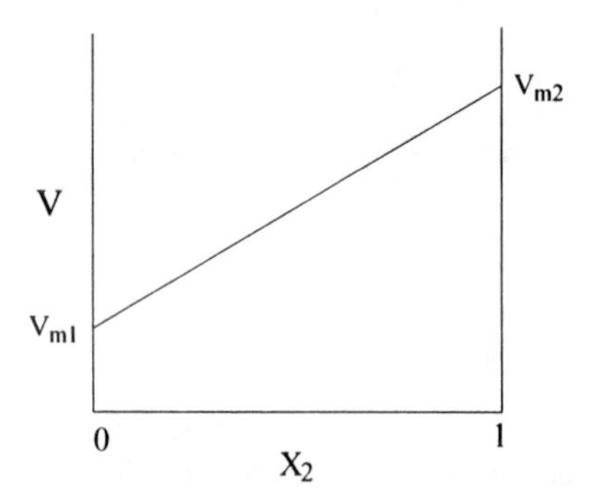

FIGURE 9.4 Total volume of an ideal binary solution. X_2 is the mole fraction of component 2.

In reality, life is not that simple. The total volume of the solution will not usually be the sum of the molar volumes weighted by their relative amounts. The volume actually occupied by the solute in the solution is called the *partial molar volume*, which can be greater than or smaller than its molar volume in the pure state. The volume of the solution will then be greater or smaller than the sum of its parts. Addition of some potassium salts to one mole of pure water results in a volume of solution that is even smaller than the initial volume of water (shrinkage occurs). These three possibilities are shown in Fig. 9.5.

9.4.1 Generalization

We have been completely arbitrary in designating component 2 with amount n_2 as the solute and n_1 as the number of moles of solvent, so we can switch designations just as arbitrarily. Thus everything we have said about component 2 in a binary solution also applies to component 1 treated as though it is the solute in solution with component 2. In a solution of completely miscible liquids, it is conventional to take the lesser

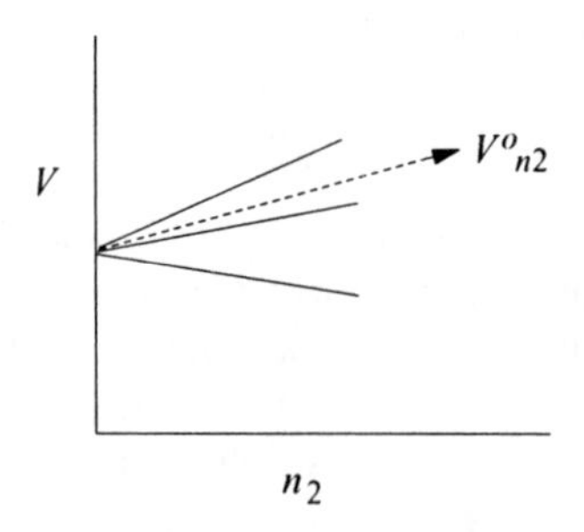

FIGURE 9.5 Volume increase (or decrease) upon adding small amounts of solute n_2 to pure solvent. Three cases are shown for $V_{n2} > V^\circ_{n2}$, $V_{n2} < V^\circ_{n2}$, and $V_{n2} < 0$.

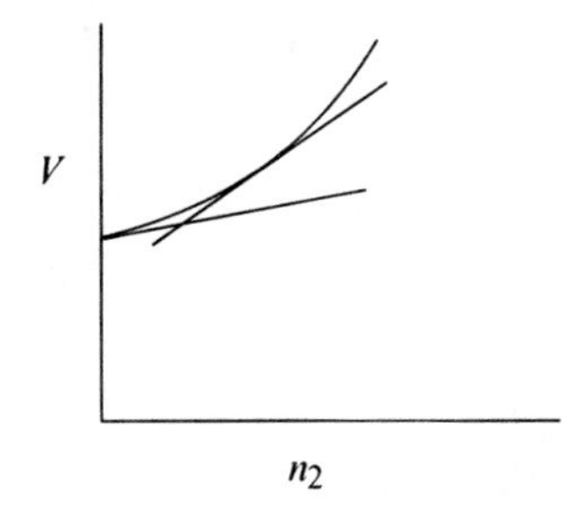

FIGURE 9.6 Partial molar volume as the slope of V vs. n_2. The lower line gives V_{m2} as $n_2 \to 0$, and the upper tangent line gives V_{m2} at a specific concentration $n_2 \neq 0$.

component as the solute and the greater component as the solvent. In many cases, the choice is obvious, for example, in KCl solutions in water, KCl is clearly the solute.

Evidently, from Fig. 9.6, the partial molar volume of one component, which we have chosen to call the solute, is the slope of one of the solid lines found by measuring the volume increment upon adding small amounts of component 2 to large amounts of component 1. In real solutions, these lines need not be straight; the partial molar volume of the solute is found in the limit as $dn_2 \to 0$.

Recognizing V_{m2} as the slope of the function of V vs. n_2 means that we have the definition

$$V_{m2} = \left(\frac{\partial V}{\partial n_2}\right)_{T,p,n_1}$$

This slope can be found anywhere on the entire curve of experimentally measured total volume vs. X_2, where $X_{n_2} = n_2/(n_1 + n_2)$. Now V_{m2} is defined as the volume change found upon adding an infinitesimal amount of component 2 to a solution of composition X_2 specified by a horizontal distance along the X_2 axis; for example, $X_2 = 0.20$ in Fig. 9.7.

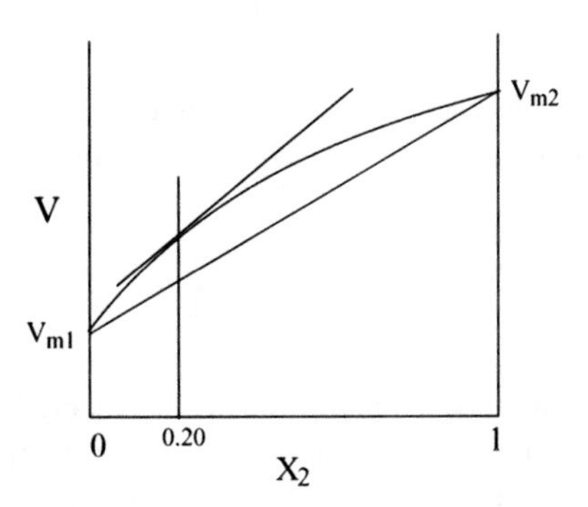

FIGURE 9.7 Volume behavior of a nonideal binary solution. X_2 is the mole fraction of component 2.

The definition of V_{m2} should come as no surprise. It comes from the condition on perfect differentials. In the case of the volume, we have

$$dV = \left(\frac{\partial V}{\partial p}\right)_{T,n_i,n_j} dp + \left(\frac{\partial V}{\partial T}\right)_{p,n_i,n_j} dT + \sum_{n_i} \left(\frac{\partial V}{\partial n_i}\right)_{T,p,n_j} dn_i$$

This generalization involving the sum

$$\sum_{n_i} \left(\frac{\partial V}{\partial n_i}\right)_{T,p,n_j} dn_i$$

gets us away from the restriction that the solution be binary, a condition we imposed at the beginning for simplicity. Now the concepts developed can be applied to any number of components. Recalling that volume is a thermodynamic property, we have

$$V = f(p, T, n_1, n_2, \ldots, n_j)$$

In view of the first two terms in the sum, the functions are sensitive to variations in both pressure and temperature hence one or both may be held constant in the phase diagrams discussed below. The variation of total volume with composition in curves like Fig. 9.7 gives rise to the term *excess volume* as the volume above the straight line expected from a simple sum of molar volumes in the pure state V_m°. The excess volume can be negative leading to a nonideal curve below the straight line in Fig. 9.7.

The volume in a real binary system corresponds to the sum in which

$$V = V_{m1}^\circ n_1 + V_{m2}^\circ n_2$$

is replaced by

$$V = \bar{V}_{m1} n_1 + \bar{V}_{m2} n_2$$

where the molar volumes $\bar{V}_{m1}$ and $\bar{V}_{m2}$ are no longer volumes in the pure phase but are *partial molar volumes*, unique to the ratio of n_1 and n_2 in the solution. In general,

$$\bar{V}_{m_i} = \left(\frac{\partial V}{\partial n_i}\right)_{T,p,n_j}$$

Each partial molar volume must be determined experimentally. There are, of course, simplified equations containing empirical constants that work more or less well for real (nonideal) solutions just as there are for nonideal gases.

The greatest generalization in this field, however, is to recognize that nothing in the arguments made is specific to volume alone. The general rule is that *partial molar quantities analogous to the partial molar volume exist for all thermodynamic state variables*. This rule gives us the partial molar energy,

$$\bar{U}_{m_i} = \left(\frac{\partial U}{\partial n_i}\right)_{T,p,n_j}$$

the partial molar enthalpy,

$$\bar{H}_{m_i} = \left(\frac{\partial H}{\partial n_i}\right)_{T,p,n_j}$$

the partial molar entropy,

$$\bar{S}_{m_i} = \left(\frac{\partial S}{\partial n_i}\right)_{T,p,n_j}$$

and the partial molar Gibbs free energy

$$\mu_{m_i} = \left(\frac{\partial \mu}{\partial n_i}\right)_{T,p,n_j}$$

This last partial molar quantity is so important that it is given a unique name and symbol. It is called the *Gibbs chemical potential* μ_{m_i}. It should be clear that in real (nonideal) systems, each of these functions has a corresponding *excess* function; the excess energy, the excess enthalpy, the excess entropy, and so on.

In generalizing thermodynamics to many-component systems, Gibbs brought about an immense expansion of the scope of the subject. All of the classical thermodynamic equations apply to partial molar quantities as well. For example, by analogy to

$$\frac{d\left(\frac{G}{T}\right)}{d\left(\frac{1}{T}\right)} = H$$

for the Gibbs free energy of ideal components (Section 6.6), we have

$$\frac{d\left(\frac{\mu}{T}\right)}{d\left(\frac{1}{T}\right)} = \bar{H}_m$$

relating the partial molar Gibbs free energy and the partial molar enthalpy. $\bar{H}_m$.

9.5 THE GIBBS PHASE RULE

Of the C mole fractions $X_i (i = 1, 2, 3, \ldots, C)$ in a many-component one-phase mixture, $C - 1$ are independent because once you know $C - 1$ of them, you can get the remaining one from $\sum_i X_i = 1$. The "last" concentration variable is not independent. If there are P coexisting phases, there are $C - 1$ independent concentration variables in each phase for a total of $P(C - 1)$. However, all P phases must be at the same chemical potential for equilibrium to exist. This makes $P - 1$ equations connecting each component in each phase,[5] and there are C components for a total of $C(P - 1)$ equations. The number of degrees of freedom is equal to the total number of independent variables $P(C - 1)$ in the system minus the number of equations $C(P - 1)$ connecting them. In the most general case of many components that may break up into many phases, we have

$$f = P(C - 1) - C(P - 1) = PC - P - CP + C = C - P$$

Adding the two remaining degrees of freedom for the system, which might be p and T, we have the celebrated Gibbs phase rule

$$f = C - P + 2$$

A problem of potentially great complexity has been reduced to simple terms. The number of degrees of freedom f of the system is equal to the number of components minus the number of phases plus 2.

9.6 TWO-COMPONENT PHASE DIAGRAMS

Two-component phase diagrams can be complicated. Without the phase rule, they would seem to have a bewildering array of unrelated behaviors. Many two-component phase diagrams, however, can be broken down into combinations of three simple diagrams that we shall label I, II, and III here. Two-component phase diagrams are characterized by one more degree of freedom than pure one-component diagrams because of the added composition variable

$$f = C - P + 2 = 2 - P + 2 = 4 - P$$

We can express composition in many ways, but the mole fraction of a two-component system is most convenient for our purposes.

$$X_B = \frac{n_B}{n_A + n_B}$$

[5]For example, $\mu_A = \mu_B = \mu_C$ connects three variables but has only two equal signs, hence only two equations.

The mole fraction of a two-component system runs from 0 to 1.0 (Fig. 9.4) where n_B is the number of moles of component B and $n_A + n_B$ is the total number of moles of both components.

9.6.1 Type 1

In a two-component one-phase Type I diagram, pressure is taken as constant. Setting $p = \text{const}$ reduces the number of degrees of freedom by 1: $f = 3 - P = 2$ for 1 phase. (Please do not confuse p and P.) The system can be represented in two dimensions as in Fig. 9.8. If one phase is present, the system can exist at a temperature and mole fraction represented by any point (T, X_B) in either area above or below the coexistance curves. The upper and lower curves represent the locus of mole fractions of the vapor (upper) and liquid (lower), respectively, in equilibrium with one another at a temperature specified on the vertical axis. A system at any temperature and mole fraction between the coexistence curves splits into two phases, a vapor (upper curve) and a liquid (lower curve).

When there are two phases in equilibrium (liquid and vapor), the number of degrees of freedom is further reduced ($f = 3 - P = 3 - 2 = 1$); and specifying X_B of the system, as the mole fraction in the liquid phase or the mole fraction in the vapor, automatically specifies the temperature on one of the two coexistence curves. These two temperatures will not be the same. The lower curve in Fig. 9.8 is the locus of points at which a trace of vapor is in equilibrium with liquid. The upper curve in Fig. 9.8 is the locus of points at which a trace of liquid is in equilibrium with vapor. The horizontal lines connecting them represent the $l \to v$ or $v \to l$ phase change at any specific temperature, T (K).

Fractional distillation is one of our most important laboratory and industrial methods of chemical purification. The separation between the upper and lower curves in a Type I diagram makes fractional distillation possible. At any temperature, the mole fraction of component B in the mixture can be read along the horizontal in Fig. 9.8. As seen from the figure, the mole fractions are different for coexisting liquid and vapor phases except at the end points of the curves. If we allow a liquid mixture of

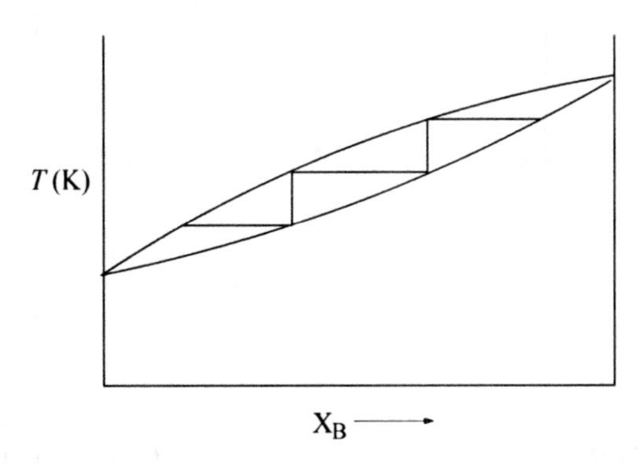

FIGURE 9.8 A Type I phase diagram. Liquid–vapor equilibriums are expressed by each of the three horizontal lines.

A and B to come to equilibrium with its vapor and then separate and condense the vapor, we shall have two solutions, one richer in A than the original solution and one richer in B. Three such equilibrations are represented in Fig. 9.8 as horizontal lines at different temperatures T, between the two T vs. X_B curves. If we start with a solution of A and B, well to the right of the figure, and allow it to come to equilibrium with its vapor, the composition of the vapor is given by the leftmost terminus of the top horizontal. Now separate that vapor from the original solution and condense it. The new liquid is richer in A than the original solution was.

A second equilibration and separation gives a solution still richer in A according to the middle horizontal in Fig. 9.8. A third repetition yields a vapor still richer in A at the lowest temperature. Now the purified concentration is given by the leftmost terminus of the three equilibration steps diagrammed in Fig. 9.8.

Of course, real distillations are not carried out by such a laborious stepwise process of equilibration and condensation. Nevertheless, a process very like this, of continual vaporization equilibration followed by condensation, does take place in the distillation columns we actually use. Real distilling columns may range in height from a hand's breadth for use in the laboratory to columns several stories tall used in the petroleum industry. By comparing the composition of the input with that of the effluent, the number of theoretical plates of a "still" can be calculated. Within normal practical considerations, the more theoretical plates, the better the still.

9.6.2 Type II

Type II phase diagrams describe liquid–liquid systems in which the components are completely miscible at some temperatures but undergo phase separation at others. Systems with a composition and temperature above the dome-shaped coexistence curve in Fig. 9.9 are completely miscible. Those with a temperature and composition below the curve split into two phases. The transition can be brought about by cooling a miscible solution so that the temperature drops along a vertical and enters the two-phase zone or it can be brought about at constant temperature by adding one or the

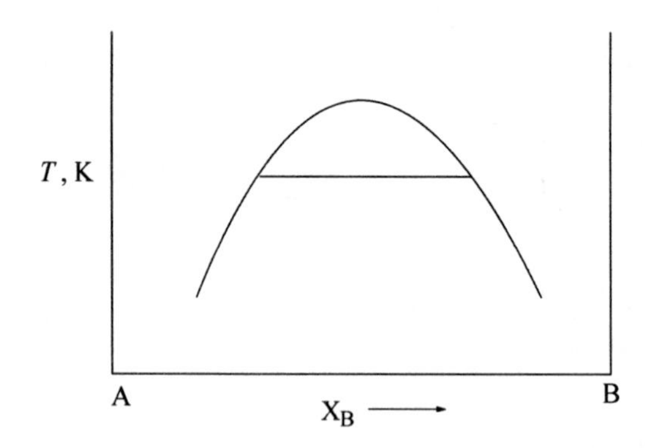

FIGURE 9.9 A Type II phase diagram. Solutions having a composition and temperature on the line split into an A-rich phase and a B-rich phase. The horizontal *tie line* is unique at each temperature.

other of the components to a miscible solution until the coexistence curve is crossed. The transition from a clear miscible system to an opaque two-phase emulsion can be quite dramatic in some cases, making it look as though water has turned into milk. Like Type I diagrams, the number of degrees of freedom is 3. We usually hold $p = \text{const}$ at 1 atm so that the rest of the phase behavior can be represented in two dimensions. On the dome-shaped coexistence curve, there is 1 degree of freedom. The horizontal under the dome is called a *tie line*. The two intersections of the tie line with the coexistence curve give the composition of the two coexisting phases. The dome need not be symmetrical. Quite a variety of shapes are possible, including one that is closed at the top *and* at the bottom, forming a closed irregular oval that is essentially an island of immiscibility in a sea of miscibility. The coexistence curve can approach the verticals representing pure A and pure B quite closely, giving rise to the folk saying "oil and water don't mix." Actually, they do mix but one phase is overwhelmingly oil-rich and the other is overwhelmingly water-rich. That is why one does not wish to drink water that has come into contact with oil or gasoline.

9.6.3 Type III

Type III solid–liquid phase diagrams are familiar as having a *eutectic point*. The locus of melting points of mixtures of A and B vs. X_B follows the two curves in Fig. 9.10. In general, mixtures of the two components have a lower melting point than either component, A or B, alone. Ordinary electrical solder is a eutectic mixture of lead and tin having a melting point that is lower than either pure Pb or pure Sn. We rely on the fact that the melting temperature of a *pure* compound is higher than that of an impure (mixed) sample of the same compound according to Fig. 9.10. Generations of pre-medical students have been judged partly on the basis of the melting point behavior of the compounds they prepare in the organic chemistry laboratory.

9.7 COMPOUND PHASE DIAGRAMS

Many complicated phase diagrams can be read quite easily as two or more simple-phase diagrams stuck together. The principle will be illustrated using a compound

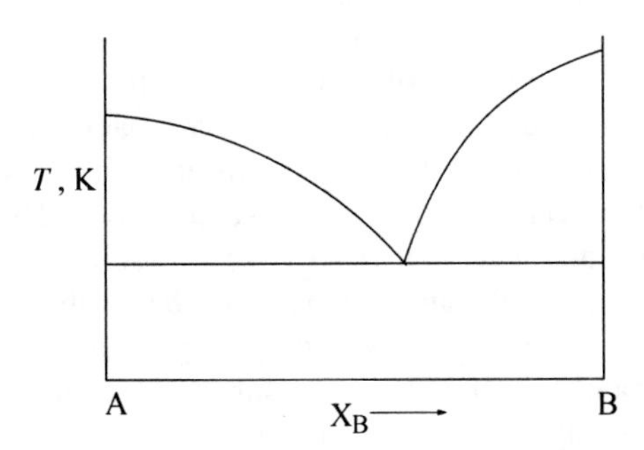

FIGURE 9.10 A Type III phase diagram. The low melting mixture is called a *eutectic mixture*.

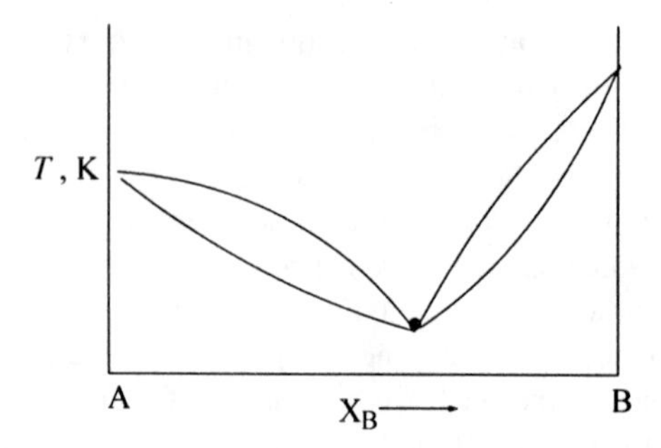

FIGURE 9.11 A compound phase diagram with a low boiling azeotrope.

phase diagram with an *azeotrope*. A phase diagram like that in Fig. 9.11 is, quite evidently, a combination of two Type I phase diagrams, one extending slightly beyond the midpoint of the composition axis and one somewhat shorter curve. Drawing isotherms as in Fig. 9.11 soon convinces us that the mixture cannot be separated by conventional fractional distillation. Even the best still, starting with a mixture to the right of the azeotropic point (•), will produce component B and azeotrope, whereas starting to the left of the point will produce pure A and azeotrope. One does not get pure A and pure B by either of these methods as one did by distilling mixtures of components having a simple Type I phase diagram. If the curve in Fig. 9.11 is turned upside down, the azeotrope is said to be *high boiling*. Ethanol and water form a high boiling azeotrope at about 95% ethanol, which is why the common laboratory ethanol is only 95% pure while most reagent grade solvents are 99+%.

9.8 TERNARY PHASE DIAGRAMS

Any mixture of three components can be represented within an equilateral triangle. All points within the triangle represent one and only one of the infinitely many possible mixtures. Because there are now three composition variables $f = C - P + 2 = 3 - P + 2$, both T and p are held constant for the three-component triangular representation on a two-dimensional surface. If only one phase is present, all solutions are permitted and the composition can take any point in the triangular 2-space. When two phases are present, the composition is restricted to the locus of points on the coexistence curve. Like the Type II phase diagram, many three-component phase diagrams are known having a dome-shaped coexistence curve. They are widely used for characterizing and selecting solvents used in industrial processes. Intersections of tie lines with the coexistence curve indicate the composition of phases in two-phase systems formed when the total composition point is below the coexistence curve.

In Fig. 9.12, solvents B and C are nearly completely immiscible, but addition of solvent A brings the composition of the two coexisting phases, the termini of the tie lines, closer together until finally there are ternary solutions of A, B, and C that are homogeneous. These solutions correspond to ternary mixtures above the coexistence dome.

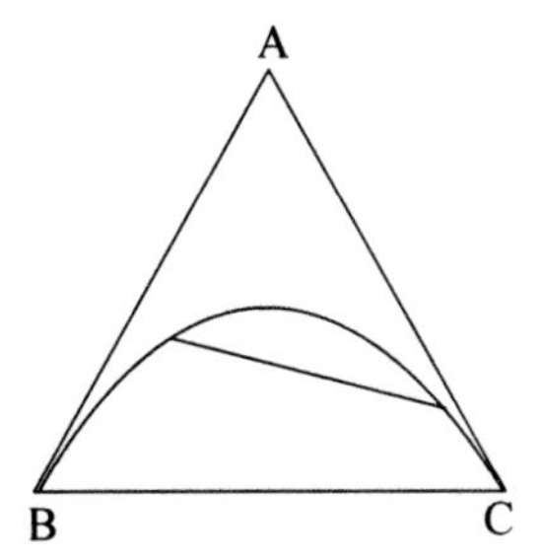

FIGURE 9.12 A ternary phase diagram with a tie line. At constant p and T, $f = 3 - 1 = 2$ within the two-dimensional triangular surface, but $f = 3 - 2 = 1$ on the coexistence curve. In general, the area under the curve is smaller at higher temperatures. Numerous, more complicated forms are known.

Addition of A may cause a milky suspension of immiscible phases to suddenly clear. Conversely, addition of C to a clear stirred solution of A and B may cause it to go milky at some concentration. The transition is sudden and sharp and can be used as the end point of a *phase titration*.

PROBLEMS AND EXAMPLES

Example 9.1 The Enthalpy of Vaporization of H_2O

Handbook values for the vapor pressures of water between 273 and 373 K are available. Plot p_{vap} vs. $-1/T$ for several temperatures in the vicinity of 298 K and use a commercial curve fitting routine to determine the enthalpy change for the vaporization of water at that temperature.

Solution 9.1 The handbook gives p_{vap} at temperatures from 273 to 373 K. We have selected six values symmetrically grouped around 298 K and presented them as $1/T\ K^{-1}$ in Table 9.1.

TABLE 9.1 Negative Inverse Temperatures and Vapor Pressures for Water.

$-1/T$	p_{vap}
−3.6600e−3	0.6110
−3.5300e−3	1.2300
−3.4100e−3	2.3400
−3.3000e−3	4.2500
−3.1900e−3	7.3800
−3.1000e−3	12.3400

The curve of p_{vap} vs. $-1/T$ for water is shown in Fig. 9.13.

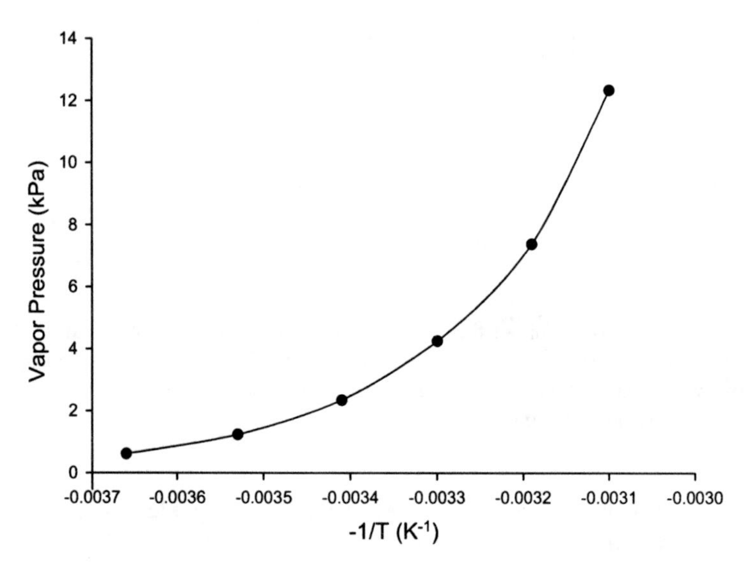

FIGURE 9.13 The liquid–vapor coexistence curve of water leading to $\Delta_{vap}H(H_2O) = 44.90\,\text{kJ mol}^{-1}$.

The vapor pressures over pure water were selected at 10 K intervals from 273 to 323 K, plotted, and submitted to a standard curve-fitting routine (SigmaPlot 11.0). A two-parameter exponential curve fit of p_{vap} vs. $-1/T$ (`Statistics → Nonlinear → Regression Wizard → Exponential Growth → Single, 2-Parameter`) gave the empirical constant b (Output `b 5400.6916`) K, where the unit of the empirical constant is kelvins in order to make the exponent $-b/T$ unitless. The curve fit is good, having a mean residual (difference between calculated and experimental values) of 1 part per thousand or 0.1% over a range of about 12 kPa. We have the relationship between the parameter b and $\Delta_{vap}H$:

$$b = \frac{\Delta_{vap}H}{R}$$

which gives us

$$\Delta_{vap}H = R(b) = 8.314\left(5.401 \times 10^3\right) = 44.90 \times 10^3 = 44.90\,\text{kJ mol}^{-1}$$

This value is "averaged out" by the curve fitting technique over the temperatures symmetrically distributed around 298 K. The handbook value of $\Delta_{vap}H$of water at 298 K is $43.99\,\text{kJ mol}^{-1}$.

Example 9.2 Ternary Phase Diagrams

The ternary phase diagram ABC in which A is completely miscible with B and C but B and C are only partially miscible in each other looks like a combination of Figs. 9.9 and 9.12 except that the tie lines are not horizontal.

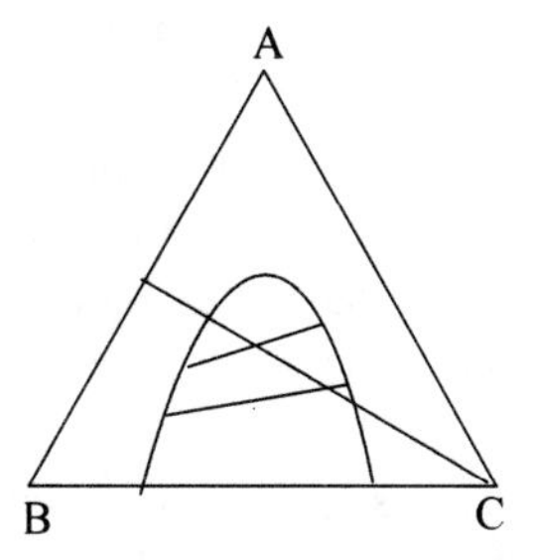

FIGURE 9.14 A ternary phase diagram in which B and C are partially miscible. Figure 9.13 might approach this form at higher temperatures.

Starting with a solution of 0.5 mol of A and 0.5 mol of B at a constant temperature corresponding to that of phase diagram Fig. 9.14, we add component C in small portions. What happens?

Solution 9.2 The phase behavior is somewhat complicated. At the first few small increments of component C, a clear homogeneous solution of A, B, and C results, corresponding to points on a straight line from the midpoint of axis AB in the direction of apex C. Soon the coexistence curve is crossed and the solution splits into two phases. The overall composition continues along the straight line toward C as increments are added, but the system now consists of two phases corresponding to the end points of the tie line. At first, a minute amount of the second phase appears; but further along in the addition process, substantial amounts of both phases are present. Their compositions always correspond to the end points of the tie line as it intersects the coexistence curve.

There comes a time when the short end of the tie line approaches the overall composition line. The amount of each phase is in the inverse ratio of the length of the tie line cut off by the overall composition line such that the AC-rich phase predominates over the AB rich phase. Ultimately the coexistence curve is crossed again and the solution clears, its composition corresponding to the points on the lower right of the overall composition curve approaching the C apex.

Problem 9.1

(a) How many components are there in a dilute solution of sodium acetate NaAc in water?

(b) A drop of HCl is added to the solution in part a. Now the anions of the weak acid Ac^- and the strong acid Cl^- are competing for the protons H^+. How many components are there in this system of Na^+, Ac^-, H^+, Cl^-, H_3O^+, and a minimal concentration of OH^-?

Problem 9.2

(a) What is the effect of a decrease in atmospheric pressure on the freezing and boiling points of water?

(b) What is the effect of a decrease in atmospheric pressure on the freezing and boiling points of benzene?

(c) Describe the behavior of a system that is carried along a horizontal line above the critical point of the phase diagram for water (Fig. 9.3).

(d) Describe the behavior of a system that is carried along a horizontal line below the critical point, but above the triple point of the phase diagram for water (Fig. 9.3).

(e) Describe the behavior of a system that is carried along a horizontal line below the triple point of the phase diagram for benzene (solid lines Fig. 9.3).

Problem 9.3

Describe the behavior of the system with a mole fraction 0.75 well above the dome-shaped coexistence curve in Fig. 9.9 as the temperature is slowly decreased.

Problem 9.4

Describe the behavior of the system with a mole fraction 0.25 well above the coexistence curve in Fig. 9.11 as the temperature is slowly decreased.

Problem 9.5

The vapor pressure of liquid benzene is given by

$$\ln P_{\mathrm{vap}} = -\frac{4110}{T} + 18.33$$

in the approximate location of the triple point (McQuarrie and Simon, 1997). The sublimation pressure of solid benzene is given by

$$\ln P_{\mathrm{sub}} = -\frac{5319}{T} + 22.67$$

What is the triple point for benzene?

Problem 9.6

Sketch a binary phase diagram resembling Fig. 9.10.

(a) Locate the points corresponding to pure components. Call them components X and Y.

(b) Locate a point corresponding to an equimolar mixture of X and Y.

(c) Locate a point corresponding to a mixture of 20% X and 80% Y.

(d) Locate the point corresponding to the compound X_2Y.

Problem 9.7

Find an area with two degrees of freedom, a coexistence curve with one degree of freedom, and a point with no degrees of freedom on the phase diagram (Fig. 9.10).

Problem 9.8

Sketch a triangular phase diagram resembling Fig. 9.12.

(a) Locate the point corresponding to a binary solution containing components in equal amounts.
(b) Locate the point corresponding (approximately) to a solution with three components in the ratio 45%, 45%, 10%.
(c) Locate the point corresponding to 33.3% for each of the three components.

Problem 9.9

Sketch the three-component phase diagram for a system in which all three of the components are miscible (soluble) in all proportions X in Y, X in Z, and Y in Z, but phase separation takes place for some of the ternary mixtures XYZ.

10

CHEMICAL KINETICS

In general, chemical thermodynamics is concerned with where a chemical system starts and where it ends up, while *chemical kinetics* is concerned with how long it takes it to get there. Many analytical techniques exist for following the rate of change of the concentration of the reactants or products of a chemical reaction, but single atom counting is not usually done. An exception is the class of *radiochemical* reactions for which counters exist to register, count, and record in computer memory every decay particle produced. For example, the decay of radium gives α particles (helium nuclei) and radon $^{222}_{86}Rn$:

$$^{226}_{88}Ra \rightarrow {}^{222}_{86}Rn + {}^{4}_{2}He$$

Each click of the counter registers one α particle, hence one decay. Here radium is the radioactive element, and both radon and helium are products. Radon is referred to as the *daughter* of the parent element radium. Radon is itself a radioactive gas, and it decays to polonium and further along a radioactive series ending eventually as lead $^{206}_{82}Pb$. Radiochemical reactions show the statistical nature of rate laws very clearly, For that reason, they are a good introduction to the entire field.

10.1 FIRST-ORDER KINETIC RATE LAWS

Suppose you have a sample consisting of a hundred billion atoms of radioactive element X. By means of a suitable counter, you observe an average of one radioactive

Concise Physical Chemistry, by Donald W. Rogers

decay per second. You have observed a *rate* of decay $-dX/dt$ in the units (number of atoms)/time, usually given as a frequency (s^{-1}). Decay causes a decrease in the number of atoms X, which accounts for the minus sign in front of the rate. If you take 200 billion atoms, you find twice as many radioactive events. That only stands to reason. Radioactive decay is a random phenomenon. Observing twice as many atoms, one expects to see twice as many events in a specified time interval. Taking k times as many atoms gives k times as many radioactive events: rate $= kX$. Now you have a *rate*, a *rate law*, and a *rate equation*. The rate of disintegration is

$$-\frac{dX}{dt} = kX$$

The rate of product production is equal but opposite in sign to the rate of decay.

Provided that the daughter is not radioactive, starting with a number of atoms X_0 at a time $t = t_0 = 0$ (an arbitrary time that you choose to start watching your counter), the average count rate gradually decreases with time. If you watch long enough,[1] the count rate will be reduced to half of what it was at t_0. This is called the *half-time* of the radioactive element.

The rate equation can be rearranged to give

$$\frac{dX}{X} = -k\,dt$$

The ratio dX/X being unitless, $-k\,dt$ must be unitless as well, so the unit of k is s^{-1} because t is in seconds. Integration between t_0 and t is straightforward:

$$\int_{t_0}^{t} \frac{1}{X} dX = -k \int_{t_0}^{t} dt$$

gives

$$\ln \frac{X}{X_0} = -k(t - t_0)$$

In exponential form

$$\frac{X}{X_0} = e^{-k(t-t_0)}$$

X_0 is the initial concentration (a constant) and t_0 is usually set to zero so that $X_0 = X(0)$ and

$$X = X_0 e^{-kt}$$

[1] About 1600 years in the case of radium.

Some chemical reactions follow the same mathematical laws as radioactive decay. The rate equation in these cases is a *first-order differential equation*, so the reactions that follow it are called *first-order reactions*. A case in point is the decomposition of a dilute solution of hydrogen peroxide H_2O_2 in contact with finely divided platinum catalyst. The probability arguments given above apply equally to radioactive decay and to first-order chemical reactions. In the case of the decomposition of a dilute solution of hydrogen peroxide in contact with Pt, the reaction is first order with a half-time of about 11 minutes.

The half-time can be determined for a first-order reaction in a simple and illustrative way. In the logarithmic form of the first-order equation above, set $X = \frac{1}{2}X_0$, which is the definition of the half-time. Now

$$\ln \frac{X}{X_0} = \ln \frac{\frac{1}{2}}{1} = -kt_{\frac{1}{2}}$$

$$kt_{\frac{1}{2}} = -\ln \frac{\frac{1}{2}}{1} = \ln \frac{1}{\frac{1}{2}} = \ln 2 = 0.693$$

and

$$k = 0.693/t_{\frac{1}{2}}$$

The half-time is found by drawing a horizontal halfway between the base line and X_0. The horizontal intersects the experimental curve at $t_{\frac{1}{2}}$. Notice that the concentration does not enter into calculation of the rate constant. Knowing either the half-time or the rate constant gives you the other one. Generally the half-time is easier to observe.

The exponential decay of X with t is seen in Fig. 10.1. The half-time is seen there as well. The linear decrease of $\ln X$ with t is seen in Fig. 10.2.

The linear plot of $\ln X$ is usually used to obtain the rate constant because it averages many experimental points and is, presumably, more reliable than any one

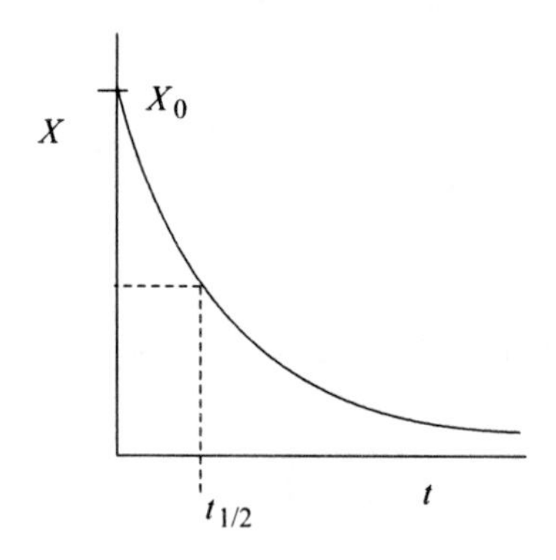

FIGURE 10.1 First-order radioactive decay.

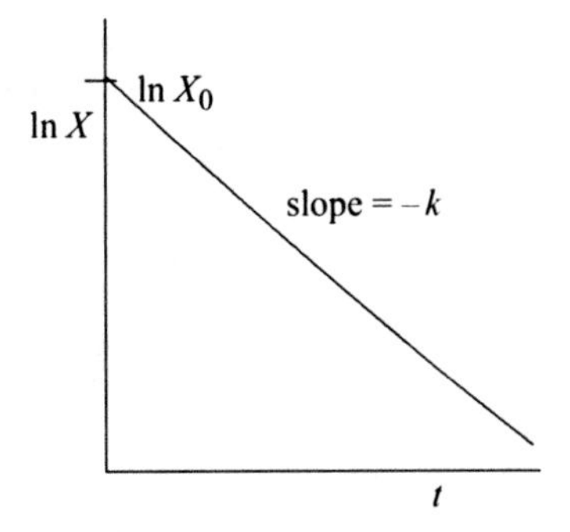

FIGURE 10.2 Logarithmic decay of a radioactive element.

measurement. From the form of the equation, ln X may be taken as the dependent variable while t may be as independent, whereupon the slope is equivalent to minus the rate constant. The constant of integration can be evaluated by setting $t = 0$,

$$C = -\ln X_0$$

enabling us to write

$$-\ln X = kt - \ln X_0$$

which is equivalent to the forms obtained by integration between limits:

$$\ln \frac{X}{X_0} = -kt$$

10.2 SECOND-ORDER REACTIONS

From the simple physical picture of a reacting mixture of reagents of two different kinds, say A and B, in which molecules collide to produce one or more products, it is logical to suppose that if either A or B is in short supply, the reaction must go slowly because fewer A–B collisions take place even though the total number of collisions may still be large. In the form of a rate law, the rate is proportional to both A and B concentrations, hence it is proportional to the *AB* product[2]:

$$\text{Rate} = -\frac{dA}{dt} = -\frac{dB}{dt} = kAB$$

[2]The identity of the molecules is given in regular type AB, but the concentration variable referring to the different species is italic *AB*.

If the reactants are mixed so that the initial concentrations A and B are equal, they will remain equal throughout, because for each molecule of one that reacts, a molecule of the other has also reacted. As a result, $AB = A^2$ throughout, giving the special case of the rate equation

$$-\frac{dA}{dt} = kAB = kA^2$$

and

$$\frac{dA}{A^2} = -k\,dt$$

Integrating

$$\int \frac{1}{A^2} dA = -k \int dt$$

gives

$$-\frac{1}{A} = -kt + C$$

To evaluate the constant of integration, let $t = 0$. The concentration A is its initial value A_0, so $-\frac{1}{A_0} = C$ and

$$\frac{1}{A} = kt + \frac{1}{A_0}$$

which leads to

$$\frac{A_0 - A}{A_0 A} = kt$$

The relation between $t_{1/2}$ and k for this restricted form of the second-order reaction is $kt_{1/2}A_0 = 1$.

An example of second-order kinetics in which the rate law $-dA/dt = kAB = kA^2$ applies is the thermal decomposition of NO_2:

$$2NO_2(g) \rightarrow 2NO(g) + O_2(g)$$

The rate is controlled by collisions of NO_2 molecules with other NO_2 molecules, hence the stipulation $A = B$ throughout is met. Although in this case the order of the reaction is the same as the stoichiometric coefficient, that need not be the case as exemplified by the thermal decomposition of N_2O_5:

$$2N_2O_5(g) \rightarrow 4NO_2(g) + O_2(g)$$

which is *first* order.

10.3 OTHER REACTION ORDERS

In general, the order of a chemical reaction is the sum of the exponents in the rate equation $a + b + c + \cdots$:

$$-\frac{dA}{dt} = kA^a B^b C^c \ldots$$

The order may be fractional or even decimal, but it is not a large number. Zero- and third-order reactions exist, but they are not common. A zero-order reaction is one in which the rate does not vary with time:

$$\frac{dA}{dt} = k$$

This rate law might be followed by a reaction depending on some constant like the surface area of a catalyst on which the reaction is taking place.

There are systematic mathematical methods for determining the order of a reaction from a data set of concentration vs. time even when the order is not simple.

10.3.1 Mathematical Interlude: The Laplace Transform

The *Laplace transform* is used to simplify differential equations. For the function $F(t)$ where t is the time and $F(t)$ is zero at $t < 0$, the Laplace transform $f(s)$ of $F(t)$ is defined as

$$f(s) = L\,[F(t)] = \int_0^\infty e^{-st} F(t)\,dt$$

The inverse Laplace transform of $f(s)$ is $F(t)$. The Laplace transform $L\,[F(t)]$ is a function of a function; it is called a *functional*. Extensive tables of Laplace transforms and inverse Laplace transforms are available (*CRC Handbook of Chemistry and Physics 2008–2009*, 89th ed.).

First we shall convert the derivative of the function $F(t)$ to a simple algebraic form. By definition,

$$L\left[\frac{dF(t)}{dt}\right] = \int_0^\infty e^{-st}\frac{dF(t)}{dt}dt$$

Integration by parts gives

$$\begin{aligned} L\left[\frac{dF(t)}{dt}\right] &= F(t)e^{-st}\Big|_0^\infty - \int_0^\infty F(t)d(e^{-st}) \\ &= -F(t=0) + s\int_0^\infty F(t)e^{-st}dt \\ &= -F(t=0) + sf(s) \end{aligned}$$

10.3.2 Back to Kinetics: Sequential Reactions

In this section we shall obtain the concentration B in the a sequence of two first-order reactions:

$$A \xrightarrow{k_1} B \xrightarrow{k_2} C$$

The differential equations are

$$\frac{dA}{dt} = -k_1 A$$
$$\frac{dB}{dt} = k_1 A - k_2 B$$
$$\frac{dC}{dt} = k_2 B$$

From the first equation, we have

$$A = A_0 e^{-k_1 t}, \qquad B_0 = C_0 = 0$$

where the subscripted 0 indicates the initial concentration. Substituting this result into the equation for $B(t)$, we obtain

$$\frac{dB(t)}{dt} = k_1 A_0 e^{-k_1 t} - k_2 B(t)$$

Now take the Laplace transform of both sides and solve for $b(s)$, the transform of $B(t)$:

$$sb(s) - B(t = 0) = \frac{k_1 A_0}{s + k_1} - k_2 b(s)$$

where $B(t = 0) = B_0 = 0$, so the remaining terms are

$$sb(s) + k_2 b(s) = \frac{k_1 A_0}{s + k_1}$$
$$b(s)(s + k_2) = \frac{k_1 A_0}{s + k_1}$$
$$b(s) = \frac{k_1 A_0}{s + k_1} \frac{1}{(s + k_2)} = k_1 A_0 \frac{1}{(s + k_1)} \frac{1}{(s + k_2)}$$

which is our solution for the Laplace transform of both sides of the rate equation. Now take the inverse Laplace transform. The Laplace transform $b(s)$ inverse transforms, of course, to $B(t)$ and the inverse transform of $\frac{1}{(s+k_1)}\frac{1}{(s+k_2)}$ is

$$\frac{1}{(s+k_1)}\frac{1}{(s+k_2)} \overset{\text{inv transform}}{\longrightarrow} \frac{1}{k_1-k_2}\left(e^{-k_1 t}-e^{-k_2 t}\right)$$

so

$$B(t) = k_1 A_0 \frac{1}{k_1-k_2}\left(e^{-k_1 t}-e^{-k_2 t}\right)$$

10.3.3 Reversible Reactions

No reaction goes to completion. Reactions we call "complete" are those in which the concentration of, for example, reactant A is negligible or not detectable in the reaction

$$\mathrm{A} \rightarrow \mathrm{B}$$

This is the case when the standard free energy of A is much larger than that of B, so that the reaction is accompanied by a significant decrease in free energy and leads to B more stable than A.

In many cases, something closer to a free energy balance exists. We write the reaction as an equilibrium characterized by an equilibrium constant K_{eq} (Chapter 7). Kinetic rate constants k_f and k_b for forward and back reactions compete to establish the equilibrium balance:

$$A \underset{k_b}{\overset{k_f}{\rightleftarrows}} B$$

$$\frac{k_f}{k_b} = K_{\text{eq}} = \frac{B}{A}$$

When A and B are mixed in concentrations A and B that are not the equilibrium concentrations, either the forward or the back reaction is faster than the other and the system is displaced so as to approach equilibrium.

The rate of depletion of A,

$$\left(-\frac{dA}{dt}\right)_{\text{obs}} = k_f A - k_b B$$

is smaller than the rate that would be found if the reaction went to completion or if we had some mechanism for removal of B immediately as it is formed. The rate of approach to equilibrium is always less than either the forward reaction or the back reaction because k_f and k_b work in opposition to one another.

For simple reversible systems,

$$A \rightleftarrows B$$

$$A + B \rightleftarrows C$$

$$A + B \rightleftarrows C + D$$

etc.

equations can be worked out that relate the observed rate constant k_{obs} for the approach to equilibrium to k_f and k_b, the elementary reactions that contribute to it (Metiu, 2006).

The concept of equilibrium as the result of opposite forward and back reactions is called the principle of detailed balance; for example, at equilibrium,

$$k_f A - k_b B = 0$$

The equations implied by the principle of detailed balance rest on the assumption that the concentrations of reactant and product vary in the simplest possible way. This assumption is often violated and must never be taken for granted.

Some enzyme-catalyzed reactions, including the famous Michaelis–Menten (Houston, 2001) mechanism, are examples of more complicated reaction mechanisms. Despite their complexity, they can often be broken down into elementary steps and equilibriums. The kinetics of complex reactions can sometimes be simplified by regarding one component of the reaction as a constant during part of the chemical process. This is the *steady-state approximation* (Metiu, 2006).

An especially important class of reaction mechanisms is that of the chain reactions in which one molecular event leads to many, possibly very many, products. The classic example for chemists is production of HBr from the elements:

$$H_2 + Br_2 \rightarrow 2HBr$$

for which we might guess the rate law to be

$$\frac{1}{2}\frac{d\,[\text{HBr}]}{dt} = k\,[\text{H}_2]\,[\text{Br}_2] \qquad \textbf{wrong}$$

but that guess would be wrong. Instead, the rate law is

$$\frac{1}{2}\frac{d\,[\text{HBr}]}{dt} = k\,[\text{H}_2]\,[\text{Br}_2]^{1/2} \qquad \textbf{right}$$

An acceptable mechanism for this reaction depends upon three elementary reaction types, *initiation*, *chain*, and *termination*:

Initiation	$Br_2 + M \xrightarrow{k_1} 2Br\cdot + M$
Chain	$Br\cdot + H_2 \xrightarrow{k_2} HBr + H\cdot$
Chain	$H\cdot + Br_2 \xrightarrow{k_3} HBr + Br\cdot$
Termination	$Br\cdot + Br\cdot \xrightarrow{k_{-1}} Br_2$

The point of this proposed mechanism is that the cyclic chain steps can continue indefinitely because the Br· *free radicals* used in the first chain step are produced in the second chain step. Initiation can be by a rather unlikely process such as collision with a high-energy molecule M in the first step or impact of a photon from a flame or spark as in chain explosions. Though initiation may not occur very often, it can have a large effect. A hydrogen–oxygen explosion is an example. A very small spark can cause a very large explosion. The HBr chain, though it may yield many molecules for one initiation step, does not, of course, go on forever. Some step such as recombination of the Br· free radicals terminates the chain or we run out of Br_2 or H_2 and the reaction stops.

If we assume that initiation and termination are rare events by comparison to the chain steps, for every Br· used up in the first step of the chain, one is produced in the second chain step, so the amount of free radical present at any time during the reaction is constant:

$$\frac{d\,[Br\cdot]}{dt} = 0$$

This is an example of the steady-state hypothesis. Making this assumption, a few lines of algebra (Houston, 2006) lead to the correct rate equation given above:

$$\frac{1}{2}\frac{d\,[HBr]}{dt} = k\,[H_2]\,[Br_2]^{1/2}$$

If, by a more complicated mechanism, two or more reactive species are produced on each step, the amount of reactive species may increase rapidly. For example, if two reactive species are produced at each step, the geometric series, 1, 2, 4, 8, 16, ... is followed. If the chain steps are fast, this kind of mechanism takes place with explosive violence. This type of mechanism is characteristic of nuclear fission bomb reactions.

Free radical and (controlled) chain reactions are also characteristic of some biochemical reactions, and they can behave in ways that are beneficial or detrimental to the organism. A hydroperoxide RCOO· free radical chain may destroy lipid molecules in a cell wall by *lipid peroxidation* with disastrous results for the cell. Free radicals derived from tocopherol (vitamin E) *antioxidants* act as sweepers in the blood, interfering with chain propagation, thereby slowing or preventing cell degradation. Free radicals have been detected in rapidly multiplying natal or neonatal cells (beneficial) but they are also found in cancers undergoing uncontrolled reproduction. The biological role of free radicals is complex and not completely understood.

10.4 EXPERIMENTAL DETERMINATION OF THE RATE EQUATION

For some reactions, determination of the rate law and the rate constant may be as simple as mixing the reagents, titrating a sample of one of the reactants or anticipated products, waiting a while, titrating another sample, and so on, to obtain a series of concentrations A, each at a time t. Treating this data set by an appropriate mathematical procedure gives the rate law and k.

Some of the most interesting contemporary research in kinetics involves tracking reaction components or intermediates by physical means, perhaps because they are of fleeting existence, being present in the reacting system for a second or less.

A straightforward approach to fairly fast reactions involves use of a *stopped-flow* reactor. One experimental design consists of two syringes connected to a mixer. To initiate the reaction, the syringes are driven simultaneously so that reactants A and B flow into the mixer, are mixed thoroughly, and flow out of the mixer into a tube where the reaction mixture is monitored by a spectrophotometer. The output from the spectrophotometer is transmitted to a microcomputer at time intervals of, say, a few milliseconds or, perhaps, microseconds. The problem is well within the capabilities of contemporary spectrophotometers and computer interfacing hardware, so the principal limiting factor is speed of mixing. Reactions with a half-time of a millisecond or less have been successfully studied in this way.

For very fast reactions, the limitation has been only the ingenuity of the researcher and *flash photolysis* studies of reactions with the astonishing half-time of $t_{1/2} = 10^{-15}$ s indicate that, for some people, this hasn't been much of a limitation (Zewail, Nobel Prize, 1999, see Zewail, 1994).

10.5 REACTION MECHANISMS

The *mechanism* by which molecules actually combine is rarely as simple as the rate law suggests. Complex reactions may involve equilibriums,

$$\mathrm{A} \rightleftarrows \mathrm{B}$$

consecutive reactions,

$$\mathrm{A} \rightarrow \mathrm{B} \rightarrow \mathrm{C}$$

parallel reactions,

$$\begin{array}{c} \mathrm{A} \rightarrow \mathrm{C} \\ \Updownarrow \\ \mathrm{B} \rightarrow \mathrm{D} \end{array}$$

or combinations of any or all of them.

Because of these complications, the terms *order* and *molecularity* must be clearly differentiated. The order is the sum of exponents in the rate equation, while the molecularity is the number of molecules taking part in a simple reaction step. The order and molecularity will be the same in only the simplest reactions. An example in which the reaction order seems at first to be unrelated to the molecularity is the important atmospheric process involving ozone

$$2O_3 \rightarrow 3O_2$$

which is *not* second order. Rather, it follows the rate law:

$$-\frac{dO_3}{dt} = k\frac{O_3^2}{O_2}$$

This can be explained by a reaction mechanism starting with a *unimolecular* step

$$O_3 \rightleftarrows O_2 + O$$

governed by the equilibrium constant

$$K_{eq} = \frac{O_2 O}{O_3}$$

The second step of the proposed reaction mechanism is *bimolecular*

$$O + O_3 \rightarrow 2O_2$$

and it is the slower step of the two. As in any sequential process, the slowest step controls the rate of the entire process. In this case there are only two steps: a relatively fast first step and a rate-controlling second step. The rate law for the bimolecular collision between O and O_3 is

$$-\frac{dO_3}{dt} = k_2 O O_3$$

Solving for O, we obtain

$$O = \frac{K_{eq} O_3}{O_2}$$

Substituting for O in the previous equation,

$$-\frac{dO_3}{dt} = k_2 O O_3 = k_2 \frac{K_{eq} O_3}{O_2} O_3 = k\frac{O_3^2}{O_2}$$

we get the observed rate law and rate constant.

Ozone is depleted by about 4% per decade in its total volume in the Earth's stratosphere and to a much larger extent over Earth's polar regions (the ozone hole). The most important process is catalytic destruction of ozone by atomic halogens due to photodissociation of chlorofluorocarbon compounds (commonly called freons) and related compounds. Since the ozone layer prevents most harmful wavelengths (270–315 nm) of ultraviolet light from passing through the Earth's atmosphere, decreases in ozone have generated concern and led to banning the production of ozone depleting chemicals. It is suspected that a variety of biological consequences such as increases in skin cancer, damage to plants, and reduction of plankton populations in the ocean's photic zone may result from the increased UV exposure due to ozone depletion (abstracted from Wikipedia).

10.6 THE INFLUENCE OF TEMPERATURE ON RATE

Variation of reaction rate with temperature is usually exponential, as you know if you have ever been tempted to heat a reaction mixture "just a little bit more." Observed exponential rate curves tempt us to write equations of the same form as the van't Hoff equation to describe the rate constant k

$$\ln k = -\frac{\Delta_a H}{R}\left(\frac{1}{T}\right) + \text{const}$$

The constant $\Delta_a H$ is the *enthalpy of activation*, often called the "energy" of activation. It can be found empirically.

The activation process can be visualized by pushing over a block of wood, as in Fig. 10.3. Position c is more stable than a. Thermodynamically, c corresponds to the products in a spontaneous reaction accompanied by a decrease in free energy. Even though the block in position a has a higher energy than c, a does not go to c spontaneously unless a force is applied to push it over, as in b. This is the *activation* process. The center of gravity of b is higher than either a or c, hence it has higher energy. Some energy—the push—must be put into the system a, and it must be driven to a less stable configuration b before it spontaneously goes to the most stable configuration c. The energy liberated in going from b to c is greater than the energy put into the push, and the process liberates energy.

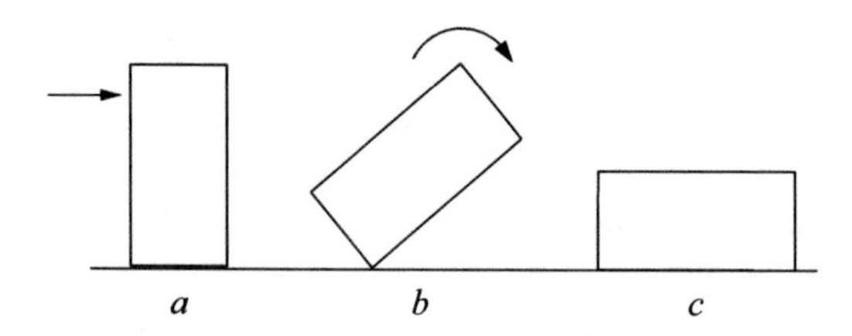

FIGURE 10.3 An activation energy barrier between an unstable position and a stable position.

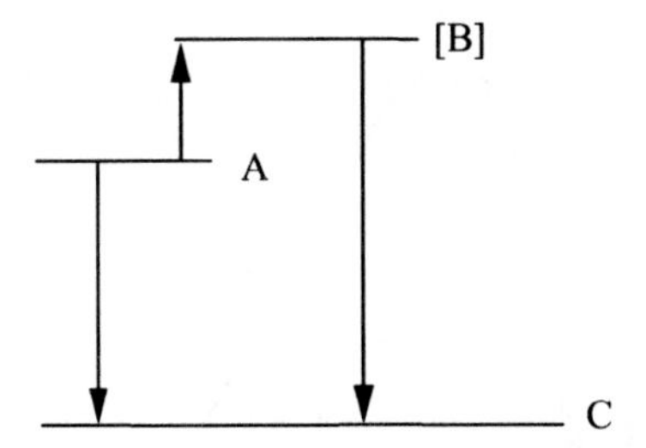

FIGURE 10.4 Enthalpy level diagram for an activated complex [B].

In a chemical system, kinetic processes can be written

$$A \rightarrow [B] \rightarrow C$$

where [B] is an *activated complex*, possibly a very fleeting intermediate like a free radical. An enthalpy diagram for the reaction involving an intermediate is given as Fig. 10.4. The enthalpy change for the reaction is given as the down arrow from level A to level C. This is the measured thermochemical value. The activation energy of the reaction is the up arrow from A to [B]. Viewed only from the enthalpy point of view, this is a nonspontaneous process. If the activation enthalpy to get to [B] is supplied, the reaction takes place and the enthalpy change from [B] to C more than repays the enthalpy debt incurred in production of [B]. The enthalpy hill from A to [B] is the *activation barrier* (Fig. 10.5).

The enthalpy to get over the activation barrier comes from ambient heat supplied to the system at temperatures greater than 0 K. Evidently, the higher the temperature, the more enthalpy is supplied to the system and the more molecules of a statistical distribution have enough enthalpy to get over the barrier. That supplies the qualitative answer to the question: Why do reactions go exponentially faster at higher temperatures than at lower temperatures? The number of activated molecules rises exponentially with temperature.

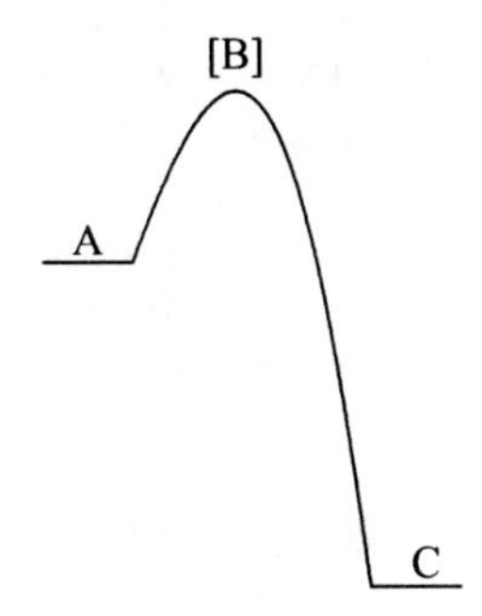

FIGURE 10.5 An activation barrier.

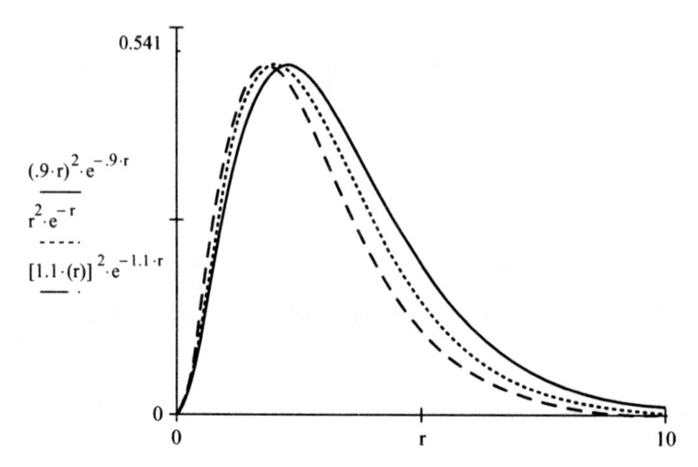

FIGURE 10.6 A Boltzmann distribution of molecular speeds. The area under the tail of the curve beyond r increases exponentially.

The quantitative relationship between the reaction rate constant and temperature is

$$k = ae^{-\Delta_a H/RT}$$

This can be appreciated by looking at a Boltzmann distribution of reactant molecules. The area under the Boltzmann curve that surpasses the activation barrier at r increases sharply (exponentially) with temperature, leading to the observed exponential increase in k (Fig. 10.6). The exponential relationship between k and T is called the *Arrhenius rate law.*

10.7 COLLISION THEORY

One can calculate the number of molecules colliding with one another in a gaseous sample at a specified temperature. Suppose Z_{AB} is the collision frequency for a collection of gas molecules capable of undergoing a bimolecular reaction. If all the molecules were activated, the rate would be directly determined by the number of collisions per unit time of molecules of kind A with those of kind B. Not all molecules will be activated, however, and we must multiply Z_{AB} by the fraction that are. This is the Boltzmann factor $e^{-\Delta_a H/RT}$.

For a gas reaction taking place between molecules that are structurally complicated, a probability factor P is introduced to account for collisions between activated molecules that are not oriented toward each other in the proper way for reaction to occur. Now

$$k = PZ_{AB}e^{-\Delta_a H/RT}$$

where the integration factor a is included in the empirical probability factor P.

For example, if a bromide ion collided with an alkyl alcohol, in all likelihood it would collide with the "wrong" part of the molecule.

$$Br^- + CH_3(CH_2)_nCH_2OH \rightarrow \text{No Reaction}$$

Only rarely would it collide with the OH group, resulting in reaction.

$$\begin{array}{c} \qquad\quad Br^- \\ \swarrow \qquad \\ CH_3(CH_2)_nCH_2OH \rightarrow OH^- + CH_3(CH_2)_nCH_2Br \end{array}$$

The probability factor P would be correspondingly low.

10.8 COMPUTATIONAL KINETICS

A critical factor determining the rate of a chemical reaction is the activation enthalpy, which is the difference $\Delta_a H$ between the reactant enthalpy and the relatively unstable *activated complex*. We can write

$$\Delta_a H = H_{ac} - H_r$$

where H_{ac} is the enthalpy of the activated complex and H_r is the enthalpy of the reactant molecule. Hehre (2006) has computed several activation enthalpies, including one for the isomerization of methyl isocyanide to acetonitrile:

$$CH_3N{\equiv}C \rightarrow CH_3C{\equiv}N$$

The postulated activated intermediate is the three-membered ring

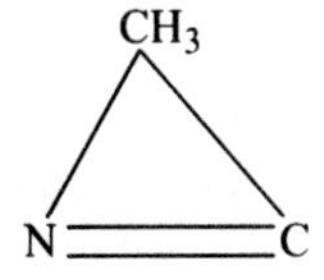

This energetic species does not lend itself to experimental study, but it can be treated computationally. Results from some computational kinetics studies are roughly comparable to measured values, but this is a difficult field because computed activation enthalpies are very sensitive to the estimated structure of the activated complex and the rate constant is an exponential function of $\Delta_a H$.

PROBLEMS AND EXAMPLES

Example 10.1 The Enthalpy of Activation

If the half-time of a first-order reaction is 20 minutes at 298 K and 4 minutes at 313 K, what is the activation enthalpy?

Solution 10.1 The half-times of the reaction at the two temperatures are inversely related to the rate constants $k = 0.693/t_{1/2}$; therefore the rate constants are in the ratio of 5:1 with temperature ratio $T_2/T_1 = 313/298$. These values are then loaded into the rate equation, which has been integrated between the appropriate limits:

$$\ln \frac{k_2}{k_1} = -\frac{\Delta_a H}{R}\left(\frac{1}{T_2} - \frac{1}{T_1}\right)$$

The ratio of the rate constants is 5.0 and the difference of the reciprocal temperatures is -1.60×10^{-4}, therefore $\Delta_a H = 83.2$ kJ mol^{-1}:

$$\ln 5 = -\frac{\Delta_a H}{R}\left(\frac{1}{313} - \frac{1}{298}\right)$$

$$1.609 = -\frac{\Delta_a H}{R}(0.00319 - 0.00336) = \frac{\Delta_a H}{R}(0.0001608)$$

$$\frac{\Delta_a H}{R} = \frac{1.609}{0.0001608} = 1.00 \times 10^4$$

$$\Delta_a H = 1.00 \times 10^4 R = 8.314 \times 10^4 \text{ J mol}^{-1} = 83.1 \text{ kJ mol}^{-1}$$

Example 10.2

The relative intensity I of fluorescence from electronically excited iodine drops off according to the time sequence below, where the time is in milliseconds.

Time, ms	0	50	100	150	200	250	300
I, relative	1	0.61	0.42	0.30	0.21	0.08	0.02

What is the rate constant and half-time for this radiative decay? Give units. Plot relative intensity as a function of time in milliseconds (Fig. 10.7).

Solution 10.2 The sequence, when sketched out, may be a first-order decay. If it is, drawing a horizontal from $I = 0.5$ (relative) cuts the curve at about $t = 70$ ms, a first estimate of the half-time. Let's try to convert to ln I vs. time to see if the plot is linear.

The reaction is first order, as demonstrated by the linear plot of ln I vs. time in Fig. 10.8 (for more detail see Houston, 2001).

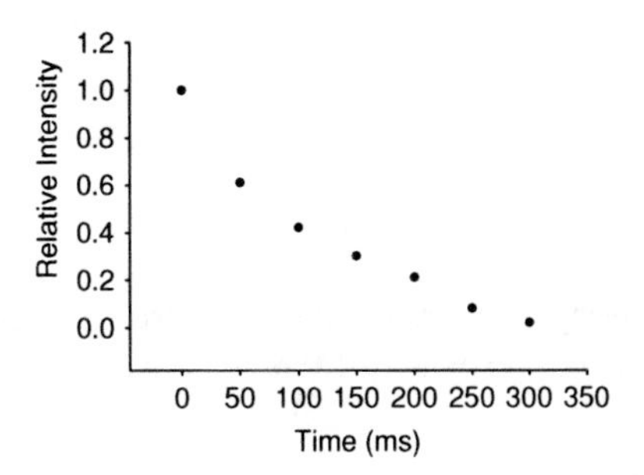

FIGURE 10.7 First-order fluorescence decline from electronically excited iodine in milliseconds.

The statistical routine of the SigmaPlot© 11.0 package gives, among other things,

Equation: Polynomial, Linear

$$f = y_0 + a^* x$$

	Coefficient	Std. Error
y_0	0.0889	0.1232
a	−0.0098	0.0007

The intercept, given here as y_0, is close to 0 as it should be and the slope, given here as a, is $-0.0098\,\text{ms}^{-1}$. This leads to a rate constant $dI/dt = -kI$ of $k = 9.8 \times 10^{-3}\,\text{ms}^{-1}$. The half-time for radiative decay is

$$t_{1/2} = \frac{\ln 2}{9.8 \times 10^{-3}\ \text{ms}^{-1}} = \frac{.693}{9.8 \times 10^{-3}\ \text{ms}^{-1}} = 71\ \text{ms}$$

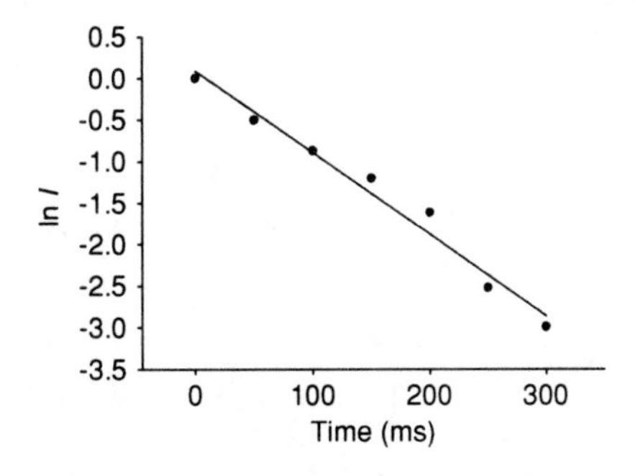

FIGURE 10.8 The natural logarithm of relative intensity vs. time for radiative decay. Within experimental error, the plot is linear (Section 10.1).

Problem 10.1

Assume that the reaction

$$\mathrm{A} \xrightarrow{k} \text{products}$$

is a simple first-order decay. If the amount of A drops from an initial value at $t = 0$ of 0.5000 mols to 0.0625 mol in exactly 1.0 hour, what is the rate constant k and half time for the reaction?

Problem 10.2

The curie (Ci) is a measure of radioactive decay frequency $1\ \mathrm{Ci} = 3.7 \times 10^{10}\ \mathrm{s}^{-1}$. A 1.50 mg sample of the β emitter technetium-99 has an activity of 5.66×10^{7} cpm (counts per minute). Recall that technetium is ${}^{99}_{44}\mathrm{Tc}$ and the β particle is an electron.

(a) What are the products of the reaction? Write the reaction.
(b) Find the frequency in units of Ci.
(c) Find the rate constant (decay constant) k, in whatever unit you choose, and find the half-life in units of years for ${}^{99}_{44}\mathrm{Tc}$.

Problem 10.3

In population ecology, an exponential growth curve

$$P(t) = Ae^{kt}$$

is often assumed for the uncontrolled growth of a newly introduced species where $P(t)$ is the population at time t and A is a parameter. A new crab was accidentally introduced into San Francisco bay several years go. Routine trapping and systematic population monitoring found 12 crabs at the beginning of the survey; two years later, 24 crabs were found at the same location by the same trapping and sampling method. What was the doubling time of the species? Find k and A. What is the estimated crab sample after 10 years from $t = 0$?

Problem 10.4

Further spot checks of the population increase of the invasive species showed the following results:

t (years)	0	2	3	5	8	10
$P(t)$	12	24	39	59	167	300

Reevaluate the parameters A and k using a linear least-squares method. What is the predicted sample at the 12 year survey?

Problem 10.5

The half-time of radioactive decay of ^{14}C is 5730 years. A wood sample from an Egyptian tomb had ^{14}C radioactivity of 7.3 ± 0.1 cpm (counts per minute) per gram of sample. Freshly harvested wood has a ^{14}C radioactivity of 12.6 ± 0.1 cpm per gram. How old is the wood from the tomb and what is the uncertainty of your answer?

Problem 10.6

Given that

$$A = A_0 e^{-k_1 t}$$

and

$$B(t) = k_1 A_0 \frac{1}{k_1 - k_2} \left(e^{-k_1 t} - e^{-k_2 t}\right)$$

how does C vary with time for the reaction

$$\mathrm{A} \xrightarrow{k_1} \mathrm{B} \xrightarrow{k_2} \mathrm{C}$$

Remember that $A + B + C = A_0$ at all times t, and the initial conditions are $A = A_0$ and $B = C = 0$.

Problem 10.7

Show that if $\dfrac{A_0 - A}{A_0 A} = kt$ for a second-order reaction, then $kt_{\frac{1}{2}} A_0 = 1$.

Problem 10.8

A sequence of irreversible first-order chemical reactions was observed for 20 hours. The sequence of reactants and products was

$$\mathrm{A} \overbrace{\rightarrow}^{k_1 = 0.6} \mathrm{B} \overbrace{\rightarrow}^{k_2 = 0.7} \mathrm{C} \overbrace{\rightarrow}^{k_3 = 0.06} \mathrm{D}\ldots$$

The first three rate constants are 0.6, 0.7, and 0.06 h^{-1} in the sequence given above, and the rate constant for decrease in component D going to another product E was 0.02. Plot the concentrations of A, B, and C. The first two equations can be found in Sections 10.1 and 10.3.2. The third equation is a rather intimidating:

$$C(t) := 0.6 \cdot 0.7 \cdot \left[\frac{e^{-0.6 \cdot t}}{(0.7 - 0.6) \cdot (0.06 - 0.6)} + \frac{e^{-0.7 \cdot t}}{(0.6 - 0.7) \cdot (0.06 - 0.7)} + \frac{e^{-0.06 \cdot t}}{(0.6 - 0.06) \cdot (0.7 - 0.06)}\right]$$

(There is a fourth and more equations that you can probably write down by extrapolation of A, B, and C. Try it!

Problem 10.9

Given the following experimental data in terms of time t and concentration of reactant NOBr in mol dm^3, find the rate law and the rate constant for the reaction:

$$NOBr(g) \rightarrow NO(g) + \tfrac{1}{2}O_2(g)$$

t	0	6	10	15	20	25
NOBr	0.025	0.019	0.016	0.014	0.012	0.011

Problem 10.10

The decomposition of NOCl to NO and Cl_2 is analogous to the reaction just studied:

$$NOCl(g) \rightarrow NO(g) + \tfrac{1}{2}Cl_2(g)$$

The second-order rate constant is 9.3×10^{-5} dm^3 mol^{-1} s^{-1} at 400 K and 1.0×10^{-3} dm^3 mol^{-1} s^{-1} at 430 K. What is the activation energy of this reaction?

11

LIQUIDS AND SOLIDS

Surface tension causes raindrops to be spherical, enables you to float a steel needle on the surface of water (if you're careful), and enables trees to draw liquid nourishment from the earth by capillary attraction. These phenomena and the beautiful hexagonal form of snowflakes are properties of liquids and solids, the condensed states of matter.

11.1 SURFACE TENSION

The surface of any liquid is enveloped by molecules that are not like other molecules because of an imbalance of forces acting on them that is different from the balanced forces in the interior (Fig. 11.1). An imbalance of forces on surface molecules causes them to be drawn in so as to form an elastic film of surface molecules enclosing the bulk molecules. The existence of such a film completely surrounding a falling rain droplet causes it to arrive at a stable structure with the minimum surface area surrounding the volume of the drop. The minimum surface area for a specified volume is that of a sphere. If there were no distortions due to gravity, air resistance, and so on, raindrops would be perfect spheres.

A model of the liquid surface film is not difficult to set up and analyze. Consider a small rectangular frame with one movable edge enclosing an elastic film such as a soap film, as in Fig. 11.2. The model is analogous to the three-dimensional expansion of a gas against a movable piston of area A except that area A is replaced by length l. The work done in the case of stretching a membrane is the product of the intensive

Concise Physical Chemistry, by Donald W. Rogers

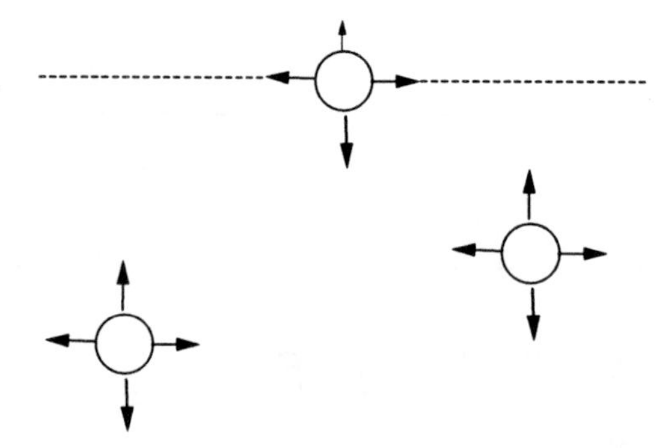

FIGURE 11.1 Intermolecular attractive forces acting upon molecules at an air–water interface.

variable force per unit length $\gamma \equiv f/l$ times the infinitesimal increase in area $d\sigma$:

$$dw = \left(\frac{f}{l}\right) d\sigma = \gamma\, d\sigma$$

The intensive variable force per unit length $\gamma = f/l$ in the two-dimensional model replaces the intensive variable pressure that is force per unit area of the piston face $p = f/A$ in the three-dimensional model of a piston and cylinder. Please do not confuse the area of the piston face A with the area of the liquid film σ.

From the physics of vibration of stretched strings, we have the force per unit length f/l defined as a *tension*. Because $d\sigma$ is an infinitesimal increment in the *surface* area of the membrane, the intensive variable $\gamma = f/l$ is called the *surface tension.*

Imagine a minute device such as that in Fig. 11.3 immersed in a soap solution and drawn up so that a film of the liquid occupies the area σ that can be expanded an amount $d\sigma$ by an upward force f on an edge of length l. Because the tension γ arises from the surfaces on either side of the liquid membrane drawn up into the framework, the work of expanding the bimembrane (a memebrane with two sides or surfaces) is twice the work for each surface, $dw = 2\gamma d\sigma$. The expansion of the surface is $d\sigma = l dh$, where dh is the increase in height of the movable edge of length l. In general, work is a force times a displacement, so $dw = 2\gamma d\sigma = 2\gamma l dh$.

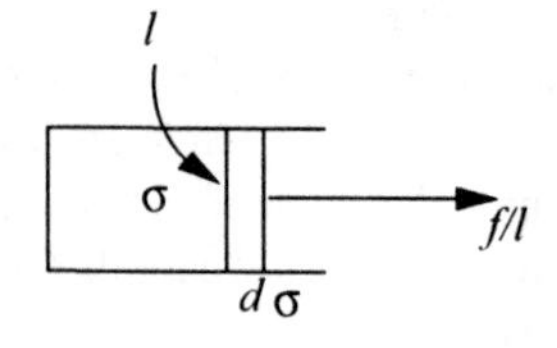

FIGURE 11.2 Stretching a two-dimensional membrane by moving an edge of length l. A real liquid would have two films, one facing you on the front surface of the page and one on the back surface.

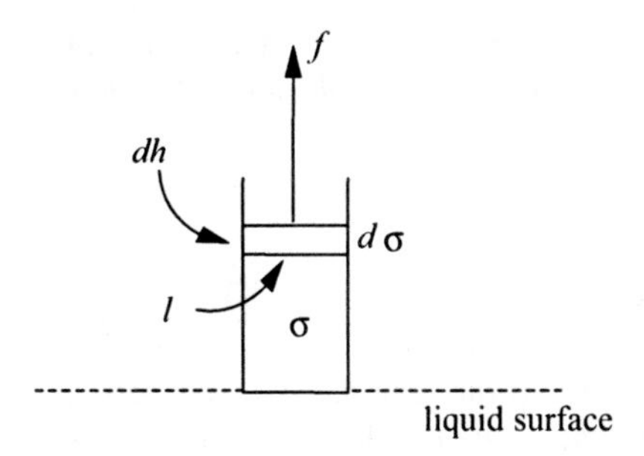

FIGURE 11.3 Stretching a two-dimensional liquid bimembrane.

Now consider capillary rise caused by surface tension in a tube of radius R. The length l of the movable edge in Fig. 11.3 is replaced by the circumference c of the tube in which the liquid rises, $c = 2\pi R$; but there is only one circular surface, so the factor 2 drops out of $2\gamma d\sigma$ only to reappear as $2\pi R$, so $dw = \gamma c dh = \gamma 2\pi R dh$. The force opposing capillary rise is mg due to lifting the mass of liquid m in opposition to gravitational acceleration g (Fig. 11.4). At equilibrium, the forces are in balance:

$$mg = \gamma 2\pi R$$

The volume of suspended liquid is that of a cylinder of height h and density $\rho = m/V$. From $V = \pi R^2 h$ for a cylinder, we get $m = \rho V = \rho\pi R^2 h$:

$$(\rho\pi R^2 h)g = \gamma 2\pi R$$

and

$$\gamma = \frac{\rho R h g}{2}$$

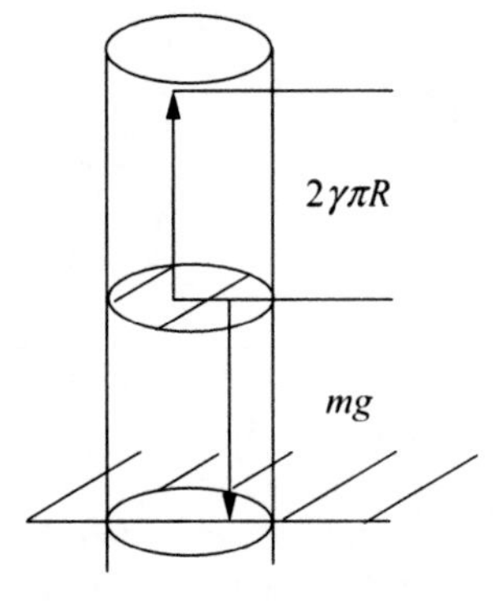

FIGURE 11.4 Capillary rise in a tube of radius R. The equilibrium height of the liquid column is determined by the balance of the capillary force and the gravitational force.

making it a simple matter to measure the surface tension of a liquid of density ρ that rises to a height h in a capillary of known radius R against the acceleration due to gravity, $g = 9.807\,\mathrm{m\,s^{-2}}$.

11.2 HEAT CAPACITY OF LIQUIDS AND SOLIDS

In 1907 Einstein showed that the heat capacity of many solids is 25 J $\mathrm{K^{-1}}$ $\mathrm{mol^{-1}}$ (in agreement with the prior law of Dulong and Petit) but that at some characteristic temperature it drops off along a sigmoidal curve to zero at 0 K (Fig. 11.5). Although his derivation is general, he used it to describe the heat capacity of diamond, an exceptional solid because of its strong tetrahedral bonding. The resulting curve, one of the most famous of twentieth-century science, was based on the new quantum theory of Max Planck, which might otherwise have gone unnoticed by many in the physics community. The success of Einstein's theory of heat capacities is shown in a reproduction of his original graph depicting the theoretical heat capacity of diamond (solid curve) compared to known experimental points.

In general, the molar heat capacity of a liquid is higher than the molar heat capacity of a gas because molecular motion in the liquid state implies distortion of the liquid structure in the vicinity of the moving molecule. This is what we would expect if we think of a liquid as an extremely nonideal gas. An example is mercury, which has the heat capacity of an ideal gas, $\frac{3}{2}R \cong 12.5\,\mathrm{J\,K^{-1}\,mol^{-1}}$ in the vapor state, but has $C_V = 23.6\,\mathrm{J\,K^{-1}\,mol^{-1}}$ in the liquid state. Notice how close liquid mercury is

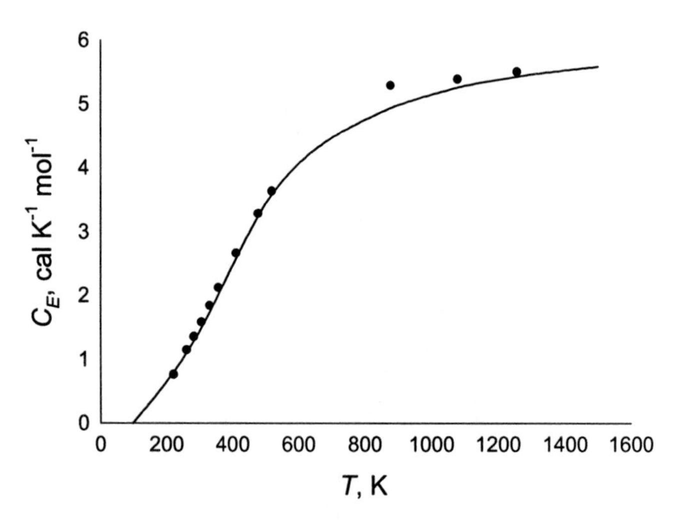

FIGURE 11.5 Heat capacity as a function of temperature. Einstein calculated the theoretical curve of the heat capacity of diamond (solid line) and compared it to known experimental points. Heat was normally measured in calories in Einstein's time. For more detail, see Rogers (2005).

in this respect to the law of Dulong and Petit, which predicts a heat capacity of 25 J K^{-1} mol^{-1} for *solid* metals. As usual, theoretical treatment of liquids is difficult because their behavior is between that of a perfectly ordered crystal and that of a perfectly random (statistical) gas.

11.3 VISCOSITY OF LIQUIDS

The viscosity of a liquid often reflects the degree of entanglement of molecules as they are moved past each other. The viscosity of heavy lubricating oil is greater than that of gasoline because the average molecular length is greater in oil than in gasoline. It takes some amount of work to push one molecule past its neighbors.

Consider a liquid flowing through a tube of radius R. At different radial distances from the center of the tube, the liquid is flowing at different rates. The greatest rate is precisely at the center of the tube, whereas the lowest rate occurs where the liquid experiences maximum frictional drag against its inner surface. It is convenient to consider the liquid flow as consisting of a large number of concentric laminar "sleeves" of thickness dr, where r is the radial distance from the center of the tube (Fig. 11.6). The viscous drag retarding the flow on one laminar sleeve relative to its neighbor sleeve is proportional to the difference in speed of adjacent lamina dv/dr times the surface area over which the lamina bear against each other. This is the circumference of the laminar sleeve times its length l:

$$\text{drag} \propto A\frac{dv}{dr} = \eta A\frac{dv}{dr} = \eta 2\pi r l\frac{dv}{dr} = f_{\text{viscous}}$$

The proportionality constant η in this equation is the *viscosity coefficient*. Because pressure is force per unit area $p = f/A$, the force driving the liquid through the tube under gravity flow is $f_{\text{grav}} = pA$, where A is the area of the end of the sleeve in

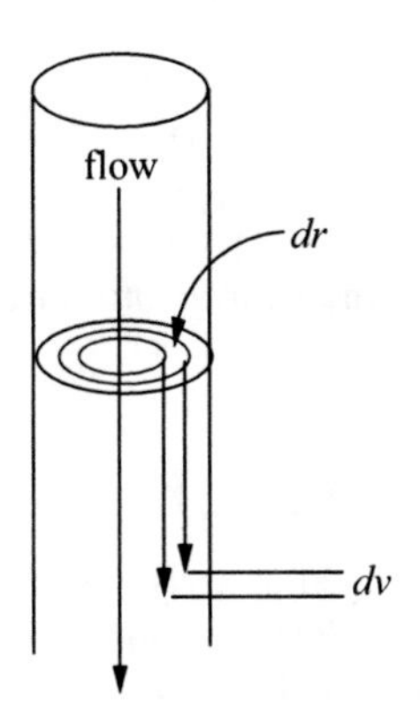

FIGURE 11.6 Approximation of laminar flow inside a tube. The difference in speed of flow between the inner and outer surfaces of a lamina is dv/dr.

question including all sleeves within it. One can think of a pencil of liquid driven by a force, f_{grav}, flowing coaxially down the tube. The gravitational force on each sleeve is $f_{grav} = pA = p\pi r^2$, where r is the radius of the sleeve. Under conditions of constant flow rate, these forces are equal and opposite:

$$f_{viscous} = -f_{grav}$$

$$\eta 2\pi r l \frac{dv}{dr} = -p\pi r^2$$

$$dv = -\frac{p}{\eta 2l} r\, dr$$

Integrating the left-hand side of the equation from velocity of flow of 0 at the inner surface of the tube to v the velocity at radius r, we get

$$\int_0^v dv = -\frac{p}{\eta 2l} \int_R^r r\, dr$$

Recall that R is the radius of the tube where $v = 0$ and $r < R$ is the radius of a moving sleeve of liquid:

$$v = -\frac{p}{\eta 4l}(r^2 - R^2) = \frac{p}{\eta 4l}(R^2 - r^2)$$

A little further analysis along the same lines leads to the Poiseulle equation for the volume of flow V per unit time:

$$V = \frac{\pi p R^4}{8\eta l}$$

or

$$\eta = \frac{\pi p R^4}{8Vl}$$

permitting determination of the viscosity coefficient η by measuring the volume of flow of a fluid through a tube of known R in unit time.

11.4 CRYSTALS

In many crystals, atoms are arranged in a very regular three-dimensional pattern of rows and columns, echelons deep. Not surprisingly, when radiation strikes a crystal, it is reflected from *planes* of atoms, ions, or molecules in the way that light is reflected from a mirror. The difference is that experimentalists use penetrating X rays to detect planes *within* the crystal. The appearance of planes within a crystal is a consequence

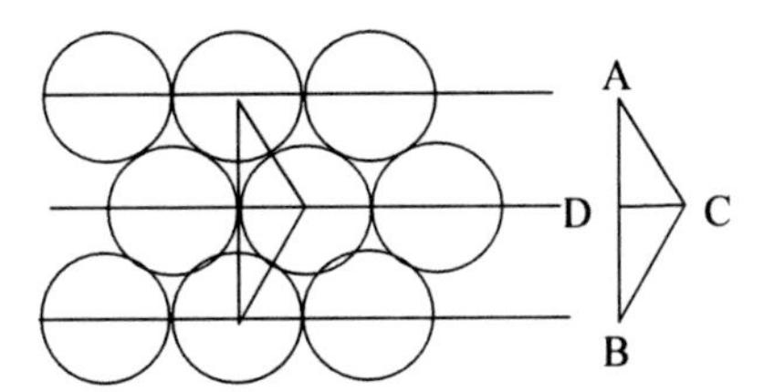

FIGURE 11.7 Close packing of marbles between two sheets.

of the regular arrangement of the atoms, ions, or molecules of the crystal, which is itself a consequence of the regular forces holding it together.

Simply dropping marbles into a box, one gets the idea of a tendency of the marbles to settle into a regular structure with layers separated by a distance that can be calculated as a function of the radii of the marbles. If you shake the box gently, so that the marbles assume a more or less compact aggregate, you may notice a repeating structural unit of a cube or hexagon.

To simplify the picture by making a two-dimensional array, think of marbles dropped into the space between two clear plastic sheets separated by a space equal to the diameter of the marbles (Fig. 11.7). Now marbles are separated into alternating rows. Repeating structural units may now be rectangles. Knowing the diameters (hence the radii) of the marbles enables one to find the distance between alternating rows.

We have extracted an isosceles triangle ABC from the marble pattern. The altitude of the triangle DC is also the radius of the marbles. The length of a side AC is twice the radius of the marbles. That gives us a right triangle ADC. The sum of the squares of the two sides of ADC is equal to the square of the hypotenuse AC. Distance DC is equal to one marble radius r, and AC is equal to $2r$ so AD is

$$\mathrm{AD} = \sqrt{\mathrm{AC}^2 - \mathrm{DC}^2} = \sqrt{(2r)^2 - r^2} = \sqrt{3r^2}$$

The distance AD is the distance between horizontal lines through the centers of the alternating rows. For example, if $r = 0.500\,\mathrm{cm}$, the distance between the horizontal lines through the centers of the marbles parallel to DC is 0.866 cm.

Distance DC permits us to take the inverse cosine $\cos^{-1}$ of the adjacent side over the hypotenuse of angle DAC to find that it is 30°. The remaining angle of the right triangle must be 60°. Now that we know all the distances and angles relating the centers of the marbles, we know all that can be known about the geometry of the marble packing everywhere the pattern in Fig. 11.7 is maintained. Dropping real marbles into a real space, one may find irregularities and fissures in the structure. This is analogous to real crystal structure as well. They show the same kind of irregularities and fissures. One more thing before going to three dimensions: If we rotate the diagram in Fig. 11.7 by 60°, we get a pattern that is identical to the one we just analyzed. There are other lines at other angles with the same geometric relationships as those we have found. The marble pattern has some *rotational symmetries*.

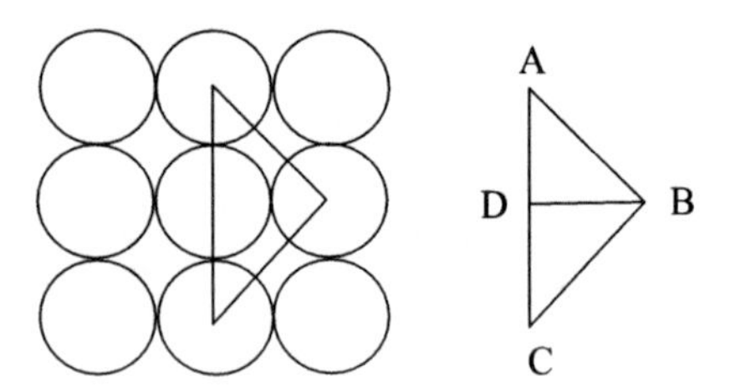

FIGURE 11.8 A less efficient packing of marbles. This packing is less efficient than close packing because the same number of marbles take up a greater space. To see this most clearly, notice that the interstitial spaces are larger than they are in Fig. 11.7.

With difficulty, marbles might be juggled into a less efficient but still regular pattern like Fig. 11.8. The vertical distance between centers for 0.500 cm marbles is now twice the radius as compared to only 0.866 r for close packing. The same number of marbles take up more space.

Suppose we had some experimental method of determining distance between layers DC. That would enable us to tell the difference between the packing pattern in Figs. 11.7 and 11.8. Assuming that atoms fall into a regular array when an element or compound crystallizes, we can picture a laminar sheet of atoms with very dense nuclei in an array similar to that Fig. 11.7 or 11.8. If marbles were packed into a three-dimensional box, the packing pattern determination would be very closely analogous to the pattern determination for marbles restricted to a plane. They would pack in a more or less regular way, and different packing patterns would be a more or less efficient with respect to space use, just as the marbles were in the simple illustrative case of two dimensions. Atoms, molecules, or ions pack in three-dimensional crystals as well. We would like to have an experimental method to solve the reverse of the problem we just solved; from an experimental value for the distance between layers of atoms, we would like to obtain the radius of the atoms themselves. The object of *X-ray crystallography* is to determine the distances and angles between atomic centers as we have done with marbles. The problem may be much more complicated, as in the example of proteins, or it may be nearly as simple as the method just described extrapolated to there dimensions, as in the case of pure metals and ionic salts.

Electromagnetic waves, from radio frequency of meter wavelengths to γrays of $\lambda \approx 0.1$ nm, have an electrical and a magnetic component. The radiation can be described mathematically as two sine waves, one electrical $\boldsymbol{\varepsilon}$ and the other magnetic $\boldsymbol{H}$, describing two oscillating vector fields oriented at right angles to one another. For *constructive interference* between two waves, the field vectors must point in the same direction. Otherwise, the radiation of one wave dims or obliterates the other by *destructive interference*.

Figure 11.9 shows that, for the radiation reflected from two adjacent horizontal lines of atoms to be in phase (to have their arrows pointed in the same direction), the difference in path must be an integer multiple of the wavelength λ. The path difference is shown as a heavy line. A path difference that is twice the wavelength of incoming radiation will interfere constructively, and one that is three times λ

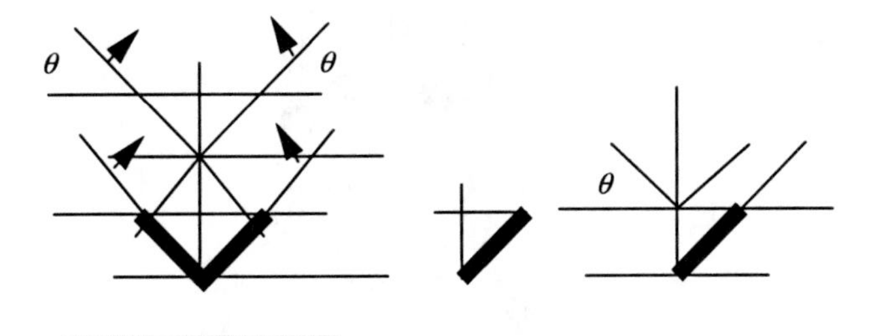

FIGURE 11.9 Bragg's law for constructive reflection.

serves as well, but one that is some uneven multiple of λ will interfere destructively. The condition demanded of wavelength for the emerging radiation to be in phase is $(1, 2, 3, \ldots,)\,\lambda = n\lambda$, where n is an integer.

From this we can extract a triangle with a hypotenuse that is one-half the wavelength of incoming radiation of wavelength λ (Fig. 11.9). The opposite side of the triangle so extracted is d, the distance between adjacent lines of atoms in the two-dimensional model. We measure θ, the angle of constructive reflection. All other angles give destructive interference and we do not see the radiation reflected. The sine of θ is its hypotenuse divided by its opposite side, which is d, so we have

$$\sin\theta = \frac{n\lambda}{2}\left(\frac{1}{d}\right)$$

or

$$d = \frac{n\lambda}{2}\left(\frac{1}{\sin\theta}\right)$$

which is *Bragg's law*.

11.4.1 X-Ray Diffraction: Determination of Interplanar Distances

The earliest and by far most common diffraction studies were on reflection and diffraction of X-radiation. When radiation is reflected from two parallel planes of a three-dimensional crystal, the reflected beams may be in phase, out of phase, or something in between. If θ is varied until the Bragg equation is satisfied, a sharp maximum in reflected radiation is recorded by photographic or other means. The (small) integer n in the Bragg equation is called the *order* of the reflection, and d is the distance between reflecting planes.

On the basis of the distance between planes found at various different incident angles, a repeating pattern is observed; from that, a repeating *unit* of crystal structure in three dimensions, the *unit cell*, is assigned. For example, if d is found to be the same in all three orthogonal directions, the cell is a simple cube; but if d values differ,

FIGURE 11.10 A face-centered cubic unit cell. The pointer indicates the face-centered Cl^- ion in the front face of the cell.

the cube may be distorted to a rhomb, tetragon, or other geometry. Let's stick to the simple cube to illustrate the principle.

From the angles within the unit cell and its orthogonal distances a, b, and c, one can find its volume V_{cell}. A measurement of the density $\rho = m/V$ of the crystalline solid gives the mass of atoms in the cell. Knowing the mass of the atoms in the cell and their *molar mass*, we can calculate how many of them there are—for example, four in the case of Na^+ and Cl^-. The *face-centered* unit cell for NaCl resembles two interpenetrating simple cubic cells (Fig. 11.10). This presents a cell with Na^+ and Cl^- ions alternating along each edge. The edge of the NaCl cell is 564 pm, but it has Na^+–Cl^-–Na^+ along each edge so the Na^+–Cl^- distance is half that dimension, $564/2 = 282$ pm. From the relative electronic structures and the fact that Na loses an electron and Cl^- gains one in the ionic bond, we can guess that the Cl^- ion will be about twice as large as Na^+. This estimate leads to an approximate ionic radius of Na^+, $r_{Na^+} = (282/3) = 94$ pm, leaving $282 - 94 = 188$ pm for r_{Cl^-}.

11.4.2 The Packing Fraction

One of the things we want to know about a crystal is how efficiently the atoms, molecules, or ions are packed. This is measured by the packing fraction, the space in the unit cell that is occupied by atoms relative to the total space in the cell. The larger the packing fraction, the greater the space occupied by atoms and the less space "wasted" in the interstices. This can be found by straightforward geometric calculation.

The idea of a unit cell packing fraction can be illustrated by arranging discs on a table top. One way is similar to the one shown for the marbles in Fig. 11.8. The repeating planar unit is a square (Fig. 11.11). A table top covered with many discs *packed in this way* can be thought of as simply a repeating pattern of square unit cells. Once we know the length of the edge of one unit cell and we know that it is square, we know the geometry of the entire table top full of discs. Chemists are not usually interested in table tops covered with discs, but we are very interested in

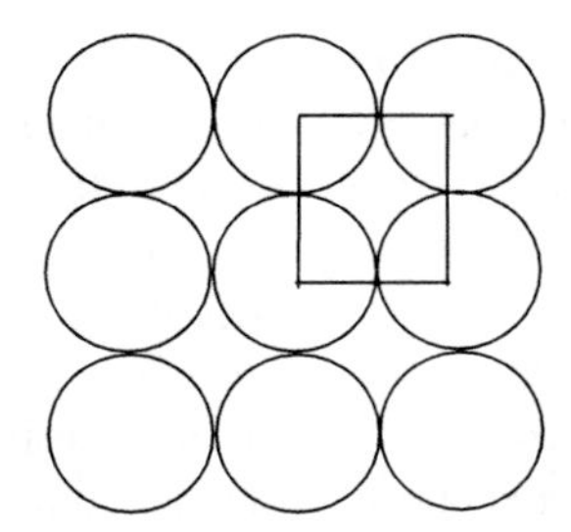

FIGURE 11.11 A two-dimensional unit cell for packing of discs.

the three-dimensional structure of atomic, molecular, and ionic solids in the regular arrangements we find in crystals. The concept of the unit cell is central in X-ray crystallography.

A crystal with the simple square atomic or molecular packing arrangement in all three directions is completely described as a repeating three-dimensional pattern of cubic unit cells. Once we know the length of the edge of the cubic unit cell, we know the geometry of the entire crystal (aside from impurities and structural imperfections).

The three-dimensional analog of Fig. 11.11 has a simple cubic unit cell (Fig. 11.12).

To simplify, let us go back to the unit cell of discs in Fig. 11.11. The total area A of discs within the cell is 4 times $A/4$, where $A = \pi r^2$ because, although there are four discs, one on each corner of the cell, only one-fourth of each disc *is actually inside* the cell. The dimension of the cell itself is $2r$ on an edge so the area of the unit cell is $(2r)^2 = 4\pi^2$. The area of the discs within the cell is πr^2, so the packing fraction is

$$P = \frac{A_{\text{disc}}}{A_{\text{cell}}} = \frac{\pi r^2}{4r^2} = \frac{\pi}{4} = 0.785$$

When this structure is taken to a three-dimensional simple cubic structure, the interstitial space at the center of the cube takes a larger proportion of the whole and the packing fraction is a rather inefficient 0.52. Simple cubic packing is not favored by many crystals. An exception is the metal polonium, ^{84}Po.

FIGURE 11.12 A simple cubic cell.

Now, let us consider a different structure. It is not likely that spheres will pack, each directly over its next lower neighbor. Instead (polonium excepted) they will slip one-half radius to the right or left to give a more efficient packing pattern with the central atom in the face of a cube or a slightly different pattern with the central atom a little back of the face so that it takes a position in the center of the cube. These are the *face-centered* and *body-centered cubic* structures fcc and bcc. You can see what is meant by face-centered by looking at the chlorine atoms only in Fig. 11.10. Metals packed as planes situated above each other so that atoms of one layer fit into the interstices of the planes above and below are *close-packed*. Close-packed spheres have a packing fraction of 0.740.

11.5 BRAVAIS LATTICES

At first thought, it would seem that there must be very many, perhaps infinitely many, unit cells in real crystals. Quite the contrary is true. As early as 1850, Bravais showed that unit cells in three dimensions can be classified into only seven systems on the basis of their rotational symmetry. These symmetries are important in X-ray crystallographic studies in which the sample is rotated in an incident beam of radiation. If a cell (crystal) is geometrically identical after rotation of 180° (π), it has at least a twofold axis of rotation. If that is all the symmetry it has about that axis, then it has one C_2 axis of symmetry. Bravais showed that the seven classifications shown in Table 11.1 permit 14 distinct crystal lattices in three dimensions.

11.5.1 Covalent Bond Radii

X-ray diffraction studies can also be carried out on *covalently* bonded molecular solids. The results can be augmented by comparison or combination with other kinds of diffraction studies in the solid, liquid, or gaseous states. It is possible to use beams of electrons or neutrons in place of X rays. These studies yield bond distances like $r_{\mathrm{C-Cl}} = 177$ pm in CCl_4. One would like to have a bond *covalent* radius for the Cl

TABLE 11.1 The Bravais Crystal Systems and Lattices[a].

Basic Types		14 Bravais Lattices
1. Cubic	$a = b = c$	P, I, F
2. Tetragonal	$a \neq c$	P, I
3. Rhombic	$\alpha,\ \beta,\ \gamma \neq 90°$	P
4. Monoclinic	$\alpha \neq 90°\ \beta,\ \gamma = 90°$	P, C
5. Hexagonal	$a \neq c$	P
6.Triclinic	$\alpha,\ \beta,\ \gamma \neq 90°$	P
7. Orthorhombic	$a \neq b \neq c$	P, I, F, C

[a]Primitive cells are designated P, body-centered cells are I, face-centered cells are F, and side-centered cells are C.

atom in CCl_4, for example, but there is no clear way of finding where, along the C–Cl bond axis, one radius leaves off and the other begins.

One approach to this difficulty is by finding the homonuclear bond distances in molecules like ethane for which the C–C bond distance is 154 pm, and Cl–Cl for which it is 200 pm. Under the assumption that homonuclear bond distances will be carried over into the heteronuclear cases, one has $(200/2 + 154/2) = 100 + 77 = 177$ pm, which successfully reproduces the experimental value for r_{C-Cl}. Now, by studying other compounds involving C and other compounds involving Cl, one can gradually build up tables of covalent bond radii such as those found in elementary textbooks (Ebbing and Gammon, 1999).

A number of points should be considered before using bond lengths or atomic radii. For one thing, combination of X-ray data with neutron diffraction data is risky because X-radiation is scattered by the electron cloud surrounding the nuclei in the molecule and neutrons are scattered by the nuclei themselves. Clearly the first is useful in studying bonds and the second is useful for structure. Also, there is no reason to expect ionic radii and covalent radii to agree with one another because of the different modes of chemical bonding involved. Tables of bond distances, ionic radii, and covalent radii should be used with some degree of reserve because of potential inconsistencies and approximations.

11.6 COMPUTATIONAL GEOMETRIES

Most present-day molecular structure–energy computer programs contain a routine that *optimizes* the geometry of the molecule under study so that each of the constituent atoms resides at the bottom of its unique potential energy well. Once knowing the complete molecular geometry, the bond lengths and angles can be calculated with great precision. The results, however, are not exactly comparable to experimental data because the energy minimum found by computational optimization for atoms in the force field of all other nuclei and electrons does not coincide with the average position of the vibrating atom. Current estimates of atomic radii and bond lengths are in good agreement overall but differ slightly according to the method used to determine them. Using the Spartan© package, the homonuclear distance $r_{H-H} = 73.6$ pm and the heteronuclear distance $r_{H-F} = 90.0$ pm are found, as compared to the experimental values of 74.2 and 91.7 pm.

11.7 LATTICE ENERGIES

Along with their geometry, we would like to know how firmly ionic crystals are held together. A quantitative measure of the energy or enthalpy holding the crystal together is its *lattice energy*. The lattice energy is the energy necessary to draw ions out of the crystal lattice and propel them into the gaseous state.

$$\text{NaI(crystal)} \rightarrow \text{Na}^+(\text{g}) + \text{I}^-(\text{g})$$

This enthalpy is the negative of the enthalpy of formation of the crystal from its gaseous ions. Please recall from Chapter 4 that this is not the standard enthalpy of formation because $Na^+(g) + I^-(g)$ is not the standard state of elemental sodium and iodine. The lattice energy for NaI and can be calculated from the standard state value $\Delta_f H^\circ(\text{NaI})$ by a procedure called the *Born–Haber cycle*, which is nothing more than an enthalpy of formation calculation like the one shown in Fig. 4.1 except that it is a little more complicated because of the nonstandard state of the products. Example 11.1 is not only an illustration of the Born–Haber cycle, but it also should serve to review and drive home the difference between standard and nonstandard states in thermochemistry.

The lattice energies of crystals show logical regularities. For example, the alkali metal iodides fall between 600 and 775 kJ mol^{-1} and decrease gradually from the lithium salt (which has a short ionic bond) to cesium iodide (which has a long one). The lattice energies of the doubly charged alkaline earth salts are very much larger than those of the alkali metals.

Lattice energies can also be calculated from a theoretical model in which the energy of the ionic crystal is supposed to be a function entirely of electrostatic forces. The model is fairly successful, but it involves some infusion of empirical data. All electrostatic attractions and repulsions can be calculated over the distances separating ions of opposite charge or the same charge. Knowing the exact geometry of the unit cell, these interionic distances a_i can be calculated precisely, not only for ions in the same unit cell but also for those in distant cells. The total Coulombic energy calculated from electrostatic theory in this way is

$$U(r) = \frac{N_A Z_+ Z_-}{r} \frac{e^2}{4\pi\varepsilon_0} \left[\frac{1}{2} \left(\sum \frac{Z_{i+}}{a_{i+}} + \sum \frac{Z_{i-}}{a_{i-}} \right) \right] + Be^{-r/\rho}$$

where the summed terms can be either positive for repulsion of like charges or negative for attraction between unlike charges. The constants B and ρ can be assigned reasonable values by an essentially empirical method (Barrow, 1996). Application to NaI gives 682 kJ mol^{-1} by comparison to the Born–Haber value of 658 kJ mol^{-1}. The difference is about 3.6%.

PROBLEMS AND EXERCISES

Exercise 11.1 The Born–Haber Cycle

Find the lattice energy (enthalpy) of NaI by the Born–Haber cycle.

Solution 11.1 Note that the small distinction between energy and enthalpy is often ignored in calculations involving large energies. We shall need several pieces of information to begin. First, we need the transition enthalpy of sodium metal in the solid form at 298 K and 1 bar, which is its standard state, to the state of the gaseous ion. This is calculated in two steps, first vaporization and then ionization, even though the steps may not be differentiated in the real process. We need not worry about the actual

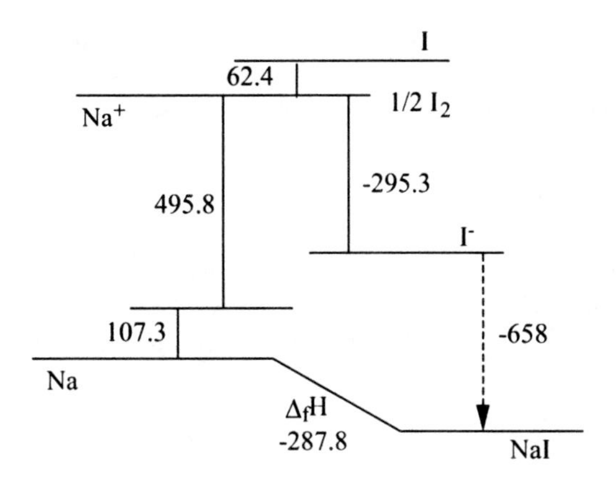

FIGURE 11.13 The Born–Haber cycle for NaI.

path the process takes because enthalpy is a state function and is path-independent. Vaporization takes up 107.3 kJ mol^{-1} and ionization takes up 495.8 kJ mol^{-1} (both endothermic). These enthalpy changes are shown as vertical lines for the left-hand side of Fig. 11.13.

Next, we sublime 0.5 mol of I_2 (endothermic 62.4 kJ mol^{-1}) and add an electron (*electron affinity*: exothermic, – 232.9 kJ mol^{-1}).

$$\tfrac{1}{2}I_2(s) \rightarrow I(g)$$

$$I(g) + e^- \rightarrow I^-$$

These two steps are shown in the top middle panel in Fig. 11.13. The enthalpy of formation of NaI is shown as the slanted line at the bottom of the diagram and after all this is done, we have the difference between Na and I in the state of their gaseous ions and the crystal NaI in its standard state. The formation of NaI from the gaseous ions is shown by the downward dotted arrow in Fig. 11.13. Adding all the enthalpies, one gets −658 kJ mol^{-1}. The reverse of this arrow is the vaporization of NaI into its gaseous ions. This is the lattice energy of the crystal, $\Delta_{\text{lattice}}H(\text{NaI}) = 658$ kJ mol^{-1}.

Problem 11.1

Calculate the surface area of a unit volume of liquid confined within

(a) a sphere and
(b) a cube

Which is larger and what is the percentage difference?

Problem 11.2

Water is unique among common liquids because it has a surface tension of 72.0 mN m^{-1}, nearly double the average of the other common liquids listed in the

CRC Handbook of Chemistry and Physics, 2008–2009, 89th ed. The unit is millinewtons per meter; be careful not to confuse the two meanings for the letter m. What is the capillary rise of water in a tube 1.000 mm in diameter?

Problem 11.3

The strongest (first-order) reflection of X rays comes when a crystal under examination is at an angle of 37.5° to the incident beam of radiation. The wavelength of the X-radiation is $\lambda = 1.54$ Å $= 154$ pm. What is the distance between the reflecting planes?

Problem 11.4

Suppose that discs of radius $r = 1$ unit are arranged on a table top in the patterns in Fig. 11.14 as they are in Figs. 11.7 (*left*) and 11.8 (*right*). Find the unit cell and packing fraction for these arrangements.

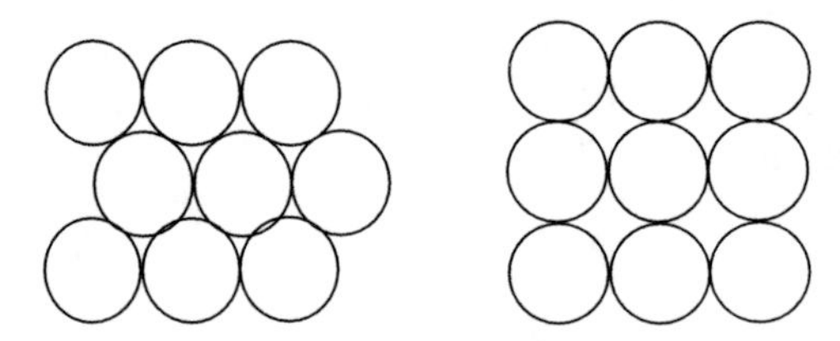

FIGURE 11.14 Close packing (*left*) and simple square unit cells (*right*).

Problem 11.5

The coordination number of an atom in a unit cell is the number of nearest neighbors. Because the entire crystal is constructed of replicated unit cells, all atoms in the system have the same coordination number. What is the coordination number of the body-centered cubic unit cell in Fig. 11.15?

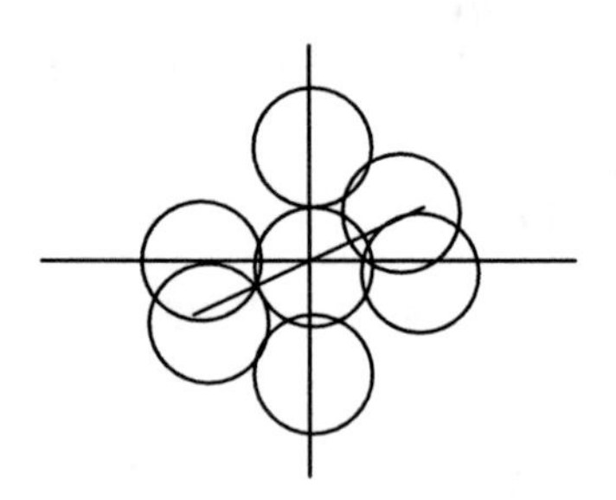

FIGURE 11.15 A body-centered primitive cubic cell.

Problem 11.6

How many atoms are there in the unit cell in Fig. 11.15?

Problem 11.7

(a) If r is set to 1.0 arbitrary length units, what is the volume of the unit cell in Fig. 11.15?
(b) What is the volume of the atoms within the unit cell in Fig. 11.15?
(c) What is the packing efficiency?

Problem 11.8

What is the lattice energy (enthalpy) of NaCl?

Take information on Na^+ from Example 11.1. The enthalpies of dissociation and ionization of Cl^- are 121.8 and -351.2 kJ mol^{-1}, respectively. The standard enthalpy of formation of NaCl is -411.2 kJ mol^{-1}.

12

SOLUTION CHEMISTRY

The concept of the ideal solution is as useful as the concept of the ideal gas, but the reference states must be very different because a model having no intermolecular interactions would not be a liquid. The next best thing is to propose a model in which no *change* in molecular forces takes place upon mixing two completely miscible liquids. In this model, mixing of two ideal liquids would be essentially the same as mixing of two ideal gases (Section 5.3.1). The enthalpy change would be zero at all proportions, and the mixing process would be entirely driven by a positive entropy change.

12.1 THE IDEAL SOLUTION

The *enthalpy* of mixing to form an ideal solution

$$\text{A(pure)} + \text{B(pure)} \rightarrow \text{AB(mixed)}$$

is zero because no change takes place in attractive forces upon going from isolated A and isolated B to AB neighbors in the mixture. On the other hand, the entropy of a system of two pure isolated liquids A and B, in which an observer knows that any randomly selected molecule will be either A or B according to which "beaker" it is drawn from, is replaced by a mixed system of greater disorder where random

Concise Physical Chemistry, by Donald W. Rogers

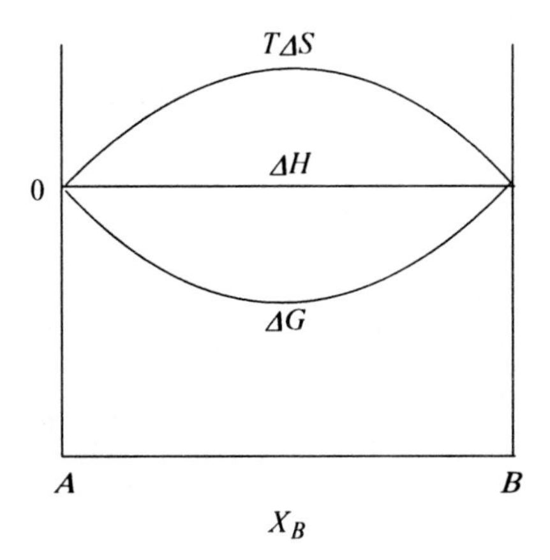

FIGURE 12.1 Entropy, enthalpy, and Gibbs free energy changes for ideal mixing at $T > 0$.

selection of a single molecule might result in either A or B. Molecular disorder has increased, hence entropy has also increased.

Uncertainty in random selection is greater for any mixture, but the maximum uncertainty is at equal concentrations $A = B$. All other ratios are less random. The enthalpy and entropy of ideal mixing are shown as the middle and upper curves in Fig. 12.1. The Gibbs free energy is $\Delta G = \Delta H - T\Delta S$, so the null result for ΔH causes the free energy change ΔG to be exactly opposite to the $T\Delta S$ curve at any mole fraction X_B and any temperature not equal to zero. Note that Fig. 12.1 could equally well have been drawn with X_A as the horizontal axis because $X_A + X_B = 1$ and $X_A = 1 - X_B$.

12.2 RAOULT'S LAW

A slightly less stringent requirement on the components of an ideal binary solution is that the partial vapor pressure of its components be a linear function of concentration, with the vapor pressure of the pure component p_A° as the slope of the function

$$p_A = X_A p_A^\circ$$

The concentration is measured in terms of the mole fraction of the component labeled A, and the variable p_A is the vapor pressure of A in the solution. This rule is called *Raoult's law*. Raoult's law is most easily understood by picking out the straight line labeled p_A in Fig. 12.2.

What is said of A is can also be said of B:

$$p_B = X_B p_B^\circ$$

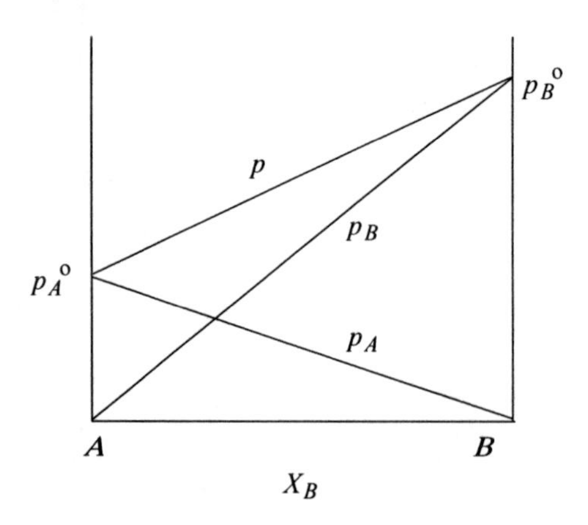

FIGURE 12.2 Partial and total pressures for a Raoult's law solution.

Normally the component with the smaller mole fraction is called the *solute*, and the one with the larger mole fraction is called the *solvent*. If we are looking at solutions over a wide range of mole fractions, this designation becomes rather arbitrary. Real solutions that come very close to satisfying the Raoult's law idealization are solutions of toluene in benzene.

The sum of two linear functions is a linear function, so the total vapor pressure p contributed by both components anywhere along the concentration axis is a linear function of the difference between the vapor pressures of the pure components

$$p = p_A^{\circ} + X_B \left(p_B^{\circ} - p_A^{\circ}\right)$$

12.3 A DIGRESSION ON CONCENTRATION UNITS

Concentration units in solution chemistry are simple but rather convoluted at times, and they can cause unnecessary errors. This section is in part a summary of concentration units and notation already used and in part an introduction to the unit of molality, which is favored in the study of chemistry in solvents other than the "universal" solvent water. Molality is also used in the study of concentrated solutions and in very precise work.

We have already been introduced to the mole fraction (Section 1.5), which is the number of moles of a selected component of the solution relative to the total number of moles of all components in the solution

$$X_i = \frac{n_i}{n_i + \sum n_j}$$

We are often interested in binary solutions in which one substance is clearly the solute and one is clearly the solvent. An example is a dilute solution of NaCl in water, in which NaCl is clearly the solute and water is the solvent. It is conventional to use n_2 to refer to the number of moles of solute while using n_1 to refer to the number of moles of solvent. The mole fractions for a binary solution are

$$X_2 = \frac{n_2}{n_1 + n_2}$$

and

$$X_1 = \frac{n_1}{n_1 + n_2}$$

12.4 REAL SOLUTIONS

Few real solutions follow Raoult's law and none follow it exactly. (This is like saying that there is no perfectly ideal gas.) Deviations may give vapor pressure curves that are consistently higher than Raoult's law as shown in Fig. 12.3, or they may be consistently lower. Rarely they may be higher at one end and lower at the other so that the vapor pressure curve crosses the Raoult's law curve to give a positive deviation at one end and a negative deviation at the other. Solutions in which attractive AB forces are weak give a positive deviation from Raoult's law because molecules leave the solution more easily than from the pure liquid. Mixtures with strong AB forces show a negative deviation.

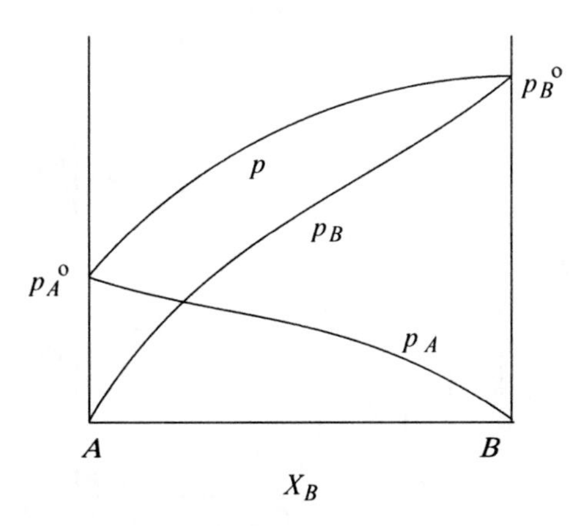

FIGURE 12.3 Consistent positive deviations from Raoult's law. In this case, intermolecular attractive forces are less strong in the solution than in the pure components A and B. The presence of B encourages A to leave the solution and go into the vapor phase.

12.5 HENRY'S LAW

The shape of the p_B curve in Fig. 12.3 suggests another model similar to Raoult's law but with a different slope for the lower (solute) end of the curve. According to *Henry's law*, the partial pressure, is the linear function found at the limit of infinitely dilute solutions (in practice, very dilute solutions). Like the ideal gas law, it is a limiting law (Rosenberg and Peticolas, 2004).

At relatively high concentrations of solute B, Henry's law is a poor approximation but it is clear from Fig. 12.4 that in the very dilute region near the vertical marker, it is a better descriptor of the partial vapor pressure of B than Raoult's law. On the contrary, at $X_B \rightarrow 1$ (toward the right in Fig. 12.4), B follows Raoult's law. Now we have a combined model: *Raoult's law* for B acting as a solvent in high concentrations and *Henry's law* for B acting as a solute in low concentrations.

12.5.1 Henry's Law Activities

We have encountered *activities* and *activity coefficients* before. A Henry's law activity of solute B in solvent A is shown as a vertical line well to the left on the X_B axis of Fig. 12.4, where X_B is small. The activity coefficient γ is the ratio of real p_B to the ideal, as determined by Henry's law. For the behavior of B shown in Fig. 12.4, the activity coefficient will be $\gamma < 1$ in all cases because real behavior is always less than ideal. It will approach $\gamma = 1$ in the limit of infinite dilution. If the partial pressure curve in Fig. 12.4 were below the Raoult's law dotted line, all Henry's law activity coefficients would be $\gamma > 1$.

The model we have described so far is for binary solutions in which the components are mutually miscible (soluble in each other in all proportions). Everything we have

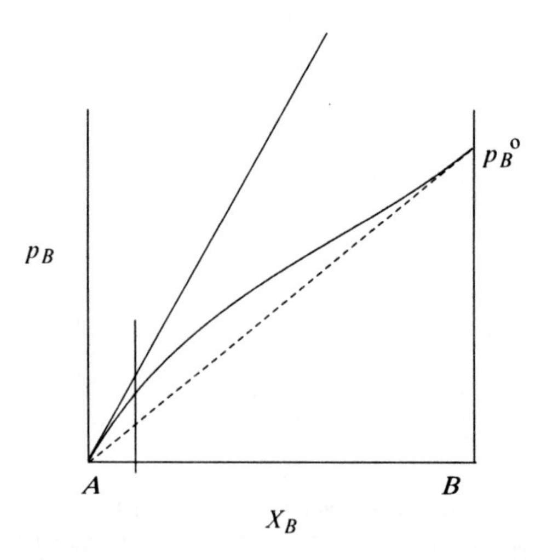

FIGURE 12.4 Henry's law for the partial pressure of component B as the solute. The solid line is a Henry's law extrapolation to infinite dilution of B. The dotted line is Raoult's law.

found for component B dissolved in solvent A is true if we change the AB ratio and start calling A the solute and B the solvent (on the right-hand side of Fig. 12.3). Therefore we can obtain Henry's law activities of both components, one from the partial pressure of very dilute solutions of A in B and the other from very dilute solutions of B in A.

12.6 VAPOR PRESSURE

Henry's original (1803) paper dealt with the solubility of CO_2 in water, for which he found that the partial pressure of p_{CO_2} confined in a closed container over water is directly proportional to the amount of CO_2 dissolved in the water. It was many years later that his law $p_{CO_2} = k\, X_{CO_2}$ where X_{CO_2} is the mole fraction of CO_2 dissolved in water, was extended to the binary solutions of mutually miscible components just discussed.

Let us go back now to Henry's point of view and look at $p_{CO_2} = k\, X_{CO_2}$ as a relationship telling us what the partial pressure of CO_2, which we shall call *component 2*, will be over a solution that we have made up to have a mole fraction of X_{CO_2}. Notice that we are looking at the same problem as Henry but from the opposite side. He was thinking about the amount of gas dissolved in a solvent (water), and we are looking at the pressure of solute leaving a solution that we have made up to some value of X_2 and going into the gas phase above the solution. At equilibrium, it's the same thing:

$$\text{Component 2(solution)} \rightleftarrows \text{Component 2(g)}$$

The critical condition for all the related *colligative properties* (Section 12.8) is a balance between the Gibbs free energy functions of two systems—for example, gas-phase CO_2 and dissolved CO_2. The Gibbs free energy relationship at equilibrium between dissolved component 2 and gas phase 2 is

$$\mu_2(\text{sol}) = \mu_2(\text{g})$$

where $\mu_2(\text{sol})$ designates the Gibbs free energy function of component 2 in solution. We know from Chapters 6 and 7 that, for an ideal gas at constant temperature, we have

$$dG = V\, dp = \frac{RT}{p} dp$$

Integrating over the difference between the Gibbs function in an arbitrary state 2 and a standard state pressure arbitrarily chosen as 1 bar, we get

$$\Delta G_2 = \int_{G_{1\text{bar}}}^{G_{p_2}} dG = RT \int_{G_{1\text{bar}}}^{G_{p_2}} \frac{dp}{p} = RT \ln \frac{p_2}{1}$$

Written in terms of Gibbs chemical potentials, this gives us

$$\mu_2(\mathrm{g}) = \mu_2^\circ(\mathrm{g}) + RT \ln \frac{p_2}{p^\circ} = \mu_2^\circ(\mathrm{g}) + RT \ln p_2$$

where $\mu_2^\circ(g)$ is the standard state free energy for an ideal gas at a partial pressure p_2 in a mixture of gases. The analogous expression for an ideal solute at mole fraction X_2 in a mixture of solutes is

$$\mu_2(\mathrm{sol}) = \mu_2^\circ(\mathrm{sol}) + RT \ln X_2$$

(We shall usually treat the special case of only one solute dissolved in one solvent.)

Equality of the two Gibbs potentials at equilibrium gives

$$\mu_2(\mathrm{sol}) = \mu_2(\mathrm{g})$$

$$\mu_2^\circ(\mathrm{g}) + RT \ln p_2 = \mu_2^\circ(\mathrm{sol}) + RT \ln X_2$$

or

$$RT \ln \frac{p_2}{X_2} = \mu_2^\circ(\mathrm{sol}) - \mu_2^\circ(\mathrm{g})$$

This leads to an expression with only constants on the right-hand side:

$$\ln \frac{p_2}{X_2} = \frac{\mu_2^\circ(\mathrm{sol}) - \mu_2^\circ(\mathrm{g})}{RT}$$

or

$$\frac{p_2}{X_2} = e^{\frac{\mu_2^\circ(\mathrm{sol}) - \mu_2^\circ(\mathrm{g})}{RT}} = e^{\mathrm{const}} = k$$

which is Henry's law, $p_2 = k X_2$. Notice that the difference between Raoult's law and Henry's law is merely in the arbitrary selection of a standard state.

12.7 BOILING POINT ELEVATION

The utility of Henry's law becomes apparent when we consider a very important class of binary solutions, the class of nonvolatile solutes in a volatile solvent—for example, sugar (sucrose) in water. The nonvolatile solute has no measurable vapor pressure, hence Raoult's law does not apply because there is no p_B° in Figs. 12.3 and 12.4. The total vapor pressure is due to the solvent:

$$p = X_1 p_1^\circ$$

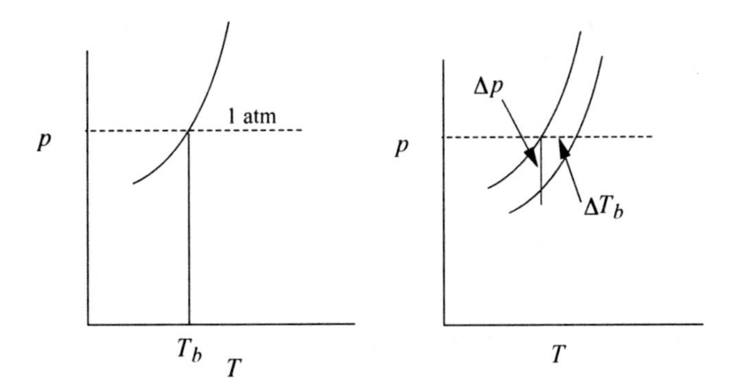

FIGURE 12.5 Boiling of pure solvent (*left*) and a solution of solvent and nonvolatile solute (*right*). The temperature of the solution must be increased by ΔT_b to restore its vapor pressure to 1 atm.

Addition of a nonvolatile solute causes a change in the vapor pressure of the solution according to the amount of solute added because $(1 - X_1) = X_2$ for a binary solution:

$$\Delta p = p_1^\circ - p = p_1^\circ - X_1 p_1^\circ = p_1^\circ(1 - X_1) = p_1^\circ X_2$$

The statement $\Delta p = p_1^\circ X_2$ is true only in the limit of very dilute solutions.

To see that addition of a small amount of nonvolatile solute causes an elevation of the boiling point of the solution relative to the pure solvent, consider a solvent that has an exponential variation of p with T as in Fig. 12.5 (*left*). Its boiling point is the temperature at which the vapor pressure is equal to the pressure of the atmosphere bearing down on the surface of the liquid (dotted line).

When a nonvolatile solute is added to the solvent, its entire vapor pressure curve is displaced downward by Δp. An increase in temperature is necessary to restore the vapor pressure to 1 atm. The vapor pressure of the solution moves along the lower exponential curve with increasing temperature until it arrives once again at 1 atm and ΔT_b, whereupon boiling recommences. It is evident that a functional relationship must exist among Δp, X_2, and ΔT_b. To find the relation between the amount of nonvolatile solute and ΔT_b we have recourse to the Clausius–Clapeyron equation

$$\ln p = -\frac{\Delta_{\text{vap}} H}{R}\left(\frac{1}{T}\right) + \text{const}$$

Differentiating, we obtain

$$d \ln p = \frac{dp}{p} = \frac{\Delta_{\text{vap}} H}{R}\left(\frac{1}{T^2}\right) dT$$

If we take dp/p as $\Delta p/p$ for a very small but finite and measurable Δp, brought about by addition of the solute, we must increase the temperature by an amount

$dT \approx \Delta T_b$ to restore the system to its original vapor pressure. These small changes are related by the equation

$$\frac{\Delta p}{p} = \frac{\Delta_{\text{vap}} H}{R} \left(\frac{1}{T^2} \right) \Delta T_b$$

where we have already shown that $\Delta p = p_1^\circ X_2$ so $\Delta p / p = X_2 p_1^\circ / p$. For experiments carried out at atmospheric pressure, p_1° is the vapor pressure at the normal boiling point of the pure solvent. The difference $\Delta p = p - p_1^\circ$ is the *vapor pressure depression*. The temperature rise ΔT_b necessary to restore p to its original value is the *boiling point elevation*.

In the limit of small amounts of solute, the vapor pressure change from that of the pure solvent is small, $p_1^\circ \approx p$, so they cancel to a good approximation and

$$\frac{X_2 p_A^\circ}{p} \approx X_2 \approx \frac{\Delta_{\text{vap}} H}{R} \left(\frac{1}{T^2} \right) \Delta T_b$$

or

$$\Delta T_b \approx \frac{R T_b^2 X_2}{\Delta_{\text{vap}} H_1}$$

Writing out X_2 in terms of grams and molar mass of the solvent M_1 (problems) leads to the boiling point elevation in practical laboratory terms as

$$\Delta T_b \approx \frac{R T_b^2 \mathrm{M}_1}{\Delta_{\text{vap}} H_1 \cdot 1000} m = K_b m$$

where K_b is called the *boiling point constant* or sometimes the *ebullioscopic* constant. The molality, m, is the number of moles per 1000 g of solvent, and M_1 is the molar mass of the solvent; that is, we are expressing the amount of solute in units of molality. Typical values for K_b range from about 0.5 to 5 K kg mol^{-1}; for example, K_b(water) = 0.51 K kg mol^{-1} and K_b (benzene) = 2.53 K kg mol^{-1}.

As we have seen, K_b is derived through Gibbs chemical potentials. A similar treatment leads to the *freezing point depression* ΔT_f and the freezing point constant K_f:

$$\Delta T_f \approx \frac{R T_f^2 \mathrm{M}_1}{\Delta_{\text{freez}} H_A \cdot 1000} m = K_f m$$

where $\Delta_{\text{freez}} H_A$ is the enthalpy of freezing of the pure solvent at 1 atm. Typical values are K_f (water) = 1.86 K kg mol^{-1} (that famous number we memorized in elementary chemistry) and K_f (benzene) = 5.07 K kg mol^{-1}. If we measure the freezing point depression brought about by dissolution of a known (small) weight of solute to produce a dilute solution, we can calculate the number of moles of solute from ΔT_f, hence its molar mass in units of kg mol^{-1}.

12.8 OSMOTIC PRESURE

Many membranes have the property that they pass small molecules but not large ones. They are *semipermeable*. The human kidney contains semipermeable membranes that separate salt and urea from the bloodstream for excretion but do not pass blood protein. If a semipermeable membrane separates two arms of a U tube and a solution of a protein is placed in one arm of the tube with the solvent in the other arm, solvent molecules will pass through the membrane into the protein solution because of the spontaneous tendency of the protein solution to evolve toward a more dilute state of higher entropy. The process is called *osmosis*. The liquid level in the protein arm of the U tube will rise until the downward pressure it exerts on the membrane due to gravity is just sufficient to balance the entropy-driven *osmotic flow* across the membrane. This *osmotic pressure*, π, follows a law similar to the ideal gas law:

$$\pi V = nRT$$

where V is the volume of a solution containing n moles of the *macromolecular* solute.

We have already made the point that liquids, contrary to gases, are subject to strong intermolecular interactions. The internal structure of the liquid state is distinct and different from that of the gaseous state, so why should a property of solutions follow what appears to be an ideal gas law? The derivation of the osmotic law to arrive at this remarkable and somewhat counterintuitive conclusion is quite distinct from ideal gas law theory, and it reveals some of the power of the condition that, at equilibrium, the Gibbs chemical potential must be the same over different macroscopic segments of the same thermodynamic system.

To apply this idea to a real system, we find the osmotic pressure on the solution in the right-hand chamber of Fig. 12.6 by a two-step process. First we shall calculate the total pressure, which is the ambient pressure p plus an arbitrary pressure π on pure solvent, through our knowledge of the variation of the Gibbs chemical potential with variation in pressure. Second we shall find the variation of the Gibbs chemical potential brought about by addition of a small amount of solute. These two steps are then summed to find the total pressure on the solution of solvent plus solute.

Step 1

Start with pure solvent and apply some pressure π at constant temperature. The Gibbs free energy of the system is the Gibbs free energy of the pure solvent at

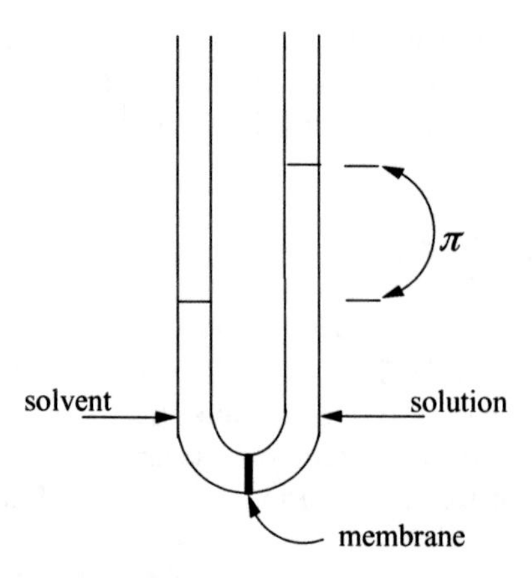

FIGURE 12.6 Osmotic pressure, π.

ambient pressure $\mu_1^\circ(p)$ plus the change resulting from the increase in pressure to $p+\pi$:

$$\mu_1^\circ(p+\pi) = \mu_1^\circ(p) + \int_p^{p+\pi} \left(\frac{\partial \mu_1}{\partial p}\right)_T dp$$

The variation of a perfect differential of a thermodynamic function is

$$d\mu = \left(\frac{\partial \mu}{\partial p}\right)_T dp + \left(\frac{\partial \mu}{\partial T}\right)_p dT$$

but the second term on the right drops out because $T = \text{const}$. We shall refer to the molar volume of solvent as $V_{m,1}$ and note that (Klotz and Rosenberg, 2008)

$$\left(\frac{\partial \mu_1}{\partial p}\right)_T = V_{m,1}$$

The Gibbs chemical potential of the pure solvent subjected to a pressure change from p to $p+\pi$ at constant $V_{m,1}$ is

$$\mu_1^\circ(p+\pi) = \mu_1^\circ(p) + \int_p^{p+\pi} V_{m,1}\, dp = \mu_1^\circ(p) + \pi V_{m,1}$$

Step 2

In the second step, add a small amount of solute. This brings about a change in the Gibbs free energy function[1]

$$\mu_1^\circ(p+\pi, X_2) = \mu_1^\circ(p+\pi) + RT \ln X_1$$

We already have an expression for the first term on the right, so

$$\mu_1^\circ(p+\pi, X_2) = \mu_1^\circ(p) + \pi V_{m,1} + RT \ln X_1$$

At equilibrium, the pressure on the solution is the same as the pressure on the pure solvent:

$$\mu_1^\circ(p) = \mu_1^\circ(p+\pi, X_2)$$

Substitution for $\mu_1^\circ(p+\pi, X_2)$ on the right gives

$$\mu_1^\circ(p) = \mu_1^\circ(p) + \pi V_{m,1} + RT \ln X_1$$

which means that

$$\pi V_{m,1} + RT \ln X_1 = 0$$

or

$$\pi V_{m,1} = -RT \ln X_1$$

In a dilute binary solution, we have

$$\ln X_1 \cong -X_2$$

so

$$\pi V_{m,1} = RTX_2$$

where $V_{m,1}$ is the molar volume of the solvent. This can be expressed in terms of the total volume as $V_{m,1} = V/n_1$, so

$$\pi V_{m,1} = \pi \frac{V}{n_1} = RTX_2$$

[1]In a binary solution, μ_1° can be expressed either as a function of X_1 or X_2.

but the mole fraction of the solute tends toward n_2/n_1 at small concentrations, so

$$\pi \frac{V}{n_1} = RT \frac{n_2}{n_1}$$

or $\pi V = n_2 RT$, which has the form, but not the content or meaning, of the ideal gas law. This is called the van't Hoff equation for osmotic pressure (as distinct from the van't Hoff equation for equilibrium).

12.9 COLLIGATIVE PROPERTIES

If we calculate molality m from the formula weight of solute—for example, NaCl—and use it in the equation for osmotic pressure, we shall be wrong. Osmotic pressure π, like Δp, ΔT_b, and ΔT_f, is a *colligative* property because it is a property of the *number of particles* in solution. Our error results from two factors: First, the number of solute particles in solution may double or triple due to ionization; and, second, the *free* solvent concentration may be reduced by solvation.

If NaCl is the nonvolatile solute in water for example, there will be an effective molality approximately two times the anticipated value because NaCl exists as Na^+ ions and Cl^- ions in aqueous solution. The number of particles in solution is the *van't Hoff i factor,* 2 for NaCl solutions, 1 for sucrose, which does not ionize, 3 for $ZnCl_2$, and so on. Van't Hoff i factors are, however, integers only at infinite dilution.

In real solutions, van't Hoff i factors show a systematic deviation from integral values due to strong solvation (hydration) of the molecules or ions. When there is a strong association between a solute molecule and solvent molecules, the solvent molecules are effectively "taken away" from the solution. The amount of free solvent is reduced and the relative amount of solute is *greater* than we conventionally calculate it to be. The measured change in colligative properties is augmented.

The freezing points of aqueous solutions of NH_3, which is not ionized to any appreciable extent, are shown in Fig. 12.7. The freezing point of water decreases with ammonia concentration according to the van't Hoff equation $\Delta T_f = -1.86\, m$ to about $m \cong 4.0$ but then the freezing point becomes more negative than theory predicts, as though the solution were more concentrated than it actually is. The effective molality (Zavitsas, 2001) is reduced to $m_1 - hm_2$, where h is a parameter called the solvation number, which gives the number of solvent molecules held so tightly by solute as to be ineffective. In water solution, h is called the *hydration number*. The solvation number h is not an integer because it is an average over many solute molecules or ions. It is not difficult to determine h; it is just an empirical parameter chosen to cause real colligative behavior to approach the van't Hoff equation. In the case of ammonia dissolved in water, the choice $h = 1.8$ leads to the function shown by open circles in Fig. 12.7. These experimental freezing points differ from those shown by solid circles only in that the molalities have been recalculated as moles of solute per kilogram of *free* water remaining after the solute has been hydrated.

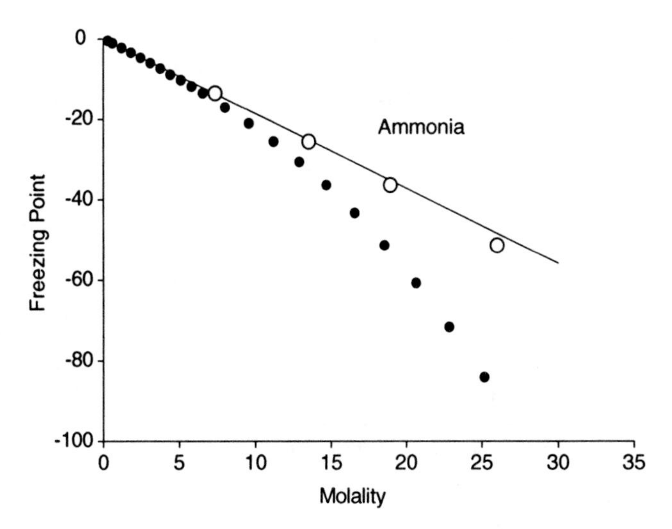

FIGURE 12.7 Lowering of the freezing point of water by ammonia (*CRC Handbook of Chemistry and Physics, 2008–2009*). The straight line has the theoretical slope of -1.86 K mol^{-1}. Open circles have had their molality corrected for hydration of NH_3.

The straight line in Fig. 12.7 was calculated from $\Delta T_f = -1.86\, m$, and the three open circles were calculated by correction of the molality for a hydration number of 1.8 by the Zavitsas method. The small deviation of the open circles from the straight line is due to failures in some of the approximations made in the derivation of K_f(for example, $\ln X_1 \cong -X_2$). They can be corrected in a more rigorous derivation (see Problems). Hydration numbers for many other solutes are known and are usually higher than 1.8, meaning that the departure of the actual colligative property from its ideal estimate is correspondingly greater. The approach to ideal behavior at small NH_3 concentrations in Fig. 12.7 shows the meaning of the term "limiting law" quite graphically. The phenomenon described here is not restricted to any one colligative property or to water as a solvent; it is general.

PROBLEMS, EXAMPLES, AND EXERCISE

Example 12.1 Partial Pressures

If we carry out experimental determinations of the partial pressure of acetone over solutions of acetone in diethyl ether, results for small values of X_2 will be something like Fig. 12.3. Some typical experimental data are given in Table 12.1. Examine the data for the pressure of acetone over diethyl ether and obtain an estimate of the activity coefficient for acetone at mole fraction $X_2 = 0.20$.

Solution 12.1 A first estimate of the Henry's law constant, which is a limiting law, can be found from the limiting slope of p_2 as a function of X_2 for the first few points

TABLE 12.1 Vapor Pressures of Acetone over Dilute Binary Solutions of Acetone in Diethyl Ether.

X_2	p_2 (kPa)[a]
0.000	0.00
0.029	2.53
0.068	5.33
0.154	9.90
0.178	10.90
0.255	14.46
0.315	16.92
0.372	19.02

[a]The units of pressure are kilopascals.

on the curve of p_2 vs. X_2. This slope is the slope of the tangent line p_2 vs. X_2 for the most dilute solutions and is approximately $2.53/0.0290 = 87$ kPa. Thus the Henry's law estimate of p_2 at $X_2 = 0.20$ is $0.20\,(87.2) = 17.4$ kPa.

The interpolated value of p_2 at $X_2 = 0.20$ is

$$\frac{0.255 - 0.200}{0.255 - 0.178} = \frac{14.5 - p_2}{14.5 - 10.9} = 0.71$$

$$p_2 = 14.5 - 0.71(3.6) = 11.9 \text{ kPa}$$

The activity coefficient is the ratio of the actual value to the Henry's law estimate $\gamma = 11.9/17.4 = 0.68$ and the activity is the effective concentration of acetone relative to the Henry's law standard $a = \gamma X_2 = 0.68$ and $(0.20) = 0.14$.

Example 12.2

What are the mole fractions of solvent and solute in a 0.1000 molar solution of NaCl?

Solution 12.2 The number of grams of NaCl in 1 liter of the solution is

$$0.1000 = \frac{0.1000\,(22.99 + 35.45)}{1.000 \text{ liter}} = \frac{5.844}{1.000 \text{ dm}^3}$$

The density $\rho = m/V$ of pure solid NaCl is 2.17 g cm^{-3}, so NaCl takes up

$$V = m/\rho = 5.844/2.17 = 2.69 \text{ cm}^3$$

This amount of the total volume of the solution is taken up by the solid NaCl, leaving 997.3 cm^3 of water in the 1.000 dm^{3}of solution. Now that we know the volume of solvent, we can calculate the mole fractions. Recall that we have 0.1000 mol of NaCl by definition and the molar mass of water is 18.02, so we have 997.3/18.02 = 55.345 mol of water. The mole fractions are

$$X_2 = \frac{n_2}{n_1 + n_2} = \frac{0.1000}{55.345 + 0.1000} = 1.804 \times 10^{-3}$$

and

$$X_1 = \frac{n_1}{n_1 + n_2} = \frac{55.345}{55.345 + 0.1000} = 0.9982$$

Comment: First, notice that the sum of the two mole fractions is exactly 1.000, as it should be because there are only two components in this solution. Second, notice that we have already made some pretty serious approximations and assumptions. We don't really have the right to assume that the volume taken up by NaCl in solution is the same as the volume in the crystalline state. Also we are assuming that the density of water is exactly 1.000 kg dm^{-3} and the volume of the solution is exactly 1.000 dm^3 without specifying the temperature. We shall have to take these approximations more seriously later.

In precise solution chemistry, we often use *molal* concentrations, the amount of solute expressed in moles *per 1000 g of solvent* in contrast to the ordinary stockroom unit of moles per liter or molarity. A 0.1000 molal solution contains 1000/18.02 = 55.494 moles of water, but so does a 0.2000 molal solution because the volume of the solute is not subtracted from that of the solvent. That is one advantage of the molal convention. Another is that the number of grams in a molal solution does not depend on the temperature, but the number of grams in a molar solution does because the volume of the solution is temperature-dependent. The mole fraction of the 0.1000 *molal* solution is

$$X_2 = \frac{n_2}{n_1 + n_2} = \frac{0.1000}{55.494 + 0.1000} = 1.800 \times 10^{-3}$$

and

$$X_1 = \frac{n_1}{n_1 + n_2} = \frac{55.494}{55.494 + 0.1000} = 0.9982$$

with a sum of 1.000 as before.

Why specify a new concentration unit when it comes out the same as the old one? As the following exercise shows, it does not come out exactly the same and sometimes it is not even close. For very dilute solutions, the difference is negligible, which is why the concentrations of solutions in working labs and elementary courses are usually given in molarity rather than molality. In medicinal and clinical chemistry, many other concentration units are used but they can all be reduced to the basic ones with a little algebra.

Exercise 12.1

(a) Calculate the mole fractions of solute and solvent in a 1.000 molar solution of sucrose in water.

(b) Calculate the mole fractions of solute and solvent in a 1.000 molal solution of sucrose in water.

Solution 12.1

(a) The number of grams of sucrose $C_{12}H_{22}O_{11}$ (M = 342.3) in 1.000 liter of the solution is $342.3/1.000\ \mathrm{dm}^3$. The density $\rho = m/V$ of pure solid sucrose is $1.580\ \mathrm{g\,cm}^{-3}$ so sucrose takes up $V = m/\rho = 342.3/1.580 = 216.6\ \mathrm{cm}^3$, leaving $783.4\ \mathrm{cm}^3$ of water in the $1.000\ \mathrm{dm}^3$of solution. We have 783.4/18.02 = 43.474 moles of water. The mole fractions are

$$X_2 = \frac{n_2}{n_1 + n_2} = \frac{1.000}{43.474 + 1.000} = 0.02248$$

and

$$X_1 = \frac{n_1}{n_1 + n_2} = \frac{43.474}{43.474 + 0.100} = 0.9775$$

for a total of 1.000.

Comment: These mole fractions depend on the assumption that the volume that sucrose molecules take up in water is the same as the molecular volume in the crystalline state. This is extremely unlikely, meaning that our mole fractions calculated in this way are probably wrong.

(b) The mole fractions of the 1.000 molal solution are

$$X_2 = \frac{n_2}{n_1 + n_2} = \frac{1.000}{55.494 + 1.000} = 0.0177$$

and

$$X_1 = \frac{n_1}{n_1 + n_2} = \frac{55.494}{55.494 + 1.000} = 0.9823$$

With a sum of 1.000 as before. Under these circumstances, the 1.000 molar and 1.000 molal solutions are very different.

Comment: There are no assumptions concerning the volume of sucrose molecules in solution therefore the molality is correct as written.

Further Comment: Sucrose molecules (or any solute molecules) may "tie up" or hydrate (solvate) water molecules, rendering them inactive in any participatory physical phenomenon like the vapor pressure over the solution, which depends on the number of solvent molecules in the solution that are free to go into the vapor phase. Other *colligative* properties (Section 12.8) are influenced in the same way by solvation.

Exercise 12.2

The freezing point of pure benzene is 5.49°C, and its freezing point constant is $K_f = 5.07°C$. A sample of a crystalline unknown was made up such that it contained 18.7 mg of unknown per 1.000 g of benzene. The freezing point of the resulting solution was found to be 4.76°C. What was the molar mass M of the unknown?

Solution 12.2 The freezing point depression was

$$\Delta_f T = 5.49 - 4.76 = 0.73°C = K_f m$$

$$m = \frac{0.73}{5.07} = 0.144 \text{ g kg}^{-1}$$

The unknown was in a concentration of 0.0187 g per 1.000 g benzene, which is 18.7 g kg^{-1}. If the solution of 18.7 g kg^{-1} is 0.144 m, then, by direct ratio, we obtain

$$\frac{18.7}{0.144} = \frac{M}{1.000}$$

$$M = 130 \text{ g mol}^{-1}$$

where M is the experimental molar mass. The unknown might be naphthalene, $C_{10}H_{10}$.

Problem 12.1

Exactly 10.00 g of NaCl was added to sufficient water to make up 100.0 g of solution.

(a) What was the weight %?
(b) How many moles of NaCl were present?
(c) What was the molality of the solution?
(d) What is the molarity of the solution?
(e) Assuming a density of 1.071 g cm^{-3} at 20°C for this solution (*CRC Handbook of Chemistry and Physics, 2008–2009*, 89th ed.), what is its volume?
(f) With this new information, find its molarity.

Problem 12.2

Acetone and chloroform are completely miscible in one another. The partial pressures of chloroform as a function of the mole fraction in acetone solutions are

X_2	p_2 (kPa)
0.00	0.00
0.20	35
0.40	82
0.67	142
0.80	219
1.00	293

Find the Raoult's law and Henry's law activity coefficients of chloroform at 0.20 mole fraction.

Problem 12.3

Derive an expression for the freezing point depression from an equality of Gibbs potentials.

Problem 12.4

Given that the enthalpy of fusion of water is about 6 kJ mol^{-1}, determine the value of the freezing point constant, K_f.

Problem 12.5

Show that

$$X_2 = \frac{\mathrm{M}_1}{1000} m$$

where X_2 is the mole fraction of dissolved solute, M_1 is the molar mass of the solvent, and m is the solute molality, that is, show that

$$\Delta T_b \approx \frac{RT_b^2 \mathrm{M}_1}{\Delta_{\mathrm{vap}} H_1 1000} m = K_b m$$

for dilute solutions.

Problem 12.6

In the section on freezing point depression, the *CRC Handbook of Chemistry and Physics, 2008–2009*, 89th ed., defines the mass % as the mass of solute divided by the total mass of the solution. Exactly 10.00 g of NaCl was added to sufficient water to make up 100.0 g of solution.

(a) What was the weight %?
(b) How many moles of NaCl were present in the solution?
(c) What was the molality of the solution?
(d) What was the molarity of the solution?
(e) Assuming a density of 1.0708 for this solution (*CRC Handbook of Chemistry and Physics, 2008–2009*, 89th ed.), what was its volume?
(f) What was its molarity?
(g) What was the mole fraction of NaCl?
(h) What was the mole fraction of water?

Problem 12.7

Exactly 10.0 g of B dissolved in 1000 g of A which has a molar mass M_A of 100 and a density of 1.000 gave an osmotic pressure of 0.0500 atm at 300 K. Find the molar mass of B.

Problem 12.8

In the section on freezing point depression, the *CRC Handbook of Chemistry and Physics, 2008–2009*, 89th ed., defines the mass % as the mass of solute divided by the total mass of the solution. Ammonia was infused into pure water until its concentration was exactly 10.00 g of NH_3 per 100.0 g of solution.

(a) What was the weight %?
(b) How many moles of NH_3 were present in 100 g of the solution?
(c) What was the molality of the solution?
(d) What was the molarity of the solution?
(e) Assuming a density of 0.9575 for this solution (*CRC Handbook of Chemistry and Physics*, 2008–2009, 89th ed.), what was its volume?
(f) What was its molarity?
(g) How many moles of water were there?
(h) If 1.8 mol of water are tied up by each mole of NH_3, how many moles of free water are there?
(i) How many grams is this?
(j) What is the molality of the solution of hydrated ammonia in water?

Problem 12.9

As part of the derivation of the osmotic pressure equation (Section 12.7), the statement was made that "In a dilute binary solution,

$$\ln x_1 \cong -x_2\text{"}$$

Verify this statement by numerical evaluation. Show that it is valid for dilute solutions only. *Hint*: Go to your calculus book and search infinite series.

Problem 12.10

A solution is made up so as to have a concentration of 1.428 g of solute per dm^3 in water. The osmotic pressure exerted by the solvent across a semipermeable membrane (Fig. 12.7) at 298 K was 0.224 atm. What is the molar mass of the solute?

13

COULOMETRY AND CONDUCTIVITY

Aqueous solutions of some substances conduct electricity well and some conduct it poorly. All aqueous solutions offer *resistance* R to flow of electrical *current* I. Good conductors offer low resistance and poor conductors offer high resistance. In what follows, we shall assume Ohm's law $V = IR$, where V is the potential difference forcing current to flow through the resistance.

13.1 ELECTRICAL POTENTIAL

A potential difference between two points separated by a distance l can be maintained by means of an electrochemical cell, commonly called a "battery" (Fig. 13.1):

$$\Delta\phi = \phi(0) - \phi(l)$$

The potential difference $V = \Delta\phi$ is called the *voltage* or *voltage drop*. We shall be concerned with the influence of a potential difference on various chemical solutions, especially salt solutions, placed between a pair of charged parallel plates. Solutions of salts in water that conduct electricity are called *electrolyte solutions*. Like other conductors, they offer some resistance to current flow.

13.1.1 Membrane Potentials

In mammalian cells, potential differences are established by a different mechanism. If a membrane is permeable to ions of one charge but impermeable (or less permeable) to

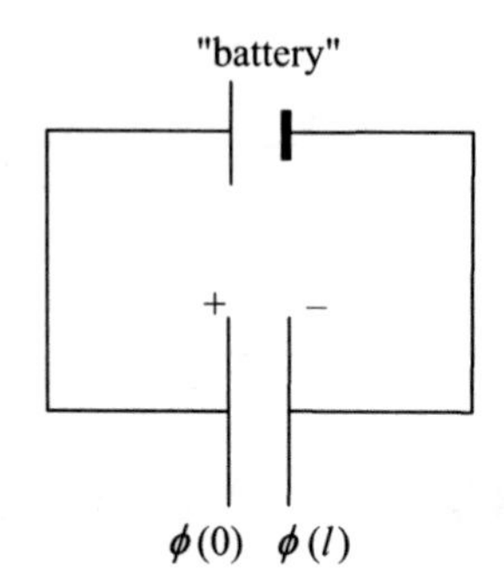

FIGURE 13.1 The potential drop between charged plates is $V = \phi(0) - \phi(l)$.

ions of the opposite charge, a potential difference is established across the membrane. Up to this point, expressions involving the Gibbs chemical potential μ have been used, assuming that there were no other potentials present; but in this section, a new potential, the *electrical potential* ϕ, is introduced. The two potentials, chemical and electrical, are additive:

$$\mu' = \mu + F\phi$$

where F is a proportionality constant (see below).

Suppose that, for a dilute solution of KCl, the potassium ion is preferentially passed through a membrane that blocks the Cl^- anion. An electrical potential builds up that is positive on the potassium-rich side of the membrane and negative on the Cl^- side. The situation is analogous to preferential passage of small molecules through a membrane in opposition to an osmotic pressure, except that now it is the electrical potential that opposes flow across the membrane. Just as in the case of osmotic flow, when the chemical potential driving the transfer is equal and opposite to the electrical potential, flow stops and an equilibrium has been established. This is the type of preferential ion flow that occurs between the interior and exterior of mammalian cells under certain conditions, and it is part of the mechanism of signal transmission along nerve networks (Fig. 13.2).

If we designate the interior of the cell α and the exterior of the cell β and take into account only the potassium ion, at equilibrium the sums of electrical and chemical potentials on either side are equal:

$$\mu_{+\alpha} + F\phi_\alpha = \mu_{+\beta} + F\phi_\beta$$

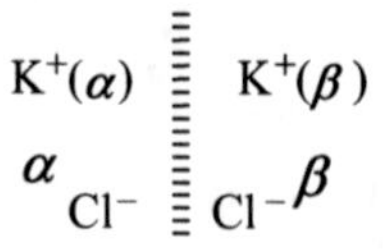

$K^+(\alpha)$ ⋮ $K^+(\beta)$
α Cl^- ⋮ Cl^- β

FIGURE 13.2 An ion-permeable membrane (schematic).

We know that, in general, $\mu = \mu^\circ + RT \ln a$; thus, in this case, we have

$$\mu^\circ_{+\alpha} + F\phi_\alpha + RT \ln a_{+\alpha} = \mu^\circ_{+\beta} + F\phi_\beta + RT \ln a_{+\beta}$$

but $\mu^\circ_{+\alpha} = \mu^\circ_{+\beta}$ because we are talking about the same ion, K^+, on either side of the membrane. Differences come, not in μ°_+, which is fixed, but in the potentials ϕ and the activities a, which are variable:

$$F\phi_\alpha + RT \ln a_{+\alpha} = F\phi_\beta + RT \ln a_{+\beta}$$

$$F\left(\phi_\alpha - \phi_\beta\right) = RT \ln \frac{a_{+\beta}}{a_{+\alpha}}$$

F is called the *Faraday constant*. It is a conversion factor from charge in coulombs C or millicoulombs mC to moles. The faraday is the charge on 1 mol of electrons, 96,485 C mol^{-1}, which may in general be regarded as 1 mol of charge.

Under the stipulation that the concentrations are rather low, the activity coefficients γ are about equal to 1.0 and we may replace the activities by molarities (or molalities) m to find the membrane potential

$$\left(\phi_\alpha - \phi_\beta\right) = \frac{RT}{F} \ln \frac{m_{+\beta}}{m_{+\alpha}}$$

Nerve cell walls are semipermeable, accounting for an equilibrium membrane potential of about 70 mV. This equilibrium potential can be disturbed by an electrical pulse from a neighbor cell, whereupon the pulse is transmitted to another neighbor cell in a sequence that is part of the mechanism for transmission of information along a nerve fiber.

13.2 RESISTIVITY, CONDUCTIVITY, AND CONDUCTANCE

If a charge suspended in a fluid medium is subjected to a potential difference $\Delta\phi$, it will move. Moving charge is called current $I = dQ/dt$, where Q is the charge in coulombs. Typically, a moving charge meets resistance R.[1] Resistance is proportional to the length l of a resistor and is inversely proportional to its cross-sectional area A:

$$R \propto \frac{l}{A} = \rho \frac{l}{A}$$

The proportionality constant $\rho = RA/l$ is a characteristic of the resistor material. It is called the *resistivity*.

These definitions prompt us to define the inverse of the resistance called the *conductance* $L \equiv 1/R$, along with the inverse of the resistivity, the *conductivity* $\kappa \equiv 1/\rho$:

$$L \equiv \frac{1}{R} = \frac{A}{\rho l} = \frac{\kappa A}{l}$$

[1]We shall not consider superconductive media here.

When a voltage V is applied to a *conductance cell* with parallel plates of area A separated by a distance l and containing an electrolyte solution, the current is

$$I = JA = \frac{\kappa A}{l} V$$

where J is the *flux* dQ/dt of charge Q passing through unit cross-sectional area A. When we combine these two equations, the current becomes $I = JA = LV$, which is another way of writing Ohm's law $V = IR$.

With these equations, it is possible to determine the conductivity of any solution from the resistance and a knowledge of l/A. It is not easy to measure l/A to the high level of accuracy required for good work in this field, but the task has already been done for solutions of KCl at several concentrations and temperatures. The results of these measurements are available (*CRC Handbook of Chemistry and Physics 2008–2009*, 89th ed). A selected value is $\kappa_{KCl} = 0.1408\ \mathrm{ohm}^{-1}\ \mathrm{m}^{-1}$ for 1.00×10^{-2} molal KCl at 298 K.

Suppose we construct a cell looking something like the one in Fig. 13.1, fill it with 1.00×10^{-2} molal KCl, and measure the resistance $R = 650$ ohms. We have

$$\kappa_{KCl} = 0.1408 = L\left(\frac{l}{A}\right) = \frac{1}{650}\left(\frac{l}{A}\right)$$

$$\frac{l}{A} = 650(0.1408) = 91.52\ \mathrm{m}^{-1}$$

We now know the geometric ratio l/A without having to make precise measurements of either l or A.

13.3 MOLAR CONDUCTIVITY

For chemical applications to solutions of varying concentrations c, an additional definition is necessary. The *molar conductivity* of an electrolyte solution is

$$\Lambda \equiv \frac{\kappa}{c}$$

where c is the concentration of the solution in $\mathrm{mol\ L}^{-1}$ or $\mathrm{mol\ dm}^{-3}$.

We shall be concerned largely with *aqueous solutions*. More than 100 years ago, F. Kohlrausch published a series of research results on conductivity of electrolytes in aqueous solutions. The conclusions he drew included what is now called Kohlrausch's law. The law is a linear connection between the molar conductivity and the square root of the electrolyte concentration:

$$\Lambda = \Lambda^{\circ} - \tilde{K}\sqrt{c}$$

where $\tilde{K}$ is an empirical constant. Kohlrausch's law leads to a graphical way of determining the parameter Λ°, a characteristic of the electrolyte alone, without

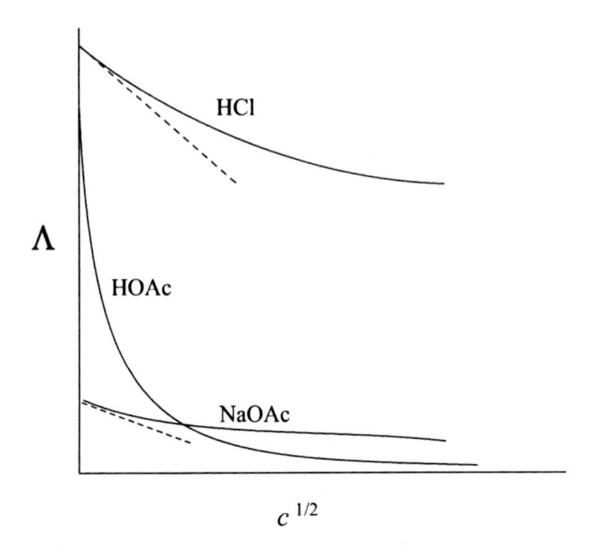

FIGURE 13.3 Kohlrausch's law for conductance of the strong electrolytes HCl and NaOAc and the weak electrolyte HOAc. Λ° is the intercept on the vertical axis.

electrostatic interference from crowding by other ions that may be of the same electrolyte or any other electrolytes present in the solution. For example, the Kohlrausch curves for the NaOAc and HCl in Fig. 13.3 can be extrapolated to a linear approximation shown by the dashed lines. The intercept of this limiting straight line with the vertical axis is Λ°. It is called the *molar conductivity at infinite dilution*. The shape of the HOAc curve, however, is such that one would not wish to find Λ° in the same way for this solute as for HCl and NaOAc because the curve is too steep near $\sqrt{c} = 0$ to do the job with acceptable accuracy.

Electrolytes in aqueous solution generally fall into two broad classes: *strong electrolytes*, which show clean linear extrapolations to Λ°, and *weak electrolytes*, which do not. A collection of experimental values for Λ° for strong electrolytes runs from about 100 to 425 ohm^{-1} and, other than high values for strong acids and bases, shows no particular regularity in itself. Kohlrausch observed, however, that many *differences* in Λ° are equal; for example,

$$\Lambda^\circ_{\mathrm{KCl}} - \Lambda^\circ_{\mathrm{NaCl}} = \Lambda^\circ_{\mathrm{KNO_3}} - \Lambda^\circ_{\mathrm{NaNO_3}} = \Lambda^\circ_{\mathrm{KOH}} - \Lambda^\circ_{\mathrm{NaOH}}$$

and so on, for a number of other Λ° values arranged pairwise in this fashion. This would be true, he said, if Λ° values were the result of contributions from each of the ions produced when the strong electrolyte goes into solution; for example,

$$\mathrm{NaCl(s)} \rightarrow \mathrm{Na^+(aq)} + \mathrm{Cl^-(aq)}$$

where (s) designates the solid compound and (aq) designates the ion in aqueous solution. If this were true, he said, the regularity in Λ° could be explained by writing the contributions to Λ° as

$$\Lambda^\circ_{KCl} = \lambda^\circ_{K^+} + \lambda^\circ_{Cl^-}$$

$$\Lambda^\circ_{NaCl} = \lambda^\circ_{Na^+} + \lambda^\circ_{Cl^-}$$

$$\vdots$$

and so on.

In this way, the difference found for

$$\Lambda^\circ_{KCl} - \Lambda^\circ_{NaCl} = \Lambda^\circ_{KNO_3} - \Lambda^\circ_{NaNO_3}$$

would be merely the difference

$$\Lambda^\circ_{KCl} - \Lambda^\circ_{NaCl} = \lambda^\circ_{Na^+} + \lambda^\circ_{Cl^-} - \lambda^\circ_{K^+} - \lambda^\circ_{Cl^-}$$
$$= \lambda^\circ_{Na^+} - \lambda^\circ_{K^+}$$

and would be the same for all pairs of sodium and potassium electrolytes because the anions, Cl^-, NO_3^-, OH^-..., would always cancel out. The same would be true for pairs of electrolytes containing a common anion, because the cation would always cancel out. In these and similar calculations, the integral charge on an ion must always be taken into account so that the expression for Kohlrausch's law of independent ion migration is

$$\Lambda^\circ = \nu_+\lambda_+ + \nu_-\lambda_-$$

where ν is the number of charges on each ion, $\nu = 1$ for the examples shown.

13.4 PARTIAL IONIZATION: WEAK ELECTROLYTES

So far we have considered mainly strong electrolytes, those for which ionization is complete in aqueous solution. This is not always the case. Ionization of weak electrolytes is partial. The usual example chosen is that of acetic acid for which ionization at ordinary concentrations is very small. Examples of the vast difference in behavior can be seen by comparing the Kohlrausch conductivity curves in Fig. 13.3 for HCl, which is completely ionized, with that of acetic acid (HOAc), which is not. Ionization of HOAc does, however, rise rapidly in dilute solutions, and Kohlrausch's law of independent migration holds at infinite dilution. At infinite dilution, the weak electrolyte HOAc *is* completely dissociated (all electrolytes are), so we have a very simple way of determining the approximate *degree of ionization* α at some finite concentration c by comparing Λ to Λ°. We simply take the ratio

$$\alpha = \frac{\Lambda}{\Lambda^\circ}$$

where α runs from 0 (the trivial case of no ionization) through small numbers at most concentrations only rising to $\alpha = 1$ for complete ionization.

The only trouble with this plan is evident from Fig 13.3. Accurate data are difficult to obtain at very low concentrations; and even if they could be found, Λ° is at the intersection of two nearly parallel lines. A way out of this dilemma is through Kohlrausch's law of independent ion migration at infinite dilution. Figure 13.3 shows that, although we cannot find an accurate value for $\Lambda^\circ_{\text{HOAc}}$ directly, we can find accurate values for $\Lambda^\circ_{\text{HCl}}$ and $\Lambda^\circ_{\text{NaOAc}}$. These, in combination with $\Lambda^\circ_{\text{NaCl}}$, yield

$$\begin{aligned}
&\Lambda^\circ_{\text{HCl}} + \Lambda^\circ_{\text{NaOAc}} - \Lambda^\circ_{\text{NaCl}} \\
&= \lambda^\circ_{\text{H}^+} + \lambda^\circ_{\text{Cl}^-} + \lambda^\circ_{\text{Na}^+} + \lambda^\circ_{\text{OAc}^-} - \lambda^\circ_{\text{Na}^+} - \lambda^\circ_{\text{Cl}^-} \\
&= \lambda^\circ_{\text{H}^+} + \lambda^\circ_{\text{OAc}^-} \\
&= \Lambda^\circ_m(\text{HOAc})
\end{aligned}$$

Armed with this new piece of information, we can calculate $\alpha = \Lambda/\Lambda^\circ$. The dissociation reaction is

$$\underset{(1-\alpha)c}{\text{HOAc}^+} \rightleftarrows \underset{\alpha c}{\text{H}^+} + \underset{\alpha c}{\text{OAc}^-}$$

where the concentrations of the individual ions at overall concentration c are controlled by the degree of dissociation αc. The undissociated HOAc $(1 - \alpha)c$ is what is left of the original HOAc after dissociation has taken place. Typical values might be 17% dissociated and 83% undissociated for $\alpha = 0.17$. The equilibrium constant for this *acid* dissociation reaction is

$$K_a = \frac{[\text{H}^+][\text{OAc}^-]}{[\text{HOAc}]} = \frac{(\alpha c)(\alpha c)}{(1-\alpha)c} = \frac{0.170(.170)}{0.83}c = 0.035\,c$$

If these dissociation data are observed for a 5.0×10^{-4} molar solution, $K_a(\text{HOAc})$ is found to be about 1.7×10^{-5} (accurate value 1.754×10^{-5}(*CRC Handbook of Chemistry and Physics, 2008–2009*)). This is the *acidity constant* of acetic acid familiar from general chemistry.

There are now computational methods of obtaining or at least approximating quantum mechanical gas-phase free energies for the components of the dissociation reaction along with the necessary computed free energies of solvent interactions.

13.5 ION MOBILITIES

In *ion mobility* studies we wish to know how fast an ion migrates in an electrical field. This is related to the charge flux $J = dQ/dt$ and to the *transport numbers* t_+ and t_-, the *proportion* of charge carried through the solution by each ion. Transport numbers, which are fractions, add up to 1 and are larger than 0.5 for an ion that carries more

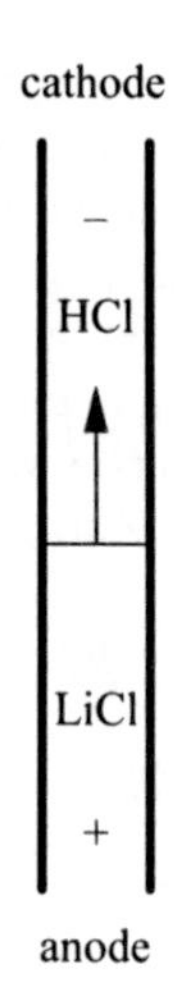

FIGURE 13.4 Moving boundary determination of the mobility of H^+. Li^+ is the follower ion.

than its "share" of charge and less than 0.5 for an ion that carries less than its share of the charge.

One way of measuring the mobility of a cation is to load a capillary tube with a solution containing the ion in question on top of a second electrolyte solution that is more dense than the measured ion solution. The cation of the more dense solution is called a *follower ion* because when a field is applied to the tube, it moves upward at the rate of the principal cation, H^+ in the case shown in Fig. 13.4. The follower ion is drawn along by H^+ to prevent a charge gap from developing between the two solutions.

Figure 13.4 shows a solution of HCl loaded into a capillary tube on top of a solution of LiCl, where Li^+ is the follower ion. Upon applying a potential difference across the cathode and anode, H^+ is attracted to the cathode (−) and Li^+ follows along. The velocity of migration is determined by the faster of the two ions. The interface between the HCl and LiCl solutions is maintained by the follower ion and can be visually observed because of the difference in refractive indices of the two solutions. The velocity is $\boldsymbol{x}$ m s^{-1} and the mobility u is the *speed* x/t per volt of potential difference:

$$u = \frac{x}{tV}\ \mathrm{m\,s^{-1}}\left(\mathrm{volt\,m^{-1}}\right)^{-1} = \frac{x}{tV}\ \mathrm{m^2\,volt^{-1}\,s^{-1}}$$

The rather odd-looking unit is meters per second per volt per meter of the resisting medium. In the present case, this is unit voltage across the conducting solution. All of the variables on the right are measurable quantities, hence the mobility can be calculated.

13.6 FARADAY'S LAWS

In the early nineteenth century, Faraday stated the laws:

1. **The mass of any substance deposited or dissolved at an electrode is proportional to the quantity of electricity passed through the solution.**
2. **One equivalent weight of any substance deposited or dissolved at an electrode requires 96,485 coulombs of electricity.**

(The modern value of 96,485 coulombs has been substituted for Faraday's original value which differed slightly.) If the quantity 96,485 coulombs is divided by the charge on the electron (measured about a century later), one obtains the Avogadro number. The explanation of this is that each ion carries an integral number of electrons (negative) or lacks an integral number of electrons (positive). In honor of one of the early geniuses of physical chemistry, the quantity 96,485 coulombs is defined as the *faraday*, F.

13.7 MOBILITY AND CONDUCTANCE

If we look at transport numbers, mobility, and ionic conductance, it should be evident that a fast-moving ion carries more of the total charge transported than a slow one. Also, transport numbers are the fraction of electrical charge carried by each ion, so they must add up to 1. In an aqueous solution of a univalent salt, $t_+ + t_- = 1.0$ and the sum of the molar (or molal) ionic conductances is $\Lambda^\circ = \lambda_+^\circ + \lambda_-^\circ$. The fraction of current carried by an ion can be given by either the transport numbers or the ratio of ionic conductivity $\lambda_\pm^\circ$ to the total molar conductivity Λ°, each at infinite dilution:

$$t_+ = \frac{\lambda_+^\circ}{\Lambda^\circ}, \qquad t_- = \frac{\lambda_-^\circ}{\Lambda^\circ}$$

In an electrolyte of faster and slower ions, the faster ion carries a greater proportion of the total charge transfer than the slower ion in proportion to their relative speeds. Therefore the ratio of transport numbers is the same as the ratio of ionic mobilities:

$$\frac{t_+}{t_-} = \frac{u_+}{u_-}$$

13.8 THE HITTORF CELL

Transport numbers can be determined directly by measuring the change in concentration brought about by transference in a compartmentalized cell called a *Hittorf cell* (Fig. 13.5). A Hittorf cell consists of three compartments: an anode compartment, a cathode compartment, and a central compartment between them. A potential is imposed on a solution in the cell containing, for example, Ag^+ ions, which are

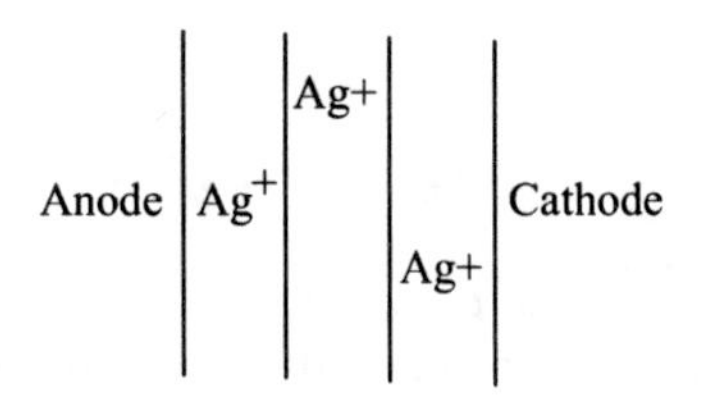

FIGURE 13.5 A three-compartment Hittorf cell.

deposited (electroplated) on the cathode. We can understand the use of the cell by taking two extreme hypothetical cases:

1. If there were no transference of charge by Ag^+ moving into the cathode compartment from the central compartment, the concentration would be reduced by one equivalent (gram atomic weight) per Faraday passed through the cell.
2. If silver ions carried *all* the charge through the Hittorf apparatus, all the Ag^+ ions lost at the cathode would be replaced by Ag^+ ions migrating into the cathode compartment from the central compartment. The concentration of Ag^+ ions would be unchanged.

In fact, the actual concentration in the cathode compartment is altered by less than one equivalent per coulomb but more than zero. The number of equivalents per coulomb transferred into the cathode compartment is the transference number of Ag^+. One can determine the transport numbers of both ions for example, Ag^+ and NO_3^-, along with their mobilities by a simple experiment (but one that requires precise analytical technique).

The conductance of a solution is the amount of charge in coulombs passing through it between electrodes at a given potential difference in volts per unit length of the resistor in units of V m^{-1}. From our idea that current flows through a solution of electrolyte due to passage of ions and from Faraday's proof that each ion carries an integral multiple of the unit charge e^-, the controlling factor must be the mobility with which ions move through the solvent, water in the case at hand. In fact the equivalent ionic conductance at infinite dilution λ° and the ionic mobility are directly proportional:

$$\lambda_+^\circ = F u_+^\circ$$

and

$$\lambda_-^\circ = F u_-^\circ$$

Taking values for Ag^+(aq) and NO_3^-(aq) (Example 13.3), we have

$$u_+^\circ = \frac{\lambda_+^\circ}{F} = \frac{62.6 \times 10^{-4}}{96{,}485} = 64.9 \times 10^{-8}\ \mathrm{m^2\,V^{-1}s^{-1}}$$

and

$$u_{-}^{\circ}=\frac{\lambda_{-}^{\circ}}{F}=\frac{70.6\times10^{-4}}{96{,}485}=73.2\times10^{-8}\,\mathrm{m^2\,V^{-1}\,s^{-1}}$$

These unusual units deserve a note. The units of speed $\mathrm{m\,s^{-1}}$ can be split off:

$$\mathrm{m^2\,V^{-1}\,s^{-1}=mV^{-1}\left(m\,s^{-1}\right)}$$

and

$$\mathrm{\left(m\,s^{-1}\right)mV^{-1}=m\,s^{-1}\frac{1}{\left(Vm^{-1}\right)}}$$

In other words, the unit is a typical speed in meters per second per unit voltage drop measured in volts per meter in the MKS system for the separation between the electrodes in the cell.

13.9 ION ACTIVITIES

One would like to have activities and activity coefficients for single ion concentrations to use just as activities, and activity coefficients were used for single molecule concentrations in Chapters 7 and 12. Unfortunately, this is not possible because single ions cannot be observed in the absence of a partner ion necessary to preserve electroneutrality. Faced with this dilemma, we make an approximation that ionic activities are equal: $a_{+}=a_{-}$. This is obviously not true, but it is the best we can do. The best we can find is an average activity, the geometric mean, $a_{\pm}\equiv\sqrt{a_{+}a_{-}}$. Other definitions follow as they did for simple molecules: $a_{\pm}=\gamma_{\pm}m$, $\gamma_{\pm}\equiv\sqrt{\gamma_{+}\gamma_{-}}$, and

$$\gamma_{\pm}\equiv\sqrt{\frac{a_{+}}{m}\frac{a_{-}}{m}}=\frac{a_{\pm}}{m}$$

Debye and Huckel have given a rather involved treatment of the electrostatic forces acting on an ion surrounded by other ions. They arrived at an expression for $\gamma_{\pm}$ in the form of the *Debye–Hückel limiting law*:

$$\ln\gamma_{\pm}=-1.172\,|Z_{+}Z_{-}|\,\sqrt{\tilde{\mu}}$$

where -1.172 is a combination of constants, $|Z_{+}Z_{-}|$ is the product of atomic charges ($|Z_{+}Z_{-}|=1.0$ in the cases we consider here), and $\sqrt{\tilde{\mu}}$ is a concentration term which is the summation of *all* the surrounding electrolytes, not just the electrolyte for which we seek $\gamma_{\pm}$. The symbol μ is almost universally used to denote both Debye–Hückel ionic strength and Gibbs chemical potential. To diminish this source of confusion, we are using the notation $\tilde{\mu}$ for the ionic strength:

$$\tilde{\mu} = \frac{1}{2}\sum_i m_i Z_i^2$$

The Debye–Hückel law is applicable only in very dilute solutions; it is a typical *limiting law*.

We know how to determine the ionization constant K_a for weak acids, say acetic acid, but we find upon close examination that the results are not quite constant with different concentrations of HOAc, and we presume that this lack is due to interference among ions. This nonideality can be expressed in terms of ionic activities:

$$K_a = \frac{a_{\mathrm{H^+}} a_{\mathrm{OAc^-}}}{a_{\mathrm{HOAc}}} = \frac{\gamma_\pm m_{\mathrm{H^+}} \gamma_\pm m_{\mathrm{OAc^-}}}{\gamma m_{\mathrm{HOAc}}}$$

In this model, nonideality is caused by charge interaction but HOAc is uncharged, so we can consider it an ideal solute and take $\gamma_{\mathrm{HOAc}} = 1.0$. The ions, however, are not ideal. This means that $\gamma_\pm \neq 1.0$, so

$$K_a = \frac{a_{\mathrm{H^+}} a_{\mathrm{OAc^-}}}{a_{\mathrm{HOAc}}} = \frac{\gamma_\pm m_{\mathrm{H^+}} \gamma_\pm m_{\mathrm{OAc^-}}}{\gamma_{\mathrm{HOAc}} m_{\mathrm{HOAc}}} = \gamma_\pm^2 \frac{m_{\mathrm{H^+}}^2}{m_{\mathrm{HOAc}}} = \gamma_\pm^2 K$$

where $m_{\mathrm{OAc^-}} = m_{\mathrm{H^+}}$ because equal numbers of product ions H^+ and OAc^- are produced in the dissociation reaction

$$\mathrm{HOAc} \rightleftarrows \mathrm{H^+} + \mathrm{OAc^-}$$

Taking logarithms of both sides, we obtain

$$\ln K_a = 2\ln\gamma_\pm + \ln K$$

where K is a measured (nonideal) value for the acid dissociation constant and K_a is the ideal value. At *infinite dilution*, $\gamma_\pm = 1.0$ and $\ln\gamma_\pm = 0.0$, so

$$\underbrace{\ln K}_{m\to 0} = \ln K_a$$

For a dilute acid in pure water, each ion concentration is αm, where α is the degree of dissociation and m is the molar concentration of acid (Section 13.4). The ionic strength is determined by the concentration of ions in the solution, but this is by no means the bulk concentration of acid. Many acids—for example, acetic acid HOAc—ionize to a very limited extent, but even ionization of pure water contributes to the ionic strength. According to the Debye–Hückel limiting law, we have

$$\ln\gamma_\pm = -1.172\sqrt{\tilde{\mu}} = -1.172\sqrt{\alpha m}$$

for the acid. The acidity (ionization) constant is

$$\ln K_a = 2\ln \gamma_\pm + \ln K$$
$$\ln K_a = 2\left(-1.172\sqrt{\alpha m}\right) + \ln K = -2.34\sqrt{\alpha m} + \ln K$$

or

$$\ln K = -2.34\sqrt{\alpha m} + \ln K_a$$

A plot of $\ln K$ vs. $\sqrt{\alpha m}$ gives an intercept of $\ln K_a$ and has a limiting slope of 2.34 as $\alpha \rightarrow 1$. As m becomes very small, the middle term of $\ln K = 2.34\sqrt{\alpha m} + \ln K_a$ drops out and the measured value of K approaches K_a.

PROBLEMS AND EXAMPLES

Example 13.1

Suppose you have access to a well-equipped electrochemistry laboratory but you do not know the value of the faraday. How would you determine the value of the faraday to four significant figures?

Solution 13.1 To work the problem *de novo* would be a pretty hard job. Remember, Faraday was a genius. When you electrolyze different solutions, however, you will soon find that many metals including silver can be plated out on an electrode by passage of current through a solution of, for example, silver nitrate. A coulomb is an ampere second. The idea is this: Measure the current flow I times the time t in seconds to force the quantity Q of electricity in coulombs through an electrolysis cell to plate the silver. Then measure the number of moles of silver that have been plated out. The fraction $I \times t$/moles Ag gives the faraday in coulombs mol^{-1}. Your first try at this method will probably be unsuccessful because the faraday is a big number. It takes more electricity than you might guess to deposit an amount of silver that is measurable to four figures of accuracy.

Many electrochemical laboratories are equipped with a variety of *potentiostats* that maintain a constant voltage across an experimental cell and *amperostats* that maintain a constant current. For our purposes, an amperostat would be most convenient because integration to find the area under an I vs. t curve would merely be multiplying constant I into the time. Both of these measurements should be possible with high accuracy. Silver has been known for many years to be plated out in 100% *current efficiency*, so if we multiply our I measurement into the time of electrolysis, both measured digitally, we get the number of coulombs passed through the cell. We can then weigh the silver electrode and find out how much it has increased in weight due to the silver deposited.

To get a rough idea of the parameters involved, suppose we have a microbalance sensitive and accurate enough to weigh to ± 0.01 mg. A silver weight of 0.2 g would be within our measurement range. If the amount of current we can maintain is about 50 milliamperes (mA), we know everything in the equation except the time. Trials show that it takes about an hour (1 h = 3600 s) to deposit 0.2 g of Ag.

$$F = \frac{I \times t}{\text{moles Ag}} = \frac{0.050 \times 3600}{\frac{0.2}{107.9}} = \frac{180}{0.00185} = 97{,}110$$

Now in our precise one-hour experiment, the real amperostat reading is 49.883 mA over the time period of 1.000 h = 3600.0 s, and the weight of silver gained by the (carefully dried) silver cathode is 0.20149 g

$$F = \frac{\text{coulombs}}{\text{mol}} = \frac{I \times t}{\text{mol}} = \frac{0.049883\,(3600.00)}{\frac{0.20101}{107.868}} = \frac{179.5788}{0.00186534} = 96{,}367$$

which is a little more than 0.1% in error.

Example 13.2

Suppose you are in the unlikely circumstance of needing to know the faraday to nine significant figures. Propose a way of calculating F from fundamental constants.

Solution 13.2 Unknown in Faraday's time, of course, unit charge is $1.602176487 \times 10^{-19}$ C and Avogadro's number is $N_A = 6.02214179 \times 10^{23}$. Since the faraday is the charge per mole of (singly charged) particles, we have

$$F = \left(6.02214179 \times 10^{23}\right) 1.602176487 \times 10^{-19} = 96485.3398\,\text{C mol}^{-1}$$

These data are from the handbook, but you will get some argument on the last two digits.

Example 13.3

(a) A Hittorf cell having 0.1000 mol of $AgNO_3$(aq) in each of its three compartments had 0.0100 faradays of charge passed through it. A reduction reaction depositing solid silver, Ag(s), took place at the cathode:

$$Ag^+(aq) + e^- \rightarrow Ag(s)$$

Analysis of the contents of the cathode compartment after the experiment showed that 0.00947 mol of Ag^+(aq) remained. What are the transport numbers of Ag^+(aq) and NO_3^-(aq)?

(b) Given the additional information that $\Lambda^{\circ}_{AgNO_3} = 133.3 \times 10^{-4}$ ohm^{-1} m^2 mol^{-1}, what are the molar ionic conductivities and mobilities of Ag^+(aq) and NO_3^-(aq)?

Solution 13.3

(a) If the Ag^+(aq) ions did not move, the amount of silver ion found in the cathode compartment after electrolysis would have been 0.0900 mols because of the removal of 0.0100 mol of Ag^+(aq) by electrodeposition. The fact that there are 0.0947 mol in the cathode compartment after deposition means that 0.0047 mol of Ag^+(aq) have been replaced by Ag^+(aq) migration from the center compartment. The transport number is the ratio of the charge transferred by Ag^+(aq) transport relative to the total charge transferred by both ions:

$$t_+ = \frac{0.0047}{0.0100} = 0.47$$

The rest of transport must have been due to NO_3^-(aq) ions migrating in the opposite direction, so $t_- = 1.00 - 0.47 = 0.53$.

(b) Transport numbers are related to ionic conductances according to the equations

$$t_+ = \frac{\lambda_+^{\circ}}{\Lambda^{\circ}} \qquad t_- = \frac{\lambda_-^{\circ}}{\Lambda^{\circ}}$$

which is reasonable because an ion that transports more than its share of charge will have a large ionic conductivity relative to the total Λ°. Given that $\Lambda^{\circ}_{AgNO_3} = 133.3 \times 10^{-4}$ ohm^{-1} m^2 mol^{-1}, we obtain

$$\lambda_+^{\circ} = t_+\Lambda^{\circ} = 0.47\left(133.3 \times 10^{-4}\right) = 62.6 \times 10^{-4} \text{ ohm}^{-1} \text{ m}^2 \text{ mol}^{-1}$$

and

$$\lambda_-^{\circ} = t_-\,\Lambda^{\circ} = 0.53\left(133.3 \times 10^{-4}\right) = 70.6 \times 10^{-4} \text{ ohm}^{-1} \text{ m}^2 \text{ mol}^{-1}$$

Problem 13.1

A solution containing 1.155×10^{-2} mol of HCl was neutralized by adding NaOH solution. The process brought about a 0.553 K temperature rise. An electrical heater of 58.7-ohm resistance brought about a 0.487 K temperature rise in 15.0 min. A potential of 6.03 volts was maintained across the resistor by a potentiostat. What is the molar enthalpy of HCl? Remember that a joule is a volt coulomb.

Problem 13.2

What is the charge on the silver ions in 1.000 dm^3 of dissolved Ag^+ in aqueous solution equilibrium with Cl^- and solid AgCl(s) at 298 K? The solubility product constant of AgCl is $K_{sp} = 1.77 \times 10^{-5}$. There is no outside source of Cl^- ion.

Problem 13.3

L-Dopa (L-(dihydroxyphenyl)alanine) is used to treat the symptoms of Parkinson's disease. It is quantitatively and exhaustively reduced in an electrolysis cell, requiring two electrons per molecule. Its molar mass is $M = 0.1972\,\text{kg mol}^{-1}$. Suppose a sample requires 57.5 μC (microcoulombs) for complete reduction. How much L-dopa was in the sample?

Problem 13.4

The resistivity of copper at 298 K is $\rho = 1.71 \times 10^{-8}$ ohm m. A potential of 1.00 mV was placed across a copper wire 1.00 mm in diameter and 1.00 m long. What was the current?

Problem 13.5

(a) The resistivity of copper at 298 K is $\rho = 1.71 \times 10^{-8}$ ohm m. Find the conductance and the conductivity of a copper wire 1.00 mm in diameter and 1.00 m long.

(b) Find the conductance and the conductivity of a copper wire 2.00 mm in diameter and .500 m long.

Problem 13.6

A conductivity cell is constructed using two platinum electrodes, precisely machined to be a square 2 cm on an edge and 1 cm apart. A 0.1000 mol dm^{-3} HCl solution was poured into the cell, and the resistance was measured and found to be 6.3882 ohms. What is the molar conductivity of HCl at this concentration? In solving this problem, be careful to express the units clearly at each step.

Problem 13.7

We can write the ionization of water in a somewhat simplified form as

$$H_2O \rightarrow H^+ + OH^-$$

The conductivity of deionized water is $\kappa = 5.50 \times 10^{-6}\ \text{ohm}^{-1}\ \text{m}^2\ \text{mol}^{-1}$.

(a) Find the concentration c of pure water in mol dm^{-3}.

(b) Find the degree of ionization α.

(c) Find the equilibrium constant K_w for the dissociation of pure water.
(d) Find the concentration of hydrogen ions H^+ in pure water.
(e) Why is it meaningful to calculate the equilibrium constant for H^+ even though we know that H_3O^+, H_5O^+, ... also exist in pure water?

Problem 13.8

The molar conductivities of NaCl are given in the handbook as

mol dm^{-3}	Conductivity	mol dm^{-3}	Conductivity
5.0000e-4	124.4400	1.0000e-3	123.6800
5.0000e-3	120.5900	0.0100	118.4500
0.0200	115.7000	0.0500	111.0100
0.1000	106.6900		

Plot these data in the form of a Kohlrausch plot. Determine the value of $\Lambda^\circ_{\mathrm{NaCl}}$ and the slope of the linear extrapolation to $\Lambda^\circ_{\mathrm{NaCl}}$.

Problem 13.9

The molar conductivities Λ° of NH_4 Cl, NaOH, and NaCl are 149.6, 247.7, and 126.4 in units of 10^4 S m^2 mol^{-1}. Find $\Lambda^\circ_{\mathrm{NH_4OH}}$.

Problem 13.10

Show that the units of conductivity κ are ohm^{-1} m^{-1}.

Problem 13.11

One meaning of the term Gibbs *free* energy is that ΔG is the maximum work that can be obtained from a chemical reaction. Conversely, the negative of ΔG is the minimum work that must be put into a chemical system to reverse its direction. Given that $\Delta G = -237$ kJ mol^{-1} for the spontaneous reaction (note the sign of ΔG)

$$H_2(g) + \tfrac{1}{2}O_2(g) \rightarrow H_2O(l)$$

and the fact that work $w =$ charge(V) where the charge is in coulombs and V is the voltage, find the minimum potential in volts necessary for the electrolysis of water to form $H_2(g) + \frac{1}{2}O_2(g)$.

14

ELECTROCHEMICAL CELLS

Advances in electrochemistry came at least a generation earlier than corresponding advances in rigorous thermochemical theory. Even prior to Faraday's numerous discoveries, Alessandro Volta's experiments on static electricity and electrical current took place largely in the eighteenth century and culminated in the first true battery of series voltaic cells. The battery was called a voltaic pile because cells were piled on top of one another. Volta's work excited great popular interest, especially when he showed that Luigi Galvani's use of a voltaic pile to activate nervous response in the dissected leg of a dead frog depended on the electrical potential of the pile and was not a property of the frog's leg. Activation of a dead animal raised the question of whether science might someday actually produce life.[1] News of Volta's experiments became something of a fad of the day. Mary Shelley's popular creation *Frankenstein* was presumably electrochemically activated.

14.1 THE DANIELL CELL

If you put a zinc rod into a solution of zinc sulfate $ZnSO_4$, there will be a difference in chemical potential between the Zn(s) in the metal rod and the Zn^{2+}(aq) ion in the *aqueous* solution. A reaction takes place, causing a minute amount of Zn to dissolve and go from the rod into the solution, leaving two electrons behind or, depending on

[1]See Gibson, D. G. et al. 2010. *Science*, *329*, 52–56 for a claim that science *has* produced life.

concentration, causing $Zn^{2+}(aq)$ to go from the solution onto the rod, taking up two electrons from the metal:

$$Zn(s) \rightarrow Zn^{2+}(aq) + 2e^-$$

or

$$Zn^{2+}(aq) + 2e^- \rightarrow Zn(s)$$

Either way, an electrical potential difference is produced between the electrode and the solution such that the sum of the chemical potential and the electrical potential, called the *electrochemical potential*, on one side balances the electrochemical potential on the other side, and equilibrium is reached.

If the same thing is done using a copper rod in contact with a $CuSO_4$ solution, an electrochemical balance is quickly reached in the same way, but the electrical potential on the Cu electrode is not the same as it was on the Zn electrode because the chemical potential of solid Cu is not the same as it is for solid Zn(s). Zn(s) is said to be more *active* than Cu(s) because it has a greater tendency to dissolve in a dilute solution of its ions than Cu(s). If a $Zn; Zn^{2+}$ *electrode* at potential ϕ is connected to a $Cu; Cu^{2+}$ electrode across a digital voltmeter and the circuit is completed by means of a neutral salt bridge, the difference in potential, called the *voltage* of the cell, can be measured. It turns out to be about 1.1 volts, depending on the concentrations of the $ZnSO_4$ and $CuSO_4$ in the electrode compartments. The entire arrangement is called a Daniell cell.[2]

If a resistance or small motor replaces the digital voltmeter, heat or work can be obtained. This is the principle of all electrochemical cells. Historically, two or more cells connected were called a *battery* of cells by analogy to gun batteries. What we call the 1.5-volt AA "battery" is really a single cell.

14.2 HALF-CELLS

There are a great variety of cells available for research and practical use. The $Cu; Cu^{2+}$ and $Zn; Zn^{2+}$ electrodes can be replaced by other metals, leading to many metal–metal ion combinations. These $M; M^{z+}$ systems, where z is the number of electrons involved in the electrode reaction, are called *half-cells* because any two of them can be combined to form a cell. The semicolon is used to denote the possibility of electron exchange and is often, but not necessarily, a physical interface such as the solid;solution interface in $Zn; Zn^{2+}$. The electrochemical standard is the *hydrogen half-cell* (Fig. 14.1). Hydrogen is allowed to pass through the cell under 1 atm

[2]Invented as early as 1837. The original Daniell cell was a copper cup containing the $CuSO_4$ in which a porous cup containing the $Zn;Zn^{2+}$ solution was suspended. The porous cup soaked with salt solution took the place of the salt bridge. The Daniell cell had many practical uses and played an important part in early electrochemistry.

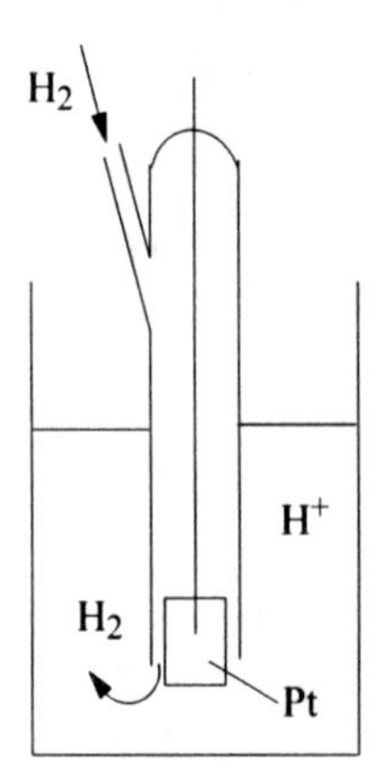

FIGURE 14.1 The hydrogen half-cell.

(or 1 bar) pressure so that the surface of the platinum electrode is alternately bathed by $H_2(g)$ and an acid solution containing $H^+(aq)$ion. The $H_2(g)$ is adsorbed onto the Pt surface (Chapter 11) so that what the $H^+(aq)$ ion "sees" appears to be a solid surface of $H_2(g)$. The electrode reaction is comparable to a metal ion reaction, except that now it is a hydrogen–hydrogen ion reaction:

$$H_2(g) \rightarrow 2H^+(aq) + 2e^-$$

Hydrogen half-cells can be connected to other half-cells, and the resulting voltage can be measured as before.

Half-cell potentials cannot be measured by conventional means. Instead, the hydrogen half-cell is chosen as a standard and is arbitrarily assigned a potential of 0.000 volts. Now cells can be constructed from the hydrogen half-cell and any other half-cell. The whole-cell potential is measured and, the contribution from $H_2(g); H^+(aq)$ being zero by definition, the entire cell potential is assigned to the other half-cell. In this way, an *electrochemical series* of many half-cell potentials is built up, going from active metals like $Zn; Zn^{2+}$ near the top to electrodes like $Cu; Cu^{2+} = +0.337$ volts farther down. The $Cu;Cu^{2+}$ is more positive than $H_2(g); H^+(aq)$ and the $Zn;Zn^{2+} = -0.763$ volts is more negative; hence the difference between them is $0.337 - (-0.763) = 1.10$ volts, which accounts for the 1.1 volts observed for the Daniell cell. All this, of course, requires a definition of the concentration of the metal (or other) ions in question. Although other conventions are sometimes used, the standard half-cell potentials are defined at ion activities of 1.00.

14.3 HALF-CELL POTENTIALS

Many half-cells have been studied in the long and rich history of electrochemistry. They are, by convention, listed as *reduction potentials* involving addition of one or

TABLE 14.1 A Few Selected Reduction Potentials.

Half-Cell	Reaction	E°(volts)
$K^+; K(s)$	$K^+ + e^- \rightarrow K(s)$	−2.925
$Zn^{2+}; Zn(s)$	$Zn^{2+} + 2e^- \rightarrow Zn(s)$	−0.763
$Cd^{2+}; Cd(s)$	$Cd^{2+} + 2e^- \rightarrow Cd(s)$	−0.403
$H^+; H_2(g)$	$2H^+ + 2e^- \rightarrow H_2(g)$	0.000
$Cl^-; AgCl(s); Ag$	$AgCl + e^- \rightarrow Ag(s) + Cl^-$	0.222
$Cu^{2+}; Cu(s)$	$Cu^{2+} + 2e^- \rightarrow Cu(s)$	0.337
$Fe^{3+}; Fe^{2+}; Pt$	$Fe^{3+} + e^- \rightarrow Fe^{2+}$	0.771
$Ag^+; Ag(s)$	$Ag^+ + 2e^- \rightarrow Ag(s)$	0.799

more electrons causing reduction of the oxidation number, one for $K^+(aq) + e^- \rightarrow K(s)$, two for $Cu^{2+}(aq) + 2e^- \rightarrow Cu(s)$, and so on (Table 14.1).

14.4 CELL DIAGRAMS

Certain conventions serve to simplify the written description of electrochemical cells. Following convention, the half-cell that is higher in the table goes on the left and the half-cell that is lower in the table goes on the right. We write the right half-cell as it appears in the table and reverse the left half-cell to give a *diagram* for the whole cell. In the case of the Daniell cell, this is

$$Zn(s); Zn^{2+}(aq) \| Cu^{2+}(aq); Cu(s)$$

where the double line represents a salt bridge. The *cell reaction* is the sum of half-cell reactions with the left-hand half-cell reversed:

$$\begin{array}{l} Zn(s) \rightarrow Zn^{2+}(aq) + 2e^- \\ \qquad\qquad Cu^{2+}(aq) + 2e^- \rightarrow Cu(s) \\ \hline Zn(s) + Cu^{2+}(aq) \rightarrow Cu(s) + Zn^{2+}(aq) \end{array}$$

The two electrons cancel upon addition. The electrode at which reduction occurs (remember that half-cell potentials are listed as reductions) is defined as the *cathode*. The other electrode is the *anode*.

The voltage of the whole cell is the sum of the half-cell potentials with the left half-cell once again reversed:

$$E^\circ = E^\circ_{\frac{1}{2}R} - E^\circ_{\frac{1}{2}L} = 0.337 - (-0.763) = 1.10 \text{ volts}$$

This is consistent with our treatment of thermodynamic potentials, for example, we have

$$\Delta G^\circ = \Delta G^\circ(\mathrm{B}) - \Delta G^\circ(\mathrm{A})$$

for the reaction

$$\mathrm{A} \rightarrow \mathrm{B}$$

The cell reaction, which involves reduction at one electrode and oxidation at the other, is called a reduction–oxidation reaction, or simply a *redox* reaction. Redox reactions are spontaneous if written according to the conventions agreed upon.

14.5 ELECTRICAL WORK

A joule is a volt coulomb, so an electrochemical cell operating at 1.0 volt does 1.0 J of work for every coulomb of electricity it produces. Don't forget that the coulomb C is an amount of charge placed in a capacitor by a specific current in a specific time $C = It$. Therefore the number of coulombs is proportional to the number of electrons driven through a resistance or motor by the cell reaction. If one electron is exchanged in the cell reaction and one mole of reactant is used up, one mole of electrons is exchanged, and one faraday of charge is produced. The work produced is 96,485 joules. In general, n electrons are exchanged and the cell potential is E volts, so

$$w = nFE = -\Delta G$$

because the amount of work done per mole of reactant consumed is the molar decrease in the Gibbs free energy. This leads to the important connection between thermodynamics and electrochemistry:

$$\Delta G = -nFE$$

14.6 THE NERNST EQUATION

For ideal solutions we have

$$\Delta G = \Delta G^\circ + RT \ln Q$$

where Q is the equilibrium quotient, not to be confused with the charge Q. Therefore

$$nFE = nFE^\circ - RT \ln Q$$

or

$$E = E^{\circ} - \frac{RT}{nF} \ln Q$$

which is the *Nernst equation*. At 298 K, RT and F can be combined to find

$$E = E^{\circ} - \frac{0.0257}{n} \ln Q$$

The meanings of ΔG and ΔG° enable us to better understand E and E°. As ΔG is the free energy change for any arbitrary equilibrium quotient Q, E represents the potential obtained from a cell under these arbitrary conditions. If the reactants and products are all in their standard states, the difference in free energy between reactants and products is ΔG°. Hence E° represents the potential of a cell in which the reactants and products in the redox reaction are in their standard states and at unit activity. This is just the cell potential from a table of standard half-cell potentials.

Knowing the concentrations of the constituents of any cell not at unit activity, one can calculate E° from the table and correct it to find the actual cell potential using the Nernst equation. For example, if the activities in the electrode reaction are

$$\mathrm{Cu(s)} + 2\mathrm{Ag}^{+}(\mathrm{aq}, 0.01\ \mathrm{m}) \rightarrow \mathrm{Cu}^{2+}(\mathrm{aq}, 0.1\ \mathrm{m}) + \mathrm{Ag(s)}$$

the equilibrium quotient is $Q = \mathrm{Cu}^{2+}(\mathrm{aq}, 0.1\ \mathrm{m})/\mathrm{Ag}^{+}(\mathrm{aq}, 0.01\ \mathrm{m})^2 = 0.1/0.01^2 = 1000$. The *standard* cell potential is $E^{\circ} = 0.799 - (0.337) = 0.462$. Applying the Nernst equation with $n = 2$, we obtain

$$E = E^{\circ} - \frac{0.0257}{n} \ln Q = 0.462 - 0.0128(6.908) = 0.373 \text{ volts}$$

Much of the older literature uses the logarithm to the base 10 in the Nernst equation. Conversion from ln to $\log_{10}$ requires multiplication of the $\ln Q$ term by 2.303 to yield

$$E = E^{\circ} - \frac{0.0257}{n} \ln Q = E^{\circ} - \frac{2.303\,(0.0257)}{n} \log Q$$

$$= 0.462 - \frac{0.0592}{2} 3 = 0.343 \text{ volts}$$

All of this assumes ideal behavior of course, as well as that $a = m$ throughout. We should be skeptical. More about this later.

14.7 CONCENTRATION CELLS

The Nernst equation suggests that we can have an electrochemical cell with half-cells that are identical in every way except for the concentrations of the ions in the electrode compartments. There will be a measurable potential that depends only on the difference in concentrations of the aqueous solutions. Such cells are called *concentration cells.*

An example would be any metal;metal ion pair with ion concentrations, say $Ag;Ag^+(aq)$ of 0.10 molar in one half-cell and 0.010 m in the other:

$$Ag(s);Ag^+(0.010\ m)|\ |Ag^+(0.10\ m);\ Ag(s)$$

The cell reaction is the half-cell potential of the cathode minus that of the anode (so as to give a positive whole-cell potential):

$$Ag^+(0.10\ m) \rightarrow Ag^+(0.010\ m)$$

that is, the direction of spontaneous change for any solution is to become more dilute. (Solutions don't spontaneously "concentrate themselves.") Q is $0.010/0.10 = 0.10$ and the Nernst equation is

$$E = 0 - 0.0257 \ln 0.10 = 0.0592\ \text{V} = 59.2\ \text{mV}$$

If the ion concentration in one of the half-cells is known and the other is unknown, the concentration of the unknown half-cell can be determined from the whole-cell potential. This is the principle behind the most widely used application of electrochemistry, the pH meter, in which a potential is measured that is linearly related through the Nernst equation to the logarithm of the hydrogen ion concentration in a half-cell. The negative logarithm (log base 10) of the hydrogen ion concentration is, as we learn in elementary chemistry, the pH. pH meters consist of an H^+-sensitive (glass) electrode and a constant voltage (calomel) electrode with suitable electronics to measure and display or record their algebraic sum. pH meters are calibrated to display the hydrogen ion concentration directly in pH units, most frequently as a digital readout.

14.8 FINDING E°

A silver electrode immersed in a solution of HCl soon becomes covered by a coating of AgCl(s). This half-cell, connected to a standard hydrogen electrode gives the cell

$$Pt(s);\ H_2(g);\ HCl(aq);\ AgCl(s);\ Ag(s)$$

The cell has no need for a salt bridge. One electrode is sensitive to H^+ and the other is sensitive to Cl^- which is precisely equal in concentration. The cell reaction is

$$AgCl(s) + \tfrac{1}{2}H_2(g) \rightleftarrows H^+(aq) + Cl^-(aq) + Ag(s)$$

The Nernst equation for this cell is

$$E = E^\circ - \frac{RT}{F} \ln \frac{a_{Ag} a_{H^+} a_{Cl^-}}{a_{AgCl} a_{H_2}^{\frac{1}{2}}}$$

We can set three of these five activities equal to 1.0 because two of them refer to solids AgCl(s)and Ag(s) and the other has an activity $a_{H_2} = 1.0$ at a hydrogen pressure of 1.0 atm (or 1.0 bar). The remaining activities are $a_{H^+} a_{Cl^-} = \gamma_\pm m_{H^+} \gamma_\pm m_{Cl^-} = \gamma_\pm{}^2 m_{HCl}{}^2$, provided that HCl is completely ionized (which it is) and there are no other sources of H^+ or Cl^-. Now the Nernst equation reads

$$\begin{aligned} E &= E^\circ - \frac{RT}{F} \ln a_{H^+} a_{Cl^-} = E^\circ - \frac{RT}{F} \ln \gamma_\pm{}^2 m_{HCl}{}^2 \\ &= E^\circ - \frac{2RT}{F} \ln \gamma_\pm - \frac{2RT}{F} \ln m_{HCl} \end{aligned}$$

or

$$E + \frac{2RT}{F} \ln m_{HCl} = E^\circ - \frac{2RT}{F} \ln \gamma_\pm$$

The right-hand side of this equation approaches E° as m approaches zero because $\gamma_\pm \to 1.0$. Plotting $E + (2RT/F) \ln m_{HCl}$ as a function of m and extrapolating to $m = 0$ is appealing, but it is not quite the way the problem is solved. Debye–Hückel theory (Section 13.9) says that $\ln \gamma_\pm = -1.171\sqrt{\tilde{\mu}}$ where, in this case, $\tilde{\mu} = m$ near infinite dilution. Let us call the term on the left E' for convenience in plotting, so that

$$E' = \left(E + \frac{2RT}{F} \ln m_{HCl} \right)$$

whereupon

$$E' = E^\circ - \frac{2RT(1.171)}{F} m^{1/2}$$

Now we need only plot E' as a function of $m^{1/2}$ to obtain an intercept of E° (Fig. 14.2). The plot will yield a straight line with a slope of $-2RT(1.171)/F$ but only in the limit of infinite dilution. The extrapolation has been carried out with considerable precision. It yields 0.22239 volts for the silver chloride–hydrogen cell and, since the standard hydrogen electrode has a half-cell potential of zero by definition, this is equal to the half-cell potential of the silver–silver chloride half-cell listed in Table 14.1.

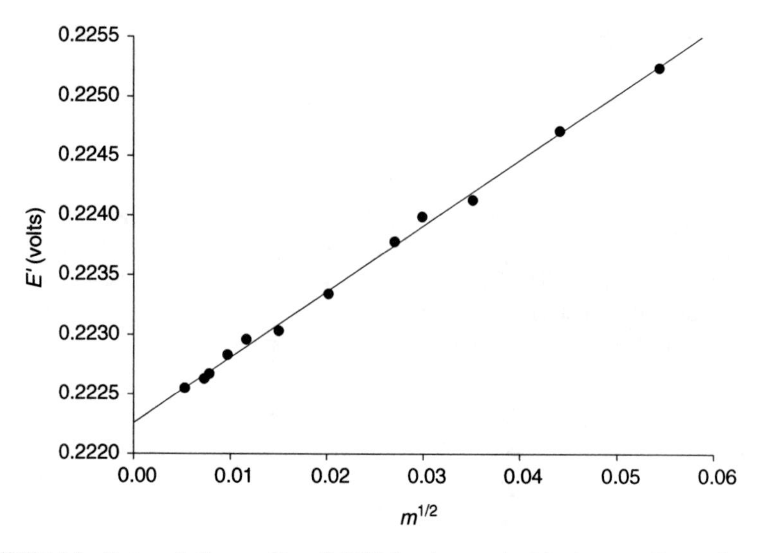

FIGURE 14.2 Extrapolation to $E^\circ = 0.2223$ for the standard hydrogen–silver–silver chloride Cell. (Standard value: 0.22239; data from Klotz and Rosenberg, 2008.) At higher concentrations the experimental points fall away from the linear function.

14.9 SOLUBILITY AND STABILITY PRODUCTS

In general chemistry we memorized the solubility product constant of silver chloride, $K_{sp} = [Ag^+][Cl^-] = 10^{-10}$, which implies that the concentration of silver ion in a solution of AgCl in pure water is $[Ag^+] = 10^{-5}$. How do you make an accurate measurement of the concentration of an ion that is 0.00001 molar (or molal)? The question becomes even more daunting for the case of copper phosphate, $K_{sp} = 10^{-37}$.

Because the Nernst equation relates the cell potential to the *logarithm* of the concentration, very small metal ion concentrations such as those in saturated solutions of sparingly soluble salts and stable complexes can be measured. For example, we can set up a cell consisting of a silver–silver ion half-cell in opposition to a silver–silver iodide half-cell which is analogous to the silver–silver chloride half-cell discussed in the previous section. The cell diagram and cell reactions are

$$\mathrm{Ag(s); AgI(s); I^-(aq)||Ag^+(aq); Ag(s)}$$
$$\mathrm{Ag^+(aq) + I^-(aq) \rightleftarrows AgI(s)}$$

The equilibrium constant for the cell reaction is

$$K_{eq} = \frac{a_{AgI}}{a_{Ag^+} a_{I^-}} = \frac{1}{K_{sp}}$$

where a_{AgI} of the solid precipitate has been set equal to 1.0 and K_{sp} is the solubility product of the AgI precipitate. The potential of this cell is 0.950 volts and, solving the Nernst equation, we find $\log K_{sp} = -0.950/0.0592 = -16.0$, so $K_{sp} \cong 10^{-16}$. Notice that we use 59.2 mV in the Nernst equation to calculate $\log_{10} K_{sp}$ (Section 14.7). Stability constants of complexes, in which metal ion concentrations are reduced to very low levels by *ligands* like EDTA, can be measured by a method analogous to the one shown for AgI.

14.10 MEAN IONIC ACTIVITY COEFFICIENTS

Back calculating from the procedure of Section 14.8 produces $\gamma_\pm$ at molalities other than zero. From

$$E = E^\circ - \frac{2RT}{F}\log\gamma_\pm - \frac{2RT}{F}\log m_{HCl}$$

$$E = E^\circ - 0.1184\log m_{HCl} - 0.1184\log\gamma_\pm$$

and

$$0.1184\log\gamma_\pm = -E + E^\circ - 0.1184\log m_{HCl}$$

$$\log\gamma_\pm = \frac{-E + E^\circ - 0.1184\log m_{HCl}}{0.1184}$$

At finite molality, E can be measured and everything on the right of the equal sign is known because E° is known.

14.11 THE CALOMEL ELECTRODE

The standard hydrogen half-cell has the advantage that its potential is 0.000 by definition, but it is bulky and presents some safety hazards in the use of hydrogen. Consequently, several half-cells that do not have these disadvantages are used as reference electrodes. The most common one is the calomel electrode (strictly, half-cell), which is easily set up, is light, and gives a constant voltage at a fixed temperature. Calomel half-cells have liquid mercury in contact with a paste of Hg_2Cl_2(s) and solid KCl in equilibrium with Cl^-(aq). Connection to external circuitry is by a Pt wire in contact with the mercury. Connection with any of a variety of other half-cells is made through a saturated KCl solution called a salt bridge. The saturated KCl solution does not change much upon drawing a small current, so the output voltage is a constant 0.2444 volts at 298 K. The electrode is designated a *saturated calomel electrode* (SCE) from the common name *calomel* for mercurous chloride (which was once used as a medicine for children!!). The cell diagram

$$\text{Pt; H}_2(1.0\text{ atm}); \text{H}^+(\text{aq})\|\text{Hg}_2\text{Cl}_2(\text{s}),\ \text{KCl}(\text{aq sat'd}); \text{Hg; Pt}$$

has a Nernst equation of

$$E_{\text{cell}} = E_{\text{calomel}} - 0.0257 \ln \text{H}^+(\text{aq})$$

because all of the activities other than $H^+(aq)$ are 1.0 or are included in E_{calomel}. This leads to

$$-\ln \text{H}^+(\text{aq}) = \frac{E_{\text{cell}} - E_{\text{calomel}}}{0.0257}$$

or, as is more commonly written,

$$\text{pH} \equiv -\log \text{H}^+(\text{aq}) = \frac{E_{\text{cell}} - E_{\text{calomel}}}{0.0592}$$

These results will not be accurate to an indefinite number of significant figures because the activity coefficient is not exactly 1.0 in real solutions. Cell potentials can be measured to four or more significant figures, but pH values should be regarded with a prudent degree of skepticism. Conversely, departure of pH values from expectation can be used to estimate activity coefficients.

14.12 THE GLASS ELECTRODE

We still have the problem of a bulky, somewhat dangerous hydrogen half-cell to determine pH in the field. For practical work, the glass electrode is used instead. Within certain limits, the potential between opposite sides of a very thin glass membrane varies in a linear way with hydrogen ion concentration. This is used in to determine pH, but the nature of the hydrogen ion sensitivity is not rigorously understood beyond knowing that there is an interchange between the aqueous hydrogen ions and the Na^+ ions in the glass on one side of the membrane that is different from the interchange on the other. Consequently, pH measurements have only relative significance. We measure pH by the potential relative to that due to a standard buffer solution for which we take the pH to be known. Glass electrodes can be made that are quite small, and the hand-held pH meter is common in many branches of laboratory and field chemistry. Other types of membranes lead to various electrodes sensitive to ions other than $H^+(aq)$—for example, $Ca^+(aq)$.

PROBLEMS AND EXAMPLES

Example 14.1 pH from the Hydrogen Half-Cell

The pH meter with a glass electrode is by no means the only way of determining pH. Suppose we reverse the method of Section 14.8 and use the half-cell potential of the

Ag;AgCl electrode in the cell

$$\mathrm{Pt(s); H_2(g,\ 1\ atm); HCl(aq = ?); AgCl(s); Ag(s)}$$

to determine the concentration of HCl(aq = ?). If the cell potential for HCl(aq = ?) is 493 mV, what is the pH of the solution in which the electrodes are immersed?

Solution 14.1 The cell reaction is

$$\mathrm{AgCl(s)} + \tfrac{1}{2}\mathrm{H_2(g)} \rightleftarrows \mathrm{H^+(aq)} + \mathrm{Cl^-(aq)} + \mathrm{Ag(s)}$$

The Nernst equation is

$$E = E^\circ - \frac{0.0592}{1} \log m_{\mathrm{HCl}} = 0.493 \text{ volts}$$

because of the unit activities of Ag(s) and AgCl(s) and $p_{\mathrm{H_2}} = 1$ atm. We know E° and we take $m_{\mathrm{H^+}} = m_{\mathrm{Cl^-}}$ because each HCl breaks up to give one of each kind of ion. We take m as the geometric mean of the ion concentrations $m_{\mathrm{HCl}} = m_{\mathrm{H^+}} m_{\mathrm{Cl^-}} = m_{\mathrm{H^+}}^2$:

$$\begin{aligned}
&0.493 = 0.222 - 0.0592 \log m_{\mathrm{H^+}}^2 \\
&0.493 - 0.222 = 2\,(0.0592)\,(-\log m_{\mathrm{H^+}}) \\
&0.271 = (0.1184)\,(-\log m_{\mathrm{H^+}}) = (0.1184)\,(\mathrm{pH}) \\
&\mathrm{pH} = 0.271/0.1184 = 2.29
\end{aligned}$$

Comment: The pH is not carried out to the same five-digit accuracy that the Ag;AgCl electrode presents because a number of approximations appear in the development of the pH equation. One is that the ionic interference between the dissociated HCl ions is not taken into account because the ions are very dilute and, second, dissolution of AgCl causes a slight imbalance between the $H^+$and Cl^- ions.

Example 14.2 A Mean Activity Coefficient

If a silver chloride–hydrogen cell with $m_{\mathrm{HCl}} = 0.1000$ produces an electrical potential of $E = 0.353$, what is the mean activity coefficient of H^+ at this molarity?

Solution 14.2 Notice that $\log m_{\mathrm{HCl}} = \log(0.1000) = -1$, so

$$\log \gamma_\pm = \frac{-E + E^\circ - 0.1184 \log m_{\mathrm{HCl}}}{0.1184}$$

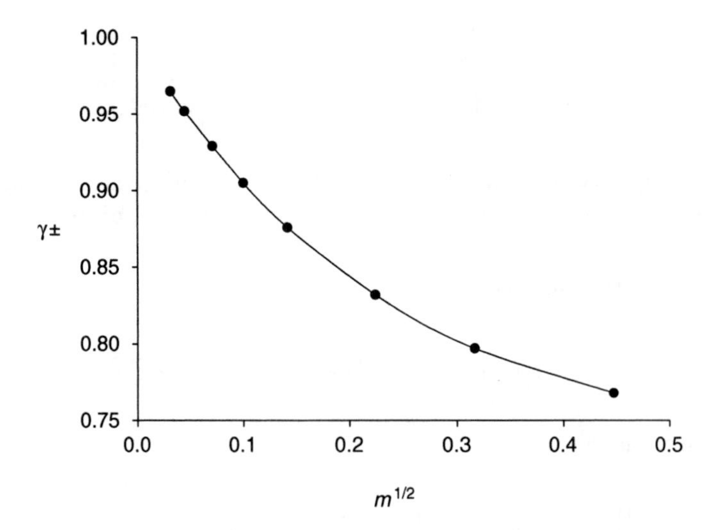

FIGURE 14.3 The mean activity coefficient of HCl as a function of $m^{1/2}$. The value $\gamma_{\pm} = -0.789$ calculated in Exercise 14.2 is the penultimate point on the curve reading from left to right. The standard value is $\gamma_{\pm} = -0.797$.

becomes

$$\log \gamma_{\pm} = \frac{-E + E^{\circ} + 0.1184}{0.1184} = \frac{-.353 + .2224 + 0.1184}{0.1184} = -0.10304$$

$$\gamma_{\pm} = -0.789$$

Experimental values of E at several specific values of m lead to specific values of $\gamma_{\pm}$ as a function of $m^{1/2}$ in Fig. 14.3. The values are less than 1.0 as predicted by the Debye–Hückel theory. The function is not linear except at low concentrations. Experimental points deviate from the linear function.

Problem 14.1

Suppose we construct a Daniell cell (Section 14.1), conventional in every way except that we replace the Zn^{2+}(aq) and Zn(rod) with Cd^{2+}(aq) and a Cd(s) rod.

(a) Which is the cathode and which is the anode, the copper cup or the Cd rod?

(b) What is the approximate voltage? (It is approximate because we don't know the exact concentrations of the Cu(aq) and Cd(aq) solutions.)

Problem 14.2

Many applications of electrochemistry have been in chemical analysis. Suppose a Zn; Zn^{2+} concentration cell is set up in which the Zn^{2+} concentration in one cell compartment is 0.0100 m but the concentration of the other Zn^{2+} half-cell is unknown. The whole-cell potential was measured and found to be 32.4 mV.

(a) Without doing a numerical calculation, is the unknown more or less concentrated than the 0.0100 m solution?
(b) What is the unknown concentration?
(c) Does your answer to part b agree with your answer to part a?

Problem 14.3

The half-cell (reduction) potential of the Ag:AgBr electrode is 0.071 volts. Use this information with information in Section 14.3 to determine the solubility product constant K_{sp} of AgBr in water.

Problem 14.4

A very important extension of electrochemical of solubility determination is generalization of the method to determine other kinds of equilibrium constant and further generalization to determination of the enthalpy, entropy, and Gibbs free energy changes for electrochemical reactions. Write or derive the equations for doing this.

Problem 14.5

Is it possible to reduce all of the Fe^{+3} ion in an aqueous solution containing Fe^{2+} and Fe^{3+} by filtering the solution through finely divided Zn(s)?

Problem 14.6

What is the standard Gibbs free energy change for the preceding reaction, and does your answer agree with the preceding answer?

Problem 14.7

A hydrogen half-cell combined with a saturated calomel electrode has a measured cell potential of 57.3 mV. What is the pH of the aqueous solution in the hydrogen half-cell? If the hydrogen electrode is a standard hydrogen electrode (SHE), what is the cell potential?

Problem 14.8

From the cell

$$Pt(s); H_2(g, 1\ atm); HCl(aq); AgCl(s); Ag(s)$$

one can obtain the following voltages at very small HCl(aq) concentrations

mol dm^{-3}	E_{cell} (volts)
0.00321	0.5205
0.00562	0.4926
0.00914	0.4686
0.0134	0.4497
0.0256	0.4182

Find the standard cell potential of the Ag(s);AgCl(s) half-cell.

Problem 14.9

It is proposed to use cerium(III) as a reducing agent in the redox titration of iron(III) according to the reaction

$$Ce^{3+}(aq) + Fe^{3+}(aq) \rightarrow Ce^{4+}(aq) + Fe^{2+}(aq) \qquad 1\ MH_2SO_4$$

The half-cell potentials written as reductions are somewhat modified in the strong acid solution:

$$Ce^{4+}(aq), Ce^{3+}(aq); Pt \text{ is } -1.44 \text{ volts} \qquad 1\ M\ H_2SO_4$$

and

$$Fe^{3+}(aq), Fe^{2+}(aq); Pt \text{ is } 0.68 \text{ volts} \qquad 1\ M\ H_2SO_4$$

Is this a plausible titration method for further development? If so, why? If not, why not?

15

EARLY QUANTUM THEORY: A SUMMARY

The first quarter of the twentieth century was a perplexing mix of discovery and wonderment as physicists progressed from the seemingly irrational quantization of energy by Max Planck (1900) to the inevitability of Werner Heisenberg's uncertainty principle (1925). For brevity, we shall refer the reader to more detailed works for the early history of quantum theory (especially Barrow, 1996; Laidler and Meiser, 1999; Atkins, 1998) and take as our starting point the observations by Louis de Broglie that small particles—in particular, electrons—have a wave nature and by Erwin Schrödinger that the spectrum of atomic hydrogen can be deduced by solving the wave equation that now bears his name.

15.1 THE HYDROGEN SPECTRUM

Some substances give off colors in the visible region of the electromagnetic spectrum when excited by an energy input. An example is sodium, which glows orange when a solution of NaCl is sprayed into a flame. In general, a multiplicity of distinct energy levels within an atom leads to a corresponding multiplicity of wavelengths characterized by its unique *electromagnetic spectrum*. Distinct colors or, more specifically, wavelengths, are given off or taken up by electrons as they emit or absorb energy of frequency ν by changing levels. Emitted radiation of different wavelengths can be separated and recorded in the form of a *line spectrum*. In most cases, the result of all these energy exchanges is of bewildering complexity, but one spectrum—that of the

Concise Physical Chemistry, by Donald W. Rogers

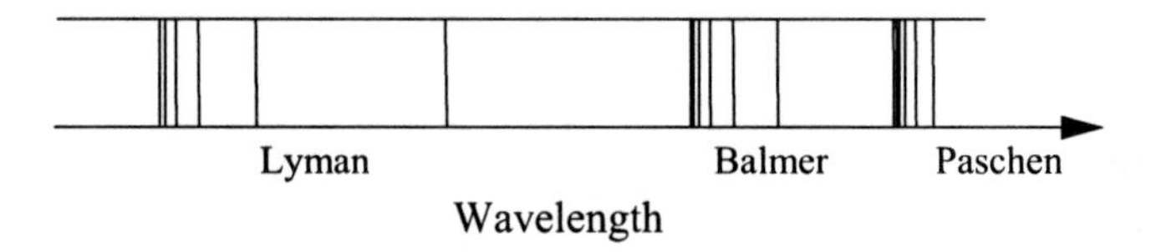

FIGURE 15.1 The hydrogen atom emission spectrum (not to scale). The wavelengths λ of the lines correspond to differences in energy between allowed levels in the hydrogen atom according to the equation $E = h\nu = hc/\lambda$, where c is the speed of light, h is Planck's constant, and ν is the frequency (Planck, 1901).

hydrogen atom—shows regular groupings that invite closer attention and theoretical explanation. Three of these groupings are shown in Fig. 15.1.

The experimental facts had been known for a half century before Schrödinger's time and had been partly explained in 1913 by Niels Bohr, who imposed a quantum number on the angular momentum of the orbital electron in hydrogen to arrive at a set of energy levels corresponding to stationary states of the atom. Transitions from one energy level to another correspond quite precisely to lines in the hydrogen spectrum, but it was not clear where the quantum number came from and it was not possible to generalize the Bohr system to more complicated atoms and molecules (Fig. 15.2).

15.2 EARLY QUANTUM THEORY

Early quantum theory showed that not only the hydrogen spectrum, but also several other important problems, can be solved by taking into account a frequency ν associated with the energy of particles according to the Planck equation $E = h\nu$. But the method, no matter how well it worked, still lacked a theoretical base and a logically consistent means of application, modification, and improvement. De Broglie (1924, 1926) pointed out that energy E of a moving particle implies momentum p. Frequency ν implies a wavelength. Therefore, if the Planck equation works (which it does),

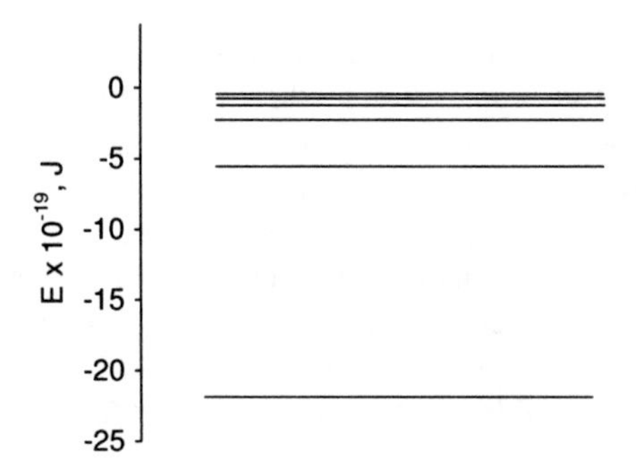

FIGURE 15.2 The first six solutions of the H atom energy calculated by Bohr (1913). The energies are negative because the electron is bound to the proton.

there should be a connection between momentum and wavelength, which he wrote as $p = h/\lambda$. At the atomic level, particles must have a wave nature. Conversely, waves have a particle nature. This was promptly experimentally verified. This mathematical equivalence between waves and particles is referred to as *wave-particle duality*.

Schrödinger (1925, 1926) reasoned that if electrons have a wavelength, they should follow a wave equation. He wrote an equation for the electron in the hydrogen atom employing a *wave function* Ψ, and he arrived at the electromagnetic spectrum of hydrogen just as Bohr had done but without making the arbitrary assumptions inherent in Bohr theory. While it should be noted that the Schrödinger equation is itself an assumption and that quantum theory is based on a set of postulates, the implications of Schrödinger's theory are far wider and the results are far more general than earlier theories. Quantum mechanics now pervades virtually all of physical science from molecular biology to string theory.

The Schrödinger equation can be written in equivalent forms with a rather intimidating array of notations, but they are all manifestations of the same thing: the postulate that a *state vector* $|\Psi\rangle$ or *state function* Ψ contains all the information we can ever have about a mechanical system such as an atom or molecule.

The *Hamiltonian function* was known from classical mechanics:

$$E = H = T + V$$

where T is the kinetic energy, written in terms of the velocities v in Cartesian 3-space $T = \frac{1}{2}m\left(v_x^2 + v_y^2 + v_z^2\right)$, and $V(x, y, z)$ is the potential energy. (Be careful not to confuse velocity v with frequency ν.)

An *operator* is a mathematical symbol telling you to do something. For example, V tells you to multiply by the potential energy (a scalar) and $\partial^2/\partial x^2$ tells you to differentiate twice with respect to x while holding some other variable constant. (It is a partial ∂x^2 because y and z are also variables.) An operator must operate on something, Ψ in the case we are interested in. If a *Hamiltonian operator* $\hat{H}$ for the hydrogen atom system is written in terms of mathematical operators V and $\hat{H} = \hat{T} - \hat{V}$ in a way that is analogous to the classical Hamiltonian for the total energy of a conservative system, we get the operator equation

$$\hat{H} = \hat{T} + \hat{V}$$

The Schrödinger equation is a special case of the Hamiltonian form where $\hat{T}$ is the operator corresponding to the *kinetic energy* of the electron:

$$\hat{T} = \frac{-\hbar^2}{2m_e}\left(\frac{\partial^2}{\partial x^2} + \frac{\partial^2}{\partial y^2} + \frac{\partial^2}{\partial z^2}\right)$$

and $\hat{V}(x, y, z)$ is the potential energy operator that describes the Coulombic attraction between the nucleus and the electron. ($\hat{V}(x, y, z)$ is usually written simply $V(x, y, z)$, where we take its operator nature as obvious.)

Atoms in their ground states are *conservative systems*. A conservative system is one that is not running down. Our solar system is a conservative system but a clock is not. This leads to the compact form

$$\hat{H}\Psi = E\Psi$$

for the *time-independent* Schrödinger equation for a conservative system, where $\Psi = \Psi(x, y, z)$ is an *amplitude* function, and E is the total energy of the system. We are interested in the time-independent form of the Schrödinger equation when we study the ground state structure and energy of atoms and molecules, which do not change over small time intervals. This equation was generalized to cover many electron atoms and molecules. As we shall see later, spectroscopic *transitions* within atoms and molecules involve the time-dependent form of the equation.

Independently, Heisenberg was developing a parallel line of quantum reasoning that led to his famous *uncertainty principle*. The Heisenberg and Schrödinger equations can be shown to be mathematically equivalent. Both Heisenberg and Schrödinger received the Nobel Prize.

Almost immediately after publication of Schrödinger's and Heisenberg's initial papers on quantum theory, Born et al. (1926) proposed that the square of the wave function $|\Psi(r)|^2\, d\tau$ (strictly, its vector inner product $\langle\Psi \mid \Psi\rangle$) describes undulations in the *probability* of finding a particle in an infinitesimal volume of space $d\tau$. For most of our purposes, we shall regard the square of the wave function as governing the *probable location* of an electron in a region of space containing many electrons.

15.3 MOLECULAR QUANTUM CHEMISTRY

As part of an emerging theory of molecular structure, the chemical bond was associated with a region of high relative electron density between two nuclei. Heitler and London (1927) took the lowest solution of the Schrödinger equation for one of the *atoms* in H_2 and combined it with an identical solution for the other *atom* in such a way as to obtain an approximate wave function for the chemical bond of the hydrogen *molecule*:

$$\psi = c_1\Psi_1 \pm c_2\Psi_2$$

The new wave function is said to be a *linear combination* of the two atomic wave functions Ψ_1 and Ψ_2, centered at their respective nuclei. By selecting appropriate values of c_1 and c_2 (which turn out to be equal), an energy minimum appears for the combined wave function. In this notation, the atomic wave functions Ψ are exact but the H_2 moleular orbitals ψ are approximate.

We are accustomed, from classical mechanics and thermodynamics, to thinking of an energy minimum as representing a stable state. And so it is with the Heitler–London energy minimum for the linear combination of atomic *orbitals* in hydrogen. The positive combination $\psi = c_1\Psi_1 + c_2\Psi_2$ leads to a stable state for two hydrogen atoms at approximately the experimental bond distance (74 pm). This is the first quantum mechanical affirmation of the two-electron bond postulated by G. N. Lewis and is familiar from general chemistry. We have the first instance of a calculated *orbital overlap* leading to stability for a molecular system. The term orbital has been substituted for the classical mechanical term orbit in recognition of the substitution of a probability function for a definite point location in space.

In the negative linear combination $c_1\psi_1 - c_2\psi_2$, we have the first example of an *antibond*. The probability function for the antibond shows that electrons are repelled from the region between the nuclei, leading to the opposite of a Lewis bond. Antibonding orbitals lie higher in energy than bonding orbitals. In some instances, infusion of energy into a molecule by incident radiation can cause electrons to be promoted from a bonding to an antibonding orbital. Only some wavelengths are absorbed. If the absorbed radiation is in the visible range (sunlight), we see the wavelengths left over after selective absorption as color.

Neither Schrödinger's nor Heisenberg's work was directed toward chemical bonding, so one cannot make the argument, "Well, they knew the answer before they set up the problem." The H—H bond in molecular hydrogen *is a consequence of the quantum nature of matter*. This point of view was developed into the *valence bond* theory by Linus Pauling (1935). Much of theoretical chemistry was to be dominated by valence bond theory (for better or for worse) for the next four decades.

Hartree (1928) is usually credited with the critical suggestion that atomic problems involving many electrons can be treated as a collection of simpler problems in which a single electron moves in an average electrostatic field created by the nucleus and all the other electrons.[1] The electrostatic attraction between a nucleus and an electron is far greater than the energy of a chemical bond. Therefore it is reasonable to suppose, as Hartree did, that one-electron wave functions would resemble Schrödinger's solutions for the hydrogen atom, being identical in the angular part and differing only in the radial probability distribution function—that is, the probable distance from the nucleus. This is the *central field approximation*.

Fock (1930) and Slater (1930), utilizing the prior concept of electron spin (Uhlenbeck and Goudsmit, 1925), recognized that the spin of an electron can be oriented in two ways, with or against its orbital motion. Therefore, electron spin must have a double-valued quantum number $m_s = \pm 1$. A double-valued quantum number demands that there be two wave functions per orbital, identical in all respects except for their different spins. The two wave functions must be *antisymmetric* just as your right and left hands are roughly identical in shape except that they are mirror images of one another (antisymmetric). Two trial functions $\varphi(1)$ and $\varphi(2)$, having opposed spins α, and β, can combine in four ways, only one of which is antisymmetric:

$$\psi = \varphi\alpha(1)\varphi\beta(1)$$

$$\psi = \varphi\alpha(2)\varphi\beta(2)$$

$$\psi = \varphi\alpha(1)\varphi\beta(2) + \varphi\alpha(2)\varphi\beta(1)$$

$$\psi = \varphi\alpha(1)\varphi\beta(2) - \varphi\alpha(2)\varphi\beta(1) \qquad \leftarrow$$

[1] Hartree himself gave credit to Bohr.

The antisymmetric wave function is the same as an expanded 2×2 determinant:

$$\begin{vmatrix} \varphi\alpha(1) & \varphi\beta(1) \\ \varphi\alpha(2) & \varphi\beta(2) \end{vmatrix} = \varphi\alpha(1)\varphi\beta(2) - \varphi\alpha(2)\varphi\beta(1)$$

To impose the condition of antisymmetry on two otherwise identical orbitals, the simple linear combinations of one-electron wave functions must be replaced by a *single-determinant* wave function of the form for the two-electron case (e.g., the helium atom):

$$\psi(1,2) = \begin{vmatrix} \varphi\alpha(1) & \varphi\beta(1) \\ \varphi\alpha(2) & \varphi\beta(2) \end{vmatrix}$$

Generalizing to several electrons led to more complicated *determinantal* wave functions. The use of *antisymmetrized orbitals*, as they were called, leads to the Hartree–Fock computational method. This method was applied to atoms and was tentatively applied to molecules larger than H_2.

15.4 THE HARTREE INDEPENDENT ELECTRON METHOD

One can treat the electrons of a many-electron atom as though they were entirely independent of one another moving under the influence of the nuclear charge Z, where Z is the atomic number, an integer greater than 1. This amounts to assuming higher-level *hydrogen-like orbitals* with principal quantum numbers $n = Z/2$ because each filled orbital contains 2 independent electrons. The resulting atomic energy will be wrong because electrons are, in fact, not independent.

This crude model can be substantially improved by taking an *effective nuclear charge* smaller than Z, motivated by our intuitive understanding that, no matter what the overall electronic probability distribution may be, there is some probability that some electrons will carry a negative charge between the nucleus and the other electrons, thereby reducing the influence of the nuclear charge on them. Any designated electron will be *shielded* from the nuclear charge by the other electrons. Simply plugging in arbitrary numbers for the effective nuclear charge $Z_{eff} < Z$ gives an improvement in agreement of the calculated energy of the system with the experimental value. It turns out that $Z_{eff} = 1.6$ is a pretty good guess in the example of helium, a two-electron problem with a nuclear charge of 2.0. Unfortunately, substitution of arbitrary numbers is theoretically meaningless. One would prefer to be able to calculate a radial electron probability distribution resulting in a theoretically derived energy and Z_{eff} for any single electron from a calculated energy distribution of all the others.

In treating helium, we shall begin with the reasonable first guess that each of its two independent one-electron wave functions resembles that of the hydrogen atom:

$$\Psi \cong \psi_1 \psi_2$$

and from the Schrödinger equation,

$$\hat{H}\Psi = E\Psi$$

we expect to find

$$\hat{h}\psi_i = \varepsilon_{ij}\psi_j$$

Mathematically, the operator $\hat{h}$ is a 2×2 transformation matrix that transforms the vector ψ_i into $\varepsilon_{ij}\psi_j$. This Hamiltonian operator is written in lowercase to show that it is approximate and not the exact operator $\hat{H}$. In the case of helium with two electrons, the state vector $|\psi\rangle$ can be expressed as a two-element column vector

$$|\psi\rangle = \begin{pmatrix} \psi_1 \\ \psi_2 \end{pmatrix}$$

We can improve upon ψ and ε by successive approximations, using an *iterative* procedure. We shall assume that we know where one electron is (we don't) and calculate its shielding effect on the other electron. We then calculate the probability density function for the location of the second electron as influenced by our initial guess as to its shielding from the first. With this information we can improve upon our initial guess as to the location of the first electron because we now have a probable location of the second electron. This iterative process can be continued until further iteration results in no further lowering of the calculated energy, whereupon we have the best estimate of the energy and the electron distribution we can get from the procedure.

The Hamiltonian operator $\hat{h}$ includes the kinetic energy operator of each electron i and the attractive electrostatic potential energy exerted on the electron by the nucleus

$$\hat{h}_i = \hat{T}_i - \frac{e^2}{r_i^2} \qquad (i = 1, 2)$$

The potential energy of one electron, call it electron 1, is influenced by the other, call it electron 2, according to their charge e and the distance between them, r_{12}. By the uncertainty principle we do not know where electron 2 is, but we do have a *probability density* function $|\psi(r_2)|^2$ governing its location in a volume element of space $d\tau_2$. Hence we have a most probable *charge distribution* function $e\,|\psi_2(r_2)|^2$. The probable potential energy of electron 1 shielded from the nucleus by electron 2 is, in suitable units, the product of the charges e divided by the distance between them

$$V_1 = e\int \frac{e\,|\psi_2(r_2)|^2}{r_{12}}d\tau_2 = e^2\int \frac{|\psi_2(r_2)|^2}{r_{12}}d\tau_2$$

The integration is over the space occupied by electron 2 because we cannot locate it as a point charge.

What can be said of the interaction between electron 1 with electron 2 can be said of electron 2 interacting with electron 1:

$$V_2 = e^2 \int \frac{|\psi_1(r_1)|^2}{r_{21}} d\tau_1$$

The operator $\hat{h}_i$ $(i = 1, 2)$ is the kinetic energy operator for electron 1 or 2:

$$\frac{-\hbar^2}{2m_e}\nabla_i^2 = \frac{-\hbar^2}{2m_e}\left(\frac{\partial^2}{\partial x^2} + \frac{\partial^2}{\partial y^2} + \frac{\partial^2}{\partial z^2}\right)$$

plus the (negative) potential energy of attraction between the nucleus and the electron, $-Ze^2/r_i^2$:

$$\hat{h}_i = \frac{-\hbar^2}{2m_e}\nabla_2^2 - \frac{Ze^2}{r_i^2}$$

With these potential and kinetic energies, we can write Schrödinger equations for electrons 1 and 2:

$$\left[\hat{T}_1 + \hat{V}_1(r_2)\right]\psi_1(1) = \hat{h}_1\,\psi_1(1) = \varepsilon_1\psi_1(1)$$

and

$$\left[\hat{T}_2 + \hat{V}_2(r_1)\right]\psi_2(2) = \hat{h}_2\,\psi_2(2) = \varepsilon_2\psi_2(2)$$

With these approximate Hamiltonians, we have two Schrödinger equations that contain the energy of the two electrons in the helium atom ε_i $(i = 1, 2)$:

$$\hat{h}\psi_i = \varepsilon_{ij}\psi_j \qquad (i = 1, 2)$$

The two equations are integrodifferential equations because they have an integral as part of $V(r)$ and second differentials in ∇^2. As written, they are *coupled* because one equation depends on the solution of the other through the potential energies. In the first equation, $V_1(r_2)$ contains the electron density of electron 2; in the second equation, $V_2(r_1)$ contains the electron density of electron 1. When we assume a wave function (any wave function) we *uncouple* these two equations. Though uncoupled and therefore solvable, the equations will no doubt lead to the wrong answer for the energies ε_i because the assumed $\psi(r)$ is only a guess. Hartree's guess was the most reasonable one he could make; he took ψ_1 and ψ_2 to be hydrogen-like orbitals.

Allowing Z to approach an effective nuclear charge Z_{eff}, we have the *Hartree equations*. Uncoupling these equations by assuming functions ψ_1 and ψ_2 was a brilliant step in quantum theory. It permits one to calculate a better approximation to V_1 and V_2, which leads to improved values of Z_{eff}, which leads to improved values of ψ_1 and ψ_2, which leads to better V_1 and V_2, and so on. This iterative process is continued until the energies do not change from one iteration to the next. When that condition has been met, the energies calculated from V_1 and V_2 are *self-consistent*. The Hartree equations are *eigenvalue* equations, hence they lead to discrete (scalar) values ε_1 and ε_2 for the two *one-electron* Schrödinger equations. The energies are not exact, but they are the best we can get from assumed hydrogen-like orbitals ψ_1 and ψ_2 by the Hartree procedure.

Conveniently, the applied mathematics of *self-consistent field* (SCF) iterations had been worked out by astronomers calculating the orbits of planets, prior to the advent of quantum mechanics. They called it the *variational method*.

15.5 A DIGRESSION ON ATOMIC UNITS

We can measure length in any unit we want: m, mm, furlongs, and so on. It is perfectly legitimate to use the radius of the first Bohr orbit as our unit of distance, replacing the meter but still related to it:

$$1\,a_0 \equiv 5.292 \times 10^{-11}\ \text{m}$$

Likewise, we can define units of mass, charge, and angular momentum:

$$1\,m_e \equiv 9.109 \times 10^{-31}\ \text{kg}$$

$$1\,e \equiv 1.602 \times 10^{-19}\ \text{C}$$

$$1\,\hbar \equiv 1.055 \times 10^{-34}\ \text{J s}$$

These definitions lead to considerable simplification in the equations of quantum chemistry (see Problems and Examples).

PROBLEMS AND EXAMPLES

Example 15.1 A Hartree Fock Solution for He

The limiting solution for a Hartree–Fock procedure to the problem of the helium atom is -2.862 hartrees, where the unit hartree (E_h) is defined as exactly half the energy of the electron in the ground state of hydrogen, $E_h \equiv 627.51\ \text{kcal mol}^{-1} = 2625\ \text{kJ mol}^{-1}$. The crude unshielded solution for the helium ionization energy is $Z^2 = 2^2 = 4$ hartrees. By our trial-and-error method, we got $Z_{eff} = 1.6$ which leads to about $(1.6)^2 = 2.56\ E_h$. The experimental value is $2.903\ E_h$.

The GAUSSIAN© program, using a restricted Hartree–Fock procedure, RHF yields

```
SCF Done: E(RHF) = -2.80778395662     A.U. after     1 cycles
```

The value after four cycles by a program named G3 (to be discussed in more detail later) gives

```
SCF Done: E(RHF) = -2.85516042615     A.U. after     4 cycles
```

A Hartree–Fock triple zeta calculation (a linear combination involving three adjustable Z parameters) gives

```
SCF Done: E(RHF) = -2.85989537425     A.U. after     3 cycles
```

The energy – 2.860 E_h agrees with the experimental value to within 1.5%.

Example 15.2

The Hamiltonian function for the ground state hydrogen atom (with angular momentum equal to zero) is

$$\begin{aligned}\hat{H} &= -\frac{\hbar^2}{2m_e}\nabla^2 + \frac{e^2}{r} = -\frac{\hbar^2}{2m_e}r^2\frac{\partial^2}{\partial r^2} - \frac{e^2}{4\pi^2\varepsilon_0 r}\\ &= -\frac{\hbar^2}{2m_e r}\frac{d}{dr}\left(r^2\frac{\partial^2}{\partial r^2}\right) - \frac{e^2}{4\pi^2\varepsilon_0 r}\end{aligned}$$

where there is no θ, ϕ term because there is no angular momentum in the ground state and $\nabla^2 = r^2(\partial^2/\partial r^2) = r^2(d^2/dr^2)$ for this one-dimensional operator in which r^2 appears because the radius vector may be oriented in any direction toward a surface element of a sphere. The surface area of a sphere goes up as the square of the radius $A = 4\pi r^2$. The $4\pi^2\varepsilon_0$ in the denominator of the potential energy arises in the same way, where $4\pi^2\varepsilon_0$ is the permittivity of free space (essentially a proportionality constant between coulombs and joules).

Allowing this operator to operate on the trial function $\phi = e^{-\alpha r}$ gives, from the Schrödinger equation ($\hat{H}\phi = E\phi$),

$$\left[-\frac{\hbar^2}{2m_e r}\frac{d}{dr}\left(r^2\frac{d^2}{dr^2}\right) - \frac{e^2}{4\pi^2\varepsilon_0 r}\right]\phi(r) = E\phi(r)$$

or

$$\frac{d^2e^{-\alpha r}}{dr^2} + \frac{2}{r}\frac{de^{-\alpha r}}{dr} + \frac{2m_e}{\hbar^2}\left(E + \frac{e^2}{4\pi^2\varepsilon_0 r}\right)e^{-\alpha r} = 0$$

We can carry out the differentiations, divide through by $e^{-\alpha r}$, and segregate the terms into those that depend on r (terms 2 and 4) and those that do not (terms 1 and 3):

$$\alpha^2 - \frac{2}{r}\alpha + \frac{2m_e}{\hbar^2}\left(E + \frac{e^2}{4\pi^2\varepsilon_0 r}\right) = 0$$

$$\alpha^2 - \frac{2}{r}\alpha + \frac{2m_e E}{\hbar^2} + \frac{2m_e}{\hbar^2}\frac{e^2}{4\pi^2\varepsilon_0 r} = 0$$

so

$$\alpha^2 + \frac{2m_e E}{\hbar^2} = \frac{2}{r}\alpha - \frac{2m_e}{\hbar^2}\frac{e^2}{4\pi^2\varepsilon_0 r}$$

The basis! function $\phi = e^{-\alpha r}$ is a negative exponential, so there must be some value of r for which the right-hand side of the equation becomes zero. But the left-hand side of the equation is a group of constants, which add up to a single constant. If this constant is zero for one value of r, it must be zero for all. Thus,

$$\alpha^2 + \frac{2m_e E}{\hbar^2} = 0$$

where $\hbar = \frac{h}{2\pi}$ The consequences of this simple equation are very important:

$$\alpha^2 = -\frac{2m_e E}{\hbar^2}$$

$$E(r) = -\frac{\hbar^2}{2m_e}\alpha^2$$

and

$$\frac{2}{r}\alpha - \frac{2m_e}{\hbar^2}\frac{e^2}{4\pi^2\varepsilon_0 r} = 0$$

$$\frac{2}{r}\alpha = \frac{2m_e}{\hbar^2}\left(\frac{e^2}{4\pi^2\varepsilon_0}\right)^2\frac{1}{r}$$

But $\alpha = m_e e^2/\hbar^2$, so

$$E(r) = -\frac{\hbar^2}{2m_e}\left(\frac{m_e e^2}{\hbar^2}\right)^2 = -\frac{m_e e^4}{2\hbar^2} - \frac{1}{2}\frac{m_e e^4}{\hbar^2} = -\frac{1}{2}$$

that is, $E(r) = \frac{1}{2}$ hartree in atomic units because the constants $m_e e/\hbar^{2^4}$ are all defined as 1 in the atomic system.

Problem 15.1

The dominant line in the emission spectrum of atomic hydrogen is a beautiful crimson line at a wavelength of 656.1 nm. The hertz (Hz) is the unit of frequency that corresponds to one cycle of the wave motion per second. The speed of electromagnetic radiation is $c = 2.998 \times 10^8\ \mathrm{m\,s^{-1}}$. Each unit of frequency represents one step of wavelength λ, so the speed of wave propagation is the number of steps per second (frequency) times the length of each step, or $c = \nu\,\lambda$. Express the crimson hydrogen line in terms of frequency in hertz and find its energy from the Planck equation. It is sometimes convenient to express frequency in wave numbers, $\tilde{\nu} = \nu/c$. Find $\tilde{\nu}$ for the crimson line in the hydrogen spectrum and use it to find λ. (This is a crosscheck because you already know λ).

Problem 15.2

Einstein's most famous equation is $E = mc^2$ and the Planck equation is $E = h\nu$. Since they both express the energy, combine them to obtain the de Broglie equation, which expresses wave particle duality.

Problem 15.3 (Classical Mechanics)

A projectile of mass m was fired upward so that it traveled along a parabolic arc with velocity v. Its initial velocity was v_0. How high will it go? Neglect wind and air resistance.

Problem 15.4

Expand determinant a so as to obtain a single number. Expand determinant b so as to obtain a single number. The job is already becoming tedious, so expand b using machine software—for example, Mathcad©. Expand c by machine. Could you have guessed the answer simply by knowing b and looking at c?

(a) $\begin{vmatrix} 2 & 3 \\ 5 & 6 \end{vmatrix}$ (b) $\begin{vmatrix} 1 & 2 & 3 \\ 4 & 5 & 6 \\ 7 & 8 & 9 \end{vmatrix}$ (c) $\begin{vmatrix} 1 & 2 & 3 \\ 4 & 5 & 6 \\ 7 & 8 & 9.1 \end{vmatrix}$

Problem 15.5

Good mathematical software contains a number of operations on matrices that will be useful in coming chapters. Use Mathcad© or similar software to calculate the sum, product, square, inversion, and product with inverse (under certain conditions equivalent to division). Choosing matrix c from between the vertical lines in the preceding problem enables one to solve most of these problems simply by inspection and then allows one to verify (test run) using the software.

Problem 15.6

According to an Einstein theory of electromagnetic radiation, the 656.1-nm radiation from atomic hydrogen in the first excited state is a particle (later named a *photon*). What is the energy, frequency, wave number, and momentum of this putative particle? Express and explain the units of the final result for each of the four steps.

Problem 15.7

Show that the operator $\nabla^2 = (1/r^2)(d/dr)r^2(d/dr)$ in spherical coordinates can be written $\nabla^2 = (d^2/dr^2) + (2/r)(d/dr)$. (This is part of the solution of the radial Schrödinger equation for the ground state of the hydrogen atom.)

Problem 15.8

Using the operator ∇^2, write the radial form of the Schrödinger equation.

Problem 15.9

Allow the operator $\nabla^2 = (d^2/dr^2) + (2/r)(d/dr)$ to operate on the function $R(r) = e^{-r}$ in spherical polar coordinates ($r, \theta,$ and ϕ). What is $\nabla^2 R(r)$? What are $(dR(r)/d\theta)$ and $(dR(r)/d\phi)$?

16

WAVE MECHANICS OF SIMPLE SYSTEMS

Because of the wave nature of the electron, solution of problems of atomic and molecular structure requires solving *wave equations*. Problems in atomic and molecular structure involve electrons that are tied to a positive nucleus or group of nuclei by electrostatic attraction. They are *bound*. Therefore we shall be concerned with mathematical *boundary conditions* imposed on the solutions of wave equations.

16.1 WAVE MOTION

Any mathematical description of a vibrating guitar string has to take into account the fact that it is tied down at both ends. This puts severe restrictions on the wave forms the string can take which are not characteristic of a free wave. A free wave can assume any wavelength, but a bound wave is restricted to waves that come to zero at either end of the string. The three bound *wavelengths* shown in Fig. 16.1 are allowed by these two boundary conditions, but an infinite number of intermediate wavelengths are not allowed. The wave of longest wavelength is the *fundamental* (one observes *half* a sine wave), and the others are called *overtones*. A fundamental and its overtones are well described by the sine function.

The wavelength λ of the first overtone is one-half the wavelength of the fundamental $\lambda = \frac{1}{2}\lambda_{\text{fundamental}}$, the next overtone has $\lambda = \frac{1}{3}\lambda_{\text{fundamental}}$, and we can imagine many overtones at $\lambda = \frac{1}{n}\lambda_{\text{fundamental}}$. This is the source of *integers n* that appear for bound waves. De Broglie's observation that particles (specifically electrons) have a

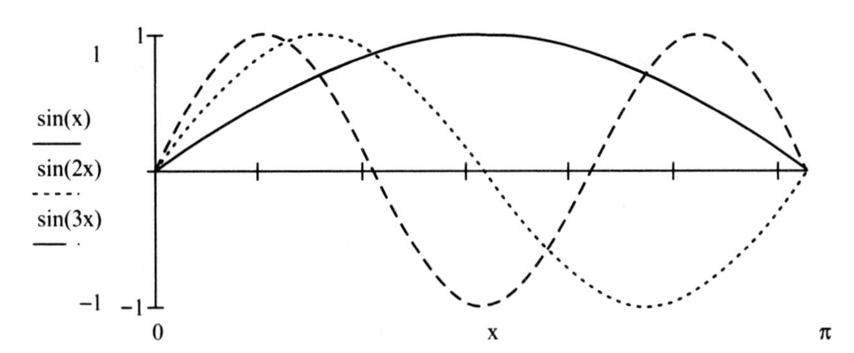

FIGURE 16.1 Graph of sin(x), sin($2x$), and sin($3x$) shown over the interval $[0, \pi]$. The fundamental mode of oscillation of a vibrating string is $\frac{1}{2}$ of a sine wave, the first overtone is a full sine wave from 0 to 2π, and the second overtone is $\frac{3}{2}$ of a sine wave.

wave nature leads to *quantum numbers n* in atomic spectra (Bohr) and to the connection between wave equations and atomic structure (Schrödinger, Hartree). Born's observation that the wave equation governs the probability of finding an electron, which may contribute to or oppose formation of a chemical bond, leads to the connection between wave equations and molecular structure, energy, and reactivity that is the basis of modern *quantum chemistry*.

16.2 WAVE EQUATIONS

The function $\sin \phi(x)$ describes the excursion away from $\phi(x) = 0$ of an infinite number of points along the variable $x = 0$ to l (Fig. 16.1). For any selected harmonic, the fraction x/λ tells us "where we are" on the sine wave. If $x = \lambda$, we are at the very end of the sine function. If $x = \lambda/2$, we are half way. In order to fully describe a wave, we need one more piece of information beyond the sine function and wavelength λ. The height of the wave is essential; a big sine wave has a large *amplitude* of oscillation A and a small sine wave has a small amplitude. The complete description of the wave is $\phi(x) = A \sin \dfrac{2\pi x}{\lambda}$.

We can obtain the second derivative of $\phi(x) = A \sin \dfrac{2\pi x}{\lambda}$:

$$\frac{d^2\phi(x)}{dx^2} = -A\frac{4\pi^2}{\lambda^2} \sin \frac{2\pi x}{\lambda} = -\frac{4\pi^2}{\lambda^2} A \sin \phi(x) = -\frac{4\pi^2}{\lambda^2}\phi(x)$$

This is typical of the wave equations that Schrödinger used:

$$\frac{d^2\phi(x)}{dx^2} = k\phi(x)$$

It is also a member of the class of *eigenvalue equations*, in which an operator $\hat{O}$ operates on the *eigenfunction* $\phi(x)$ to give an *eigenvalue* k times the same function:

$$\hat{O}\phi(x) = k\phi(x)$$

In this case, the operator is d^2/dx^2, and the eigenvalue is $k = -4\pi^2/\lambda^2$.

The eigenfunction may be represented as a vector:

$$\phi(x) = \begin{pmatrix} \xi_1 & (x) \\ \xi_2 & (x) \end{pmatrix}$$

The effect of the operator is to stretch or contract the *eigenvector* by an amount equal to the eigenvalue, or to change its direction, or both.

Waves in two dimensions (x, y) such as those of a vibrating plate or membrane are described by

$$\frac{\partial^2\phi(x, y)}{\partial x^2} + \frac{\partial^2\phi(x, y)}{\partial y^2} = -\frac{4\pi^2}{\lambda^2}\phi(x, y)$$

and the wave equation in 3-space is

$$\frac{\partial^2\phi(x, y, z)}{\partial x^2} + \frac{\partial^2\phi(x, y, z)}{\partial y^2} + \frac{\partial^2\phi(x, y, z)}{\partial z^2} = -\frac{4\pi^2}{\lambda^2}\phi(x, y, z)$$

16.3 THE SCHRÖDINGER EQUATION

By the de Broglie equation $p = h/\lambda$, it follows that $\lambda^2 = h^2/p^2$. Substituting for λ^2 in a three-dimensional wave equation, we obtain

$$\frac{\partial^2\phi(x, y, z)}{\partial x^2} + \frac{\partial^2\phi(x, y, z)}{\partial y^2} + \frac{\partial^2\phi(x, y, z)}{\partial z^2} = -\frac{4\pi^2}{\lambda^2}\phi(x, y, z) = -\frac{4\pi^2 p^2}{h^2}\phi(x, y, z)$$

The momentum of a moving particle is its mass times its velocity, $p = mv$. This leads to $p^2 = m^2v^2 = 2m(\frac{1}{2}mv^2) = 2mT$, where T is the classical kinetic energy $\frac{1}{2}mv^2$. Now, with a slight change in symbol for notational simplicity, we obtain

$$\frac{\partial^2\Psi}{\partial x^2} + \frac{\partial^2\Psi}{\partial y^2} + \frac{\partial^2\Psi}{\partial z^2} = \left(\frac{\partial^2}{\partial x^2} + \frac{\partial^2}{\partial y^2} + \frac{\partial^2}{\partial z^2}\right)\Psi = -\frac{8\pi^2 mT}{h^2}\Psi$$

The bracketed operator $(\partial^2/\partial x^2 + \partial^2/\partial y^2 + \partial^2/\partial z^2)$ operates on the wave function $\Psi(x, y, z)$. It is given the shorthand notation ∇^2. Because its eigenvalue contains a group of constants times the kinetic energy T, ∇^2 is called a *kinetic energy operator*. The total energy of a classical system is the sum of the kinetic energy and the potential energy, $E = T + V$, so the kinetic energy is the difference $T = E - V$ and

$$\nabla^2\Psi = -\frac{8\pi^2 m}{h^2}(E - V)\Psi$$

which is one form of the famous Schrödinger equation.

Other notational conveniences include the definition of "h bar," $\hbar \equiv h/2\pi$, which gives

$$\nabla^2\Psi = -\frac{2m}{\hbar^2}(E - V)\Psi$$

and

$$-\frac{\hbar^2}{2m}\nabla^2\Psi + V\Psi = E\Psi$$

The kinetic energy operator plus the potential energy operator $(-\frac{\hbar^2}{2m}\nabla^2\Psi + V\Psi)$ is defined as the *Hamiltonian operator* $\hat{H}$ by analogy to the classical Hamiltonian function $H = T + V$ (Chapter 15). The kinetic and potential energy operators are $\hat{T}$ and $\hat{V}$, so $\hat{H} = \hat{T} + \hat{V}$. (Even though $\hat{V}$ is an operator, it is often written V.) These notational changes give the concise form of the Schrödinger equation:

$$\hat{H}\Psi = E\Psi$$

where the total energy E is a *scalar* eigenvalue. It is distinguished from the operators by not having a circumflex notation. Clearly energy is a scalar because it has magnitude but not direction.

16.4 QUANTUM MECHANICAL SYSTEMS

A *system* is a collection of mechanical entities governed by physical laws. If we know the *state* of a system, we know every physical property it can have. It is astonishing but true that all this information can be known by specifying a very few fundamental variables (degrees of freedom) and a small number of postulates.

The wave function satisfies all of the properties of a vector; therefore it is written as a vector $|\Psi\rangle$ or $\langle\Psi|$. We shall use either the vector form $|\Psi\rangle$ or the functional form Ψ, according to which is more convenient. When it is useful to specify the variables as degrees of freedom, we shall do that: $|\Psi(x_1, x_2, \ldots)\rangle$ or $\Psi(x_1, x_2, \ldots)$, but usually the simpler notation $|\Psi\rangle$ or Ψ is preferred.

One of the postulates of quantum mechanics is as follows:

If a system is in a state **s** described by $|\Psi\rangle$ an eigenvector of the operator $\hat{\mathbf{A}}$ (or, equivalently if Ψ is an eigenfunction of $\hat{\mathbf{A}}$), the corresponding *observable* a is an eigenvalue:

$$\hat{\mathbf{A}}\,|\Psi\rangle = a\,|\Psi\rangle$$

or, equivalently,

$$\hat{\mathbf{A}}\Psi = a\,\Psi$$

and experiments on s in state $|\Psi\rangle$ will *always* yield the observable a.

In the case of the atomic and molecular systems that concern us here, the general operator $\hat{\mathbf{A}}$ is the Hamiltonian operator, $\hat{H}$, and the observable a is the energy level E_i of the state $|\Psi_i\rangle$. Usually there are many states leading to many energies E_i. The set$\{E_i\}$ represents a *spectrum* or multiplicity of energy levels, one for each eigenvector:

$$\hat{H}\,|\Psi_i\rangle = E_i\,|\Psi_i\rangle$$

An operator operating on a vector produces another vector. The operator $\hat{H}$ operating on $|\Psi\rangle$ produces the vector $\left|\hat{H}\,|\Psi\right\rangle$. The inner product of $\langle\Psi|$ premultiplied into $\left|\hat{H}\,|\Psi\right\rangle$ is

$$\langle\Psi|\,\hat{H}\,|\Psi\rangle = \langle\Psi|\,E\,|\Psi\rangle = E\,\langle\Psi\mid\Psi\rangle$$

The inner product of two vectors $\langle\Psi\mid\Psi\rangle$ is a scalar, which is why the eigenvalue E can be moved out of the brackets on the right. The *operator* $\hat{H}$ cannot be factored out. This leads to an expression for the energy

$$E = \frac{\langle\Psi|\,\hat{H}\,|\Psi\rangle}{\langle\Psi\mid\Psi\rangle}$$

The Born probability postulate leads to a further simplification. The inner product $\langle\Psi\mid\Psi\rangle$ amounts to the integral over all space of the wave function squared Ψ^2 or the product $\Psi * \Psi$ if the wave function is complex. The sum of the probabilities of finding an electron over all space must be 1.0 (certainty) because the electron has to be somewhere. With the inner product $\langle\Psi\mid\Psi\rangle = 1.0$ we have the energy of an eigenstate as

$$E_i = \langle\Psi_i|\,\hat{H}\,|\Psi_i\rangle$$

16.5 THE PARTICLE IN A ONE-DIMENSIONAL BOX

Facility in working quantum mechanical energy problems can be gained by going from simple problems to more complicated ones. The problem usually chosen as the starting point is the *particle in a one-dimensional box* of length l:

$$x = 0 \quad \text{|}\_\_\_\_\_\_\bullet\_\_\_\_\_\_\text{|} \quad x = 1$$

Since there is only one dimension, the wave function is $\Psi(x)$. The Schrödinger equation in this case is

$$-\frac{\hbar^2}{2m}\nabla^2\Psi(x) + V\Psi(x) = E\Psi(x)$$

We stipulate that the potential energy is zero inside the box $V = 0$ and that it is infinite outside the box. These conditions mean that the particle cannot escape from the box. We note that the one-dimensional operator is $\nabla^2 = \dfrac{d^2\Psi(x)}{dx^2}$, so

$$-\frac{\hbar^2}{2m}\frac{d^2\Psi(x)}{dx^2} = E\Psi(x)$$

inside the box, or

$$\frac{d^2\Psi(x)}{dx^2} = -\frac{2mE}{\hbar^2}\Psi(x)$$

which is a typical wave equation and is an eigenvalue problem.

From our knowledge of wave equations (Section 16.2), we know that a function like $\Psi(x) = A\sin\dfrac{2\pi x}{\lambda}$ will satisfy this equation. The second derivative of $\Psi(x)$ is

$$\frac{d^2\Psi(x)}{dx^2} = -A\frac{4\pi^2}{\lambda^2}\sin\frac{2\pi x}{\lambda} = -\frac{4\pi^2}{\lambda^2}\Psi(x)$$

but it is also true that $\dfrac{d^2\Psi(x)}{dx^2} = -\dfrac{2mE}{\hbar^2}\Psi(x)$, so

$$\frac{4\pi^2}{\lambda^2} = \frac{2mE}{\hbar^2}$$

We also know that bound waves must have wavelengths that go up as half-integers of the limits placed on the oscillatory excursion—that is, the length of a vibrating guitar string (Fig. 16.1) or the length l of a box. Wavelengths allowed by the boundary conditions are $\lambda = \frac{1}{n}\lambda_{\text{fundamental}}$ (Section 16.1), where $\lambda_{\text{fundamental}} = 2l$. It follows that

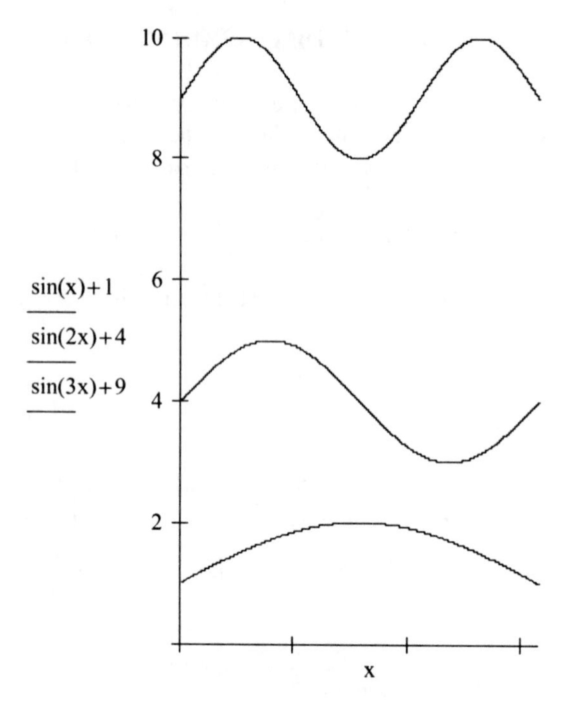

FIGURE 16.2 Wave forms for the first three wave functions of the particle in a box. The waves are drawn at the lowest three energy levels: $n = 1, 2, 3$; ($n^2 = 1, 4, 9$).

$\lambda = \frac{2}{n}l$. A little algebra goes from

$$-\frac{4\pi^2}{\left(\frac{2}{n}l\right)^2} = -\frac{2mE}{\hbar^2}$$

to the energy *spectrum* (Fig. 16.2):

$$E = \frac{n^2\hbar^2\pi^2}{2ml^2} = \frac{n^2h^2}{8ml^2}$$

Except for the lowest one, each wave function has internal values of x or *nodes* at which $\Psi(x) = 0$. The Born probability density $|\Psi(x)|^2$ of finding the particle precisely at one of these nodes is zero. Not counting the nodes at the extremities of the wave function, the number of internal nodes goes up as 0, 1, 2, ... Each internal node in the wave function $\Psi(x)$ yields an internal zero point shown as a minimum in the probability function in Fig. 16.3. There are $n - 1$ internal nodes.

By elaboration of the methods used for the particle in a one-dimensional box, the problem can be solved in two dimensions to produce solutions for the vibratory

$$n := 1, 2 .. 3$$

$$P(n, x) := (\sin(n \cdot x))^2 + n^2$$

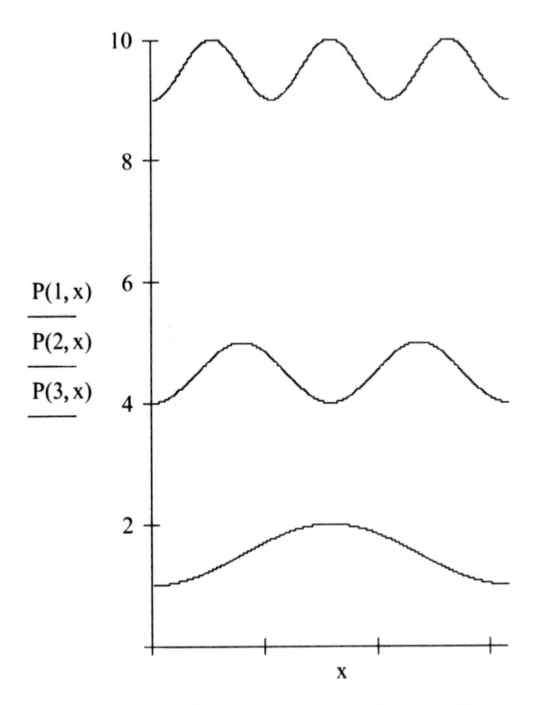

FIGURE 16.3 A Mathcad © sketch of the born probability densities at the first three levels of the particle in a box. The lowest wave function has no internal nodes, the second wave function has one, and the third has two.

nodes of a square plate or membrane. It can be solved in 3-space to produce solutions to the vibratory motions in a solid cube. Other geometries are possible such as a circle, rectangle, parallelepiped, cylinder, and so on. Each problem lends insight and is recommended to the interested reader.

16.6 THE PARTICLE IN A CUBIC BOX

In the case of a particle confined to the interior of a cube, we have a three-dimensional wave equation of the form

$$-\frac{\hbar^2}{2m}\nabla^2\Psi(x, y, z) + V\Psi(x, y, z) = E\Psi(x, y, z)$$

We simplify this equation immediately by setting $V\Psi(x, y, z) = 0$, so the middle term drops out. The kinetic energy operator looks complicated:

$$-\frac{\hbar^2}{2m}\frac{d^2\Psi(x, y, z)}{dx^2} - \frac{\hbar^2}{2m}\frac{d^2\Psi(x, y, z)}{dy^2} - \frac{\hbar^2}{2m}\frac{d^2\Psi(x, y, z)}{dz^2} = E\Psi(x, y, z)$$

until we make the entirely reasonable observation that, in a cube, there is no reason to prefer particle motion in any one direction over motion in any other. The energy E_x supplied by particle motion Ψ_x is the same as E_y supplied by Ψ_y and E_z supplied by Ψ_z. This being the case, we can split one complicated equation into three simple ones:

$$-\frac{\hbar^2}{2m}\frac{d^2\Psi(x)}{dx^2} = E_x\Psi(x)$$

$$-\frac{\hbar^2}{2m}\frac{d^2\Psi(y)}{dy^2} = E_y\Psi(y)$$

$$-\frac{\hbar^2}{2m}\frac{d^2\Psi(z)}{dz^2} = E_z\Psi(z)$$

Each of these equations has already been solved with the results

$$\Psi(x) = A\sin\frac{2\pi x}{\lambda}$$

$$\Psi(y) = A\sin\frac{2\pi y}{\lambda}$$

$$\Psi(z) = A\sin\frac{2\pi z}{\lambda}$$

and

$$E_x = \frac{n_x^2 h^2}{8ml^2}$$

$$E_y = \frac{n_y^2 h^2}{8ml^2}$$

$$E_z = \frac{n_z^2 h^2}{8ml^2}$$

In the energy equations, the length of one edge of the cube l is in the denominator. If the geometry had been other than a cube, there would be different dimensions—perhaps a, b, and c for a parallelepiped. This would have made the solutions slightly more complicated, but it would not have made any significant change in the result.

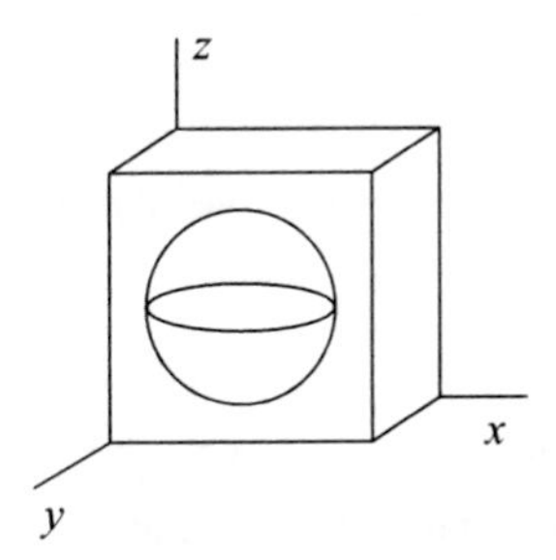

FIGURE 16.4 The ground state orbital of a particle confined to a cubic box.

16.6.1 Orbitals

The ground state probability density has a maximum in the center when viewed from each of the three dimensions (Fig. 16.4). Thus there is a high probability density at the center of the cube diminishing symmetrically in all directions (spherical symmetry). This is the geometry of an s atomic *orbital*.

16.6.2 Degeneracy

If the particle is excited to the $n = 2$ state in the x direction while its motion in both the y and z directions remain at the lowest energy, the resulting orbital has an internal node in the x direction but no nodes in the y and z directions (Fig. 16.5). This gives the geometry of the first p orbital which we denote p_x. What was said for the x direction can be said for the y and z directions, so we have three orbitals with energies that are identical. Different orbitals with the same energy are said to be *degenerate* (Fig. 16.6). Perhaps we remember from elementary chemistry that the p orbitals of hydrogen are three-fold degenerate.

16.6.3 Normalization

One can also find the constant A in $\Psi(x) = A \sin \dfrac{2\pi x}{\lambda}$ by *normalizing* the wave function. Normalization requires setting the integral of Ψ^2 over all space equal to 1.0

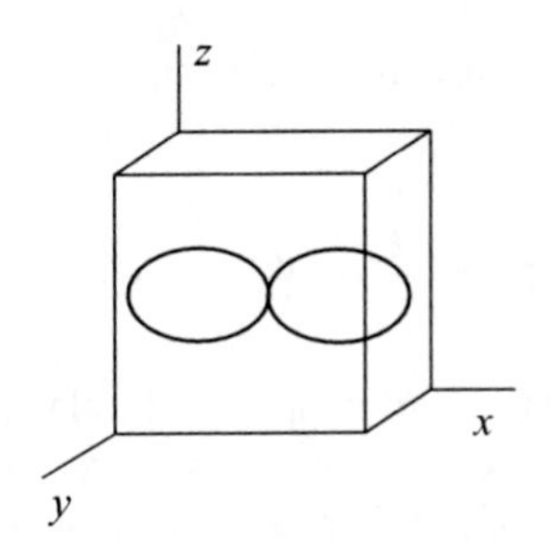

FIGURE 16.5 The first excited state of a particle confined to a cubic box.

FIGURE 16.6 Degenerate energy levels in a one-dimensional box. Degeneracy increases with n in the sequence 1, 4, 9, ...

by Born's principle. Normalization yields $A = \sqrt{\frac{2}{l}}$ (see Problems) for the particle in a one-dimensional box. Normalization constants, sometimes rather messy ones, appear as premultipliers of wave functions.

Further elaboration of the method [see many textbooks, including Levine (2000) and Barrow (1996)] yields solutions for the harmonic oscillator (one dimensional vibrator) and the rigid rotor. These systems have the same kind of energy level spectrum (Section 16.5) as the particle in a box except that the spacings are different. The spacing is in equal steps for vibration of the harmonic oscillator but not for rotation.

16.7 THE HYDROGEN ATOM

The kinetic energy operator of an electron moving in the vicinity of a proton (H^+ nucleus) is the same as a particle in a box:

$$\hat{T} = \frac{\hbar}{2m_e}\nabla^2$$

Many texts replace the mass of the electron m_e by μ designating the reduced mass of the electron and proton rotating about their center of gravity. The effect of this correction is very small for the proton–electron pair, and we shall ignore it until we reach comparable problems of molecular rotation where more nearly equal masses are involved.

The potential energy is not zero in this problem; rather, it is

$$V = \frac{e^2}{r}$$

according to the electrostatic attraction between the nucleus and the electron. This leads to

$$\nabla^2\Psi(r,\theta,\phi) = \frac{2m_e}{\hbar^2}(E(r,\theta,\phi) - V(r,\theta,\phi))\Psi(r,\theta,\phi) = 0$$

where we have expressed $\Psi(r,\theta,\phi)$ in spherical polar coordinates r, θ, and ϕ. The problem is simpler this way because of its spherical symmetry. Routine algebraic methods exist for conversion of problems in Cartesian coordinates to spherical polar coordinates (Barrante 1998).

Upon writing this equation out in full, it becomes rather intimidating, as the reader will see by scanning several textbooks. Fortunately, expressed in r, θ, and ϕ, it breaks up into three separate equations just as the particle in a cubic box did. The equations are

$$\frac{\partial}{\partial r}r^2\frac{\partial R(r)}{\partial r}+\frac{2m_e r^2}{\hbar^2}\left(\frac{e^2}{r}+E\right)R(r)=R(r)\beta$$

$$\frac{1}{\sin\theta}\frac{\partial}{\partial\theta}\sin\theta\frac{\partial\Theta(\theta)}{\partial\theta}-\frac{m_e^2}{\sin^2\theta}=-\beta\,\Theta(\theta)$$

and

$$\Phi(\phi)=-\frac{1}{\sqrt{2\pi}}e^{im_e\phi}$$

Each of these equations involves only one variable $R(r)$, $\Theta(\theta)$, and $\Phi(\phi)$. The first $R(r)$ is called the *radial equation* and the second two are often lumped together as the *spherical harmonics* $Y(\theta,\phi)=\Theta(\theta)\,\Phi(\phi)$. The name is apt; they describe oscillations taking place on the surface of a sphere with the boundary constraints that no wave can have a discontinuity or seam anywhere on the sphere.

The wave function or orbital for the electron in the ground state of the hydrogen atom is $\Psi(r)=e^{-\alpha r}$, where r is the radial distance between the proton and the electron, and α contains several constants. Solution for α by any one of several methods gives

$$E=-\frac{1}{2}\left(\frac{me^4}{(4\pi\varepsilon_0)^2\hbar^2}\right)=-\frac{me^4}{32\pi^2\varepsilon_0^2\hbar^2}$$

This purely quantum mechanical result is in precise quantitative agreement with Bohr's semiclassical result for the ground state of hydrogen. When higher energy states are considered, quantum numbers appear just as they do in Bohr's result.

In addition to the principal quantum number n, a more general solution for the spherical harmonics $Y(\theta,\phi)=\Theta(\theta)\,\Phi(\phi)$ leads to a quantum number l for the $\Theta(\theta)$ equation and m for the $\Phi(\phi)$ equation. There are also restrictions on the quantum numbers in the complete solution. For example, $n\leq l+1$, so that l can take on a zero value even though n cannot.

16.8 BREAKING DEGENERACY

For the simple one-electron system of hydrogen, the $2s$ and $2p$ orbitals are degenerate, as are the more complicated $3s$, $3p$, and $3d$ orbitals. When we start putting electrons into these orbitals to *build up* the atomic table by the *aufbau* principle, however, some of these degeneracies are lost. For example, the probability density function of the $2p$ orbital is small near the nucleus, where, in contrast, the s orbital has a

FIGURE 16.7 Reduced degeneracy in the energy levels for hydrogen-like atoms. Some of the degeneracy of Fig. 16.6 has been lost.

maximum probability. Consequently, the energy of an electron drifting about in a distant p orbital is less negative than that of a tightly held s electron. This causes the sp degeneracy of the hydrogen atom to be broken up in beryllium, boron, carbon, and the heavier elements. The $2s$ level is lower than the $2p$ levels in beryllium because of its greater average radial distance from the nucleus and because of s shielding.

We see this when we traverse the series H $\rightarrow$ He $\rightarrow$ Li $\rightarrow$ Be $\rightarrow$ B in the atomic table. The $1s$ level readily accommodates H and He, but the electron in Li is excluded from the $1s$ level which is "filled" by two electrons.[1] It *could* go into either the $2s$ or one of $2p$ levels, but in fact it goes into the $2s$ level which has a lower energy. This preference is even more remarkable in beryllium. The valence electron in Be has good reason to go into the $2p$ orbital to escape charge and spin correlation with the electron already in the $2s$ orbital, but it refuses this safe haven and goes into the lower energy $2s$ alternative. We know from ionization potentials that the electron in Be ionizes from the $2s$ orbital, not the $2p$, and that Be is a typical metal, not as metallic as Li perhaps, but not a metalloid like boron. Atoms with orbital configurations similar to hydrogen but with reduced degeneracy are called *hydrogen-like* atoms (Fig. 16.7).

16.8.1 Higher Exact Solutions for the Hydrogen Atom

The first six eigenfunctions of the hydrogen atom are shown in Table 16.1. The geometric parameter $a = 52.9\,\text{pm} = 5.29 \times 10^{-11}$ m is $1/\alpha$, the first orbital radius in Bohr theory. The integer Z is the atomic number. It is 1 for hydrogen but 2 or 3 for the higher hydrogen-like atoms shown here. The orbitals fall into two different groups, s and p. More complete listings include $d, f, g, \ldots$ orbitals of increasing complexity. Note that the terms function and orbital are used as synonyms. The s eigenfunctions are $1s$, $2s$, and $3s$. They do not contain sin or cos parts because they do not depend upon angular location. They are *spherically symmetrical*. Table 16.2 shows the orbitals reduced to their simple functional form. Let us concentrate on the simple polynomial form of the s orbitals by ignoring the premultiplying constants. The first orbital is simply a negative exponential. From the monotonic decrease of e^{-r}, we expect zero roots for the $1s$ function, but from the polynomials $(1 - r)$ and $\left(1 - r + r^2\right)$ we expect two and three roots, respectively. A Mathcad© graph of the $3s$ radial wave function in Fig. 16.8 shows that the polynomial part produces two nodes and one asymptotic approach when the function is plotted on the vertical axis against r as the independent variable.

[1]The universal restriction of two electrons with opposed spins per orbital is called the Pauli exclusion principle.

TABLE 16.1 The First Six Wave Functions for Hydrogen.

$$\psi_{1s} = \frac{1}{\pi^{1/2}}\left(\frac{1}{a}\right)^{3/2} e^{-Zr/a}$$

$$\psi_{2s} = \frac{1}{4\cdot(2\pi)^{1/2}}\left(2-\frac{Zr}{a}\right)^{3/2} e^{-Zr/2a}$$

$$\psi_{2p_z} = \frac{1}{4\cdot(2\pi)^{1/2}}\left(\frac{Z}{a}\right)^{5/2} re^{-Zr/2a}\cos\theta$$

$$\psi_{2p_x} = \frac{1}{4\cdot(2\pi)^{1/2}}\left(\frac{Z}{a}\right)^{5/2} re^{-Zr/2a}\sin\theta\cos\phi$$

$$\psi_{2p_y} = \frac{1}{4\cdot(2\pi)^{1/2}}\left(\frac{Z}{a}\right)^{5/2} re^{-Zr/2a}\sin\theta\sin\phi$$

$$\psi_{3s} = \frac{1}{81(3\pi)^{1/2}}\left(\frac{Z}{a}\right)^{3/2}\left(27-18\frac{Zr}{a}+2\frac{Z^2r^2}{a^2}\right)^{3/2} e^{-Zr/3a}$$

TABLE 16.2 The First Three *s* Wave Functions for Hydrogen (Simplified Form).

1*s*	e^{-r}
2*s*	$(2-r)e^{-r} = 2e^{-r} - re^{-r}$
3*s*	$(27-18r+2r^2)e^{-r} = 27e^{-r} - 18re^{-r} + 2r^2e^{-r}$

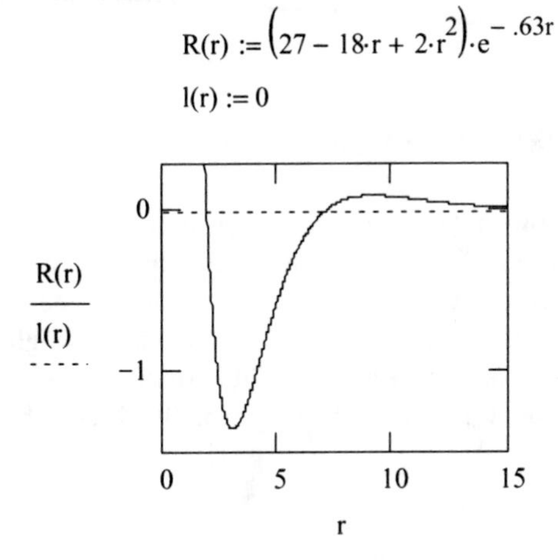

FIGURE 16.8 Roots of the radial 3*s* wave function of atomic hydrogen as a function of distance *r*. Three nodes correspond to zeros of the function at about $r = 2,\ 7,$ and ~ 15. The function *l*(*r*) simply draws a horizontal line.

$$a := .529$$

$$S(r) := \frac{1}{81\sqrt{3\pi}\,a^{1.5}}\left[27 - 18\cdot\frac{r}{a} + 2\cdot\left(\frac{r}{a}\right)^2\right]e^{\frac{-r}{3a}}$$

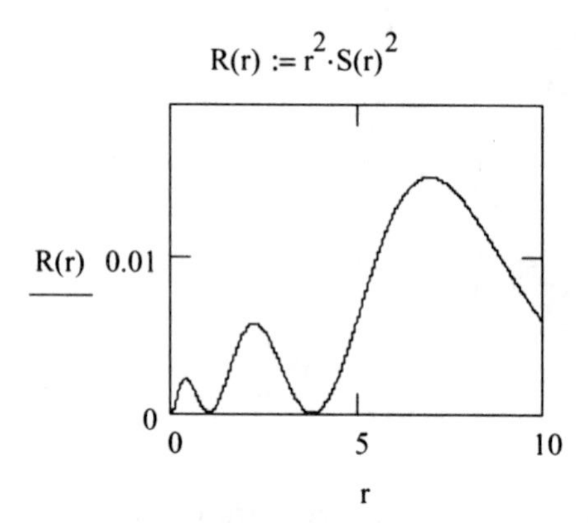

FIGURE 16.9 The radial probability density for an electron in the 3*s* orbital of hydrogen.

To find the probability density function for the lowest three orbitals, we square the wave function. At each node the probability density of finding the electron at the radial distance r is zero. Between the nodes are maxima called *antinodes*. For the 3*s* orbital, this leads to Fig. 16.9 in which there are the antinodes at about $r = 0.5$, 2, and 7. The radii of maximum probability densities correspond to the "shells" in early atomic theory.

16.9 ORTHOGONALITY AND OVERLAP

It is not difficult to show that the probability density of the product of the p_z orbital times the p_x orbital is zero. Their *overlap* is zero. Unlike *s* orbitals which are positive (+) everywhere, the 2*p* orbitals have one angular nodal plane. Therefore they pass from − to + and back again as the radius vector passes through either 0 or π. The *p* orbital has a *plane of inversion* as one of its *symmetry elements*. The *p* orbitals have a positive *lobe* and a negative lobe (Fig. 16.10). If a *p* orbital and an *s* orbital

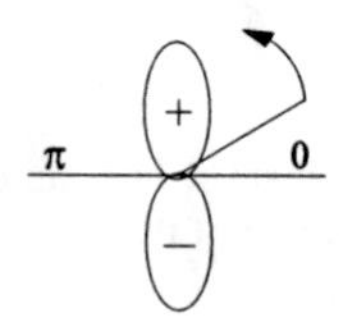

FIGURE 16.10 The radial node of the 2*p* atomic orbital.

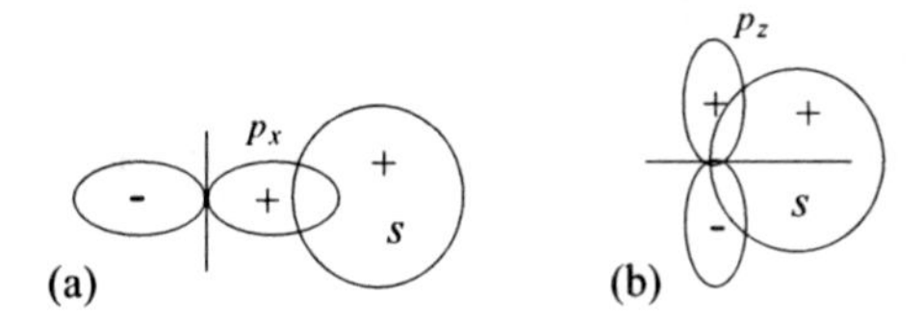

FIGURE 16.11 Favorable sp_x and unfavorable sp_z overlap of orbitals depending upon orbital symmetry. Diagram a depicts bond formation but orbital overlap exactly cancels in diagram b.

are combined as basis functions (vectors) to describe a chemical bond, the result may be favorable, as in Fig. 16.11a, resulting in enhancement of the probability density between bonded atoms. The overlap of an s positive orbital and a p_x positive orbital is positive. Or it may be unfavorable, depending on the orbital symmetry, as in Fig. 16.11b where any $+\,+$ overlap between one lobe of the $2p_z$ orbital with the $1s$ orbital is canceled by the negative $+\,-$ overlap of the other. In analyzing Fig. 16.11, we are asking the question of whether p orbitals can or cannot be used as basis vectors to describe a favorable overlap, thus a chemical bond. The answer is *yes* for the p_x orbital and no for the p_z. A natural question is: What is so special about the p_x orbital? Couldn't we just rotate the p_z orbital a quarter turn into a favorable overlap?

The answer is that there is *nothing* special about the p_x orbital and yes, we can rotate the p_z orbital into a position of favorable overlap, but in so doing, we rotate the p_x orbital *out* of its position of favorable overlap so we haven't really changed anything. We are merely saying that if one of the p orbitals is favorable, the other two orthogonal orbitals are not. The orbitals don't care how we label them.

16.10 MANY-ELECTRON ATOMIC SYSTEMS

If we regard each electron as obeying an orbital (function) that is independent of all other electrons except that it moves in the potential field of the nuclei and their *average* electrostatic forces, we get a wave function for the entire system that is approximated by a product of many one-electron orbitals called the *Hartree product*:

$$\psi_{\text{Hartree}} = \psi_1(r_1)\psi_2(r_2), \ldots, \psi_N(r_N) = \prod_{i=1}^{N} \psi_i(r_i)$$

This product produces N integrodifferential equations:

$$\hat{h}_i \psi_i(r_i) = \varepsilon_i \psi_i(r_i), \qquad i = 1, 2, \ldots, N$$

where the operator $\hat{h}_i$ includes the kinetic energy and the potential energy of attraction along with the potential energy of interelectronic repulsion V_i:

$$\hat{h}_i = \left[-\tfrac{1}{2}\nabla_i^2 - \frac{Z}{r_i} \right] + V_i(\psi_i(r_j)), \qquad j \neq i$$

The operator $\hat{h}_i$ is approximate because the position of each individual electron j_i is uncertain. The potential energy of repulsion between electrons V_i is included as an average by integrating over the Born probability function for each electron $j \neq i$ and then summing the results:

$$V_i(\psi_i(r_j)) = \sum_{j \neq i} \int \frac{|\psi_j(r_j)|^2}{r_{ij}} \, d\tau$$

The N equations are then solved using approximations to $\psi_1, \psi_2, \ldots, \psi_n$ and entering an iterative procedure. Each of the solutions ε_i for an atomic system is the energy of an atomic orbital 1*s*, 2*s*, 2*p*, and so on. Each energy coincides with a distinct set of self-consistent one-electron wave functions $\psi_1, \psi_2, \ldots, \psi_n$. The *radial wave functions* lead to electron probability density functions with 1, 2, 3, ... maxima at specific distances from the nucleus corresponding to the 1*s*, 2*s*, 2*p*, and so on, "shells" of electron density familiar from elementary chemistry.

PROBLEMS

Problem 16.1

The integral

$$\int_{-\infty}^{\infty} x e^{-x^2} dx$$

is an "improper integral because both of its limits are infinite (one would do to make it improper). Evaluate this integral, that is, find *y*(*x*) for

$$y(x) = \int_{-\infty}^{\infty} x e^{-x^2} dx$$

Verify your answer using a numerical integration computer routine such as Mathcad©.

Problem 16.2

Show that the first two orbitals of a particle in a one-dimensional box of length 1 unit are orthogonal, that is,

$$\int_0^1 \psi_1 \psi_2 \, d\tau = 0$$

for

$$\psi_1 = \sqrt{2} \sin(\pi x)$$

and

$$\psi_2 = \sqrt{2} \sin(2\pi x)$$

Problem 16.3

If the operator $\hat{A} = d^2/dx^2$, find the eigenfunction $\phi(x)$ and the eigenvalue a for the eigenvalue equation

$$\hat{A}\phi(x) = a\phi(x)$$

Problem 16.4

According to a theory of Niels Bohr (1913) for an electron to move in a stable classical orbit, the centrifugal force mv^2/r pulling away from the nucleus must be exactly balanced by the electrostatic force of attraction between the negative electron and the positive proton e^2/r^2.

(a) Write an expression for the velocity of the electron.
(b) Calculate the velocity of the electron. To get your answer in SI units, use $e^2/4\pi\,\varepsilon_0$ for the charge on the electron, where $4\pi\,\varepsilon_0 = 1.113 \times 10^{-10}\ \mathrm{C^2\ s^2\ kg^{-1}\ m^{-3}}$ is called the *permittivity* in a vacuum.
(c) Give units.
(d) Suppose the reader forgets to take a square root in the last step and arrives at the result $v = 4.784 \times 10^{12}$. How can he immediately know that something has gone wrong?

Problem 16.5

Normalize the eigenfunction $\Psi(x) = A \sin \frac{2\pi x}{\lambda}$ for the particle in a one-dimensional box of dimension a.

Problem 16.6

Suggest a simple classical (macroscopic) mechanical system for which the probability function varies in a regular way with position. What does the probability function look like?

Problem 16.7

Carry out the "little algebra" to go from

$$-\frac{4\pi^2}{\left(\frac{2}{n}l\right)^2} = -\frac{2mE}{\hbar^2}$$

to

$$E = \frac{n^2 h^2}{8m\ell^2}$$

Problem 16.8

What is the probability of finding an electron within one Bohr radius of the nucleus of the hydrogen atom?

Problem 16.9

The normalized wave function for a particle in a one-dimensional box of unit length is

$$\Psi(x) = \sqrt{2} \sin \frac{2\pi x}{\lambda}$$

where x tells you where you are on the excursion of the particle in the one-dimensional x-space. What is the probability of finding a particle in the first quarter of its excursion?

17

THE VARIATIONAL METHOD: ATOMS

Exact orbital solutions for the hydrogen atom cannot be replicated for other atoms or molecules, even small ones, but they are of inestimable value in the mathematical treatment of larger systems. The energy of nuclear attraction is very much larger than the binding energies that dominate chemistry, so one can regard the chemical bond as a perturbation on nuclear attraction. Thus the orbitals developed for H are the foundation stones for larger systems of ions, atoms, and the interdependent groups of atoms that we call molecules.

17.1 MORE ON THE VARIATIONAL METHOD

A precise expression of the variational equation for approximate wavefunctions is

$$\langle E_0 \rangle = \frac{\int_{-\infty}^{\infty} \phi_0^*(\tau) \hat{H} \phi_0(\tau)\, d\tau}{\int_{-\infty}^{\infty} \phi_0^*(\tau) \phi_0(\tau)\, d\tau}$$

In this notation, the subscripted 0 indicates the ground state energy and wave functions, the angle brackets around $\langle E_0 \rangle$ indicate an expectation value, the asterisk (*) designates the complex conjugate of the function, and τ is a variable indicating that the integrals are to be taken over all space.

Concise Physical Chemistry, by Donald W. Rogers

In our treatment we shall unencumber this expression from some of its notation by writing the expectation value of the energy simply as

$$E = \frac{\int \phi \hat{H} \phi \, d\tau}{\int \phi \phi \, d\tau}$$

In order to evaluate this energy, we need a trial function ϕ, which we are free to choose, and the Hamiltonian $\hat{H}$, which is determined by the system.

17.2 THE SECULAR DETERMINANT

In very many cases, the wave function will not be expressed as a single function at all but as a sum of functions. To illustrate, let us take the sum of two additive terms (strictly, basis vectors in a vector space.)

$$\phi = c_1 u_1 + c_2 u_2$$

For a variational treatment of the sum as an approximate wave function, we need the integrals

$$E = \frac{\int \phi \hat{H} \phi \, d\tau}{\int \phi \phi \, d\tau} = \frac{\int c_1 u_1 + c_2 u_2 \hat{H} c_1 u_1 + c_2 u_2 \, d\tau}{\int (c_1 u_1 + c_2 u_2)(c_1 u_1 + c_2 u_2) \, d\tau}$$

$$= \frac{c_1^2 \int u_1 \hat{H} u_1 \, d\tau + c_1 c_2 \int u_1 \hat{H} u_2 \, d\tau + c_1 c_2 \int u_2 \hat{H} u_1 \, d\tau + c_2^2 \int u_2 \hat{H} u_2 \, d\tau}{c_1^2 \int u_1^2 + 2 c_1 c_2 \int u_2 u_1 + c_2^2 \int u_2^2 \, d\tau}$$

$$E = \frac{c_1^2 H_{11} + 2 c_1 c_2 H_{12} + c_2 H_{22}}{c_1^2 S_{11} + 2 c_1 c_2 S_{12} + c_2^2 S_{22}}$$

The integration problem has been broken into sums of smaller integrals. Each of the smaller integrals above has been given a separate symbol. The integrals in the numerator are denoted H_{ij} and the integrals in the denominator are denoted S_{ij}. The assumption that

$$\int u_1 \hat{H} u_2 \, d\tau = \int u_2 \hat{H} u_1 \, d\tau$$

has been made in order to simplify the numerator. Operators $\hat{H}$ for which this is true are called *Hermitian* operators. The equivalent substitution of S_{12} for S_{21} has been made in the denominator.

One can multiply through by the denominator to find

$$\left(c_1^2 S_{11} + 2c_1c_2S_{12} + c_2^2 S_{22}\right) E = c_1^2 H_{11} + 2c_1c_2H_{12} + c_2H_{22}$$

Our objective is to differentiate the energy with respect to each of the minimization parameters c_1 and c_2 so as to find the simultaneous minimum with respect to both of them. This will be the minimum energy obtainable from the sum of functions $\phi = c_1u_1 + c_2u_2$. Differentiating first with respect to c_1, we obtain

$$(2c_1S_{11} + 2c_2S_{12})\,E + \frac{\partial E}{\partial c_1}\left(c_1^2 S_{11} + 2c_1c_2S_{12} + c_2^2 S_{22}\right) = 2c_1H_{11} + 2c_2H_{12}$$

Differentiating with respect to c_2, we get

$$(2c_1S_{12} + 2c_2S_{22})\,E + \frac{\partial E}{\partial c_2}\left(c_1^2 S_{11} + 2c_1c_2S_{12} + c_2^2 S_{22}\right) = 2c_1H_{12} + 2c_2H_{22}$$

In order to find the minimum, we set

$$\frac{\partial E}{\partial c_1} = \frac{\partial E}{\partial c_2} = 0$$

This causes the two $\partial E/\partial c$ terms to drop out:

$$\frac{\partial E}{\partial c_1}\left(c_1^2 S_{11} + 2c_1c_2S_{12} + c_2^2 S_{22}\right) = \frac{\partial E}{\partial c_2}\left(c_1^2 S_{11} + 2c_1c_2S_{12} + c_2^2 S_{22}\right) = 0$$

We are left with a pair of simultaneous equations:

$$(2c_1S_{11} + 2c_2S_{12})\,E = 2c_1H_{11} + 2c_2H_{12}$$

and

$$(2c_1S_{12} + 2c_2S_{22})\,E = 2c_1H_{12} + 2c_2H_{22}$$

These two equations are somewhat more conformable to computer solutions if we divide by 2 and write them in the equivalent form:

$$(H_{11} - ES_{11})\,c_1 + (H_{12} - ES_{12})\,c_2 = 0$$
$$(H_{12} - ES_{12})\,c_1 + (H_{22} - ES_{22})\,c_2 = 0$$

The differences $H_{11} - ES_{11}$, and so on, are scalar energies, so we are left with nothing more than a simple simultaneous equation pair like the ones we solved in high school:

$$ax + by = p$$
$$cx + dy = q$$

The only difference is that the case of $p = q = 0$ is considered an unusual case in elementary presentations, whereas it is the case of interest here. We wish to solve the simultaneous equations in the H and S integrals for the solution set c_1 and c_2.

Alas, we cannot find unique values for c_1 and c_2 because equations with $p = q = 0$ are *inhomogeneous*. Without an extra piece of information, we can only get the ratio of c_1 to c_2. This is not as much of a limitation as it may seem because for larger sets of N equations, we can get the ratios of $N - 1$ coefficients to each other, where N may be very large. In other words, we can get almost all of the information we want. We can obtain the ratios of nonzero solutions we seek if the second equation is a *linear combination* of the first. For a linearly dependent equation pair, it must be possible to multiply one equation by a constant k and obtain the other. In this case, the equations are linearly related if $ka = c$ and $kb = d$. We can assure ourselves that this is true and that the ratio of c_1 to c_2 exists if we stipulate that the determinant of the coefficients be zero

$$\begin{vmatrix} a & b \\ c & d \end{vmatrix} = \begin{vmatrix} ka & kb \\ c & d \end{vmatrix} = kad - ckb = cd - cd = 0$$

In the case of the simultaneous equations we have

$$(H_{11} - ES_{11})c_1 + (H_{12} - ES_{12})c_2 = 0$$
$$(H_{12} - ES_{12})c_1 + (H_{22} - ES_{22})c_2 = 0$$

The stipulation for a linear pair is

$$\begin{vmatrix} H_{11} - ES_{11} & H_{12} - ES_{12} \\ H_{12} - ES_{12} & H_{22} - ES_{22} \end{vmatrix} = 0$$

This is called the *secular determinant*. When expanded, it leads to a quadratic equation with two roots, E_1 and E_2, a pair of energy estimates. Upon imposing the normalization condition, which is the extra piece of information necessary to complete the solution, we take the lower of the two roots as the ground state energy for the system.

The same problem for a sum of N terms demands that the general secular determinant be zero. This leads to N roots:

$$\begin{vmatrix} H_{11} - ES_{11} & H_{12} - ES_{12} & \dots & H_{1N} - ES_{1N} \\ H_{21} - ES_{21} & H_{22} - ES_{22} & \dots & H_{2N} - ES_{2N} \\ \vdots & \vdots & \dots & \vdots \\ H_{N1} - ES_{N1} & H_{N2} - ES_{N2} & \dots & H_{NN} - ES_{NN} \end{vmatrix} = 0$$

Many readers will be familiar with the use of this mathematical formalism in Hückel molecular orbital theory.

17.3 A VARIATIONAL TREATMENT FOR THE HYDROGEN ATOM: THE ENERGY SPECTRUM

Starting with the radial Hamiltonian

$$\hat{H} = \left(-\tfrac{1}{2}\nabla_1^2 - \frac{2}{r_1}\right) = -\frac{1}{2r^2}\frac{d}{dr}r^2\frac{d}{dr} - \frac{2}{r}$$

and the exact wave function for the hydrogen atom $\phi(r) = e^{-\alpha r}$, the variational method leads to

$$E = \frac{h^2\alpha^2}{8\pi^2 m_e} - \alpha e^2$$

where m_e is the mass of the electron (McQuarrie, 1983). This result is similar to the exact solution for the particle in a cubic box $E = h^2n^2/8ml^2$ except that there is a potential energy term $-\alpha e^2$. We can expect some similarities between the two systems. One similarity is that in each there is a *spectrum* of specific energy levels, each corresponding to a specific quantum number. Another is that there is no zero energy because the lowest quantum number is 1, not 0.

We would like to carry out a systematic search for the minimum energy, so we set the first derivative of E with respect to α equal to zero. The derivative goes to zero at a minimum, maximum, or inflection point. If our trial function is reasonably good,[1] only the minimum is found at this level of calculation. (For more complicated systems, maxima, saddle points, and the like may be found.)

The derivative is

$$\frac{dE}{d\alpha} = \frac{2\hbar^2\alpha}{m_e} - e^2 = 0$$

[1] As chemists, we bring centuries of empirical evidence to bear on the decision of what is or is not a good atomic or molecular wave function. A negative exponential is reasonable for the hydrogen atom.

This amounts to finding the wave function $\phi(r) = e^{-\alpha r}$ that has a specific α selected from among all possible values of α. The minimization gives

$$\alpha = \frac{m_e e^2}{\hbar^2}$$

From this, the minimum energy can be found to be [Problem 7.11, McQuarrie (1983)]

$$E = -\frac{1}{2}\left(\frac{m_e e^4}{\hbar^2}\right) = -2.180 \times 10^{-18} \text{ J}$$

which is the ground state of H originally found by Bohr (Fig 15.2). The spectrum of allowed energies can be calculated from the quantum numbers n:

$$\varepsilon_\text{H} = -\frac{1}{2}\left(\frac{m_e e^4}{\hbar^2}\right)\left(\frac{1}{n^2}\right), \qquad n = 1, 2, 3, \ldots$$

The energies are negative for a stable system. The resulting energy level differences can be used to determine the lines in the hydrogen spectrum in agreement with Fig. 15.1. This should come as no surprise because we started out with the exact ground state orbital.

At this point it is reasonable to define an atomic unit of energy, the *hartree* $E_\text{h} = \left(m_e e^4/\hbar^2\right)$, which is twice the energy of the ground state of the hydrogen atom ε_H. (Please do not confuse the measured energy ε_H with the *unit* of energy E_H.)

$$\begin{aligned}\varepsilon_\text{H} &\equiv 2E_\text{h} = 2(2.180 \times 10^{-18} \text{ J}) = 4.359 \times 10^{-18} \text{ J} = 2625 \text{ kJ mol}^{-1} \\ &= 627.51 \text{ kcal mol}^{-1}\end{aligned}$$

The process of searching out energy minima in this way is frequently referred to as *minimization* for obvious reasons. It is also called *optimization* because in cases where the orbital is not known, the optimum energy and the best possible values of any parameters in the orbital expression are found. These parameters are sometimes called *optimization parameters* to distinguish them from known constants. One arrives at different optimization parameters by different optimization procedures, but universal constants like Planck's constant h do not change.

17.3.1 Optimizing the Gaussian Function

Let us try a function that is similar to, but not identical to, the exact function. A variational calculation can be carried out with the approximate *Gaussian* function $\phi(r) = e^{-\alpha r^2}$. It is of the same form as the lowest hydrogen orbital except that the numerator has an r^2, where r should be. This function can be optimized by calculus

```
# gen

hatom gen

0 2
h

1 0
S        1 1.0
0.280000 1.0
  ****
```

FILE 17.1 Gaussian gen input for the hydrogen atom.

and algebra, but we choose to do it now by running the Gaussian© computer program, using a gen input. Selecting a value of $\alpha = 0.28$, one finds

$$E_{\psi} = -0.4244\ E_{\mathrm{h}}$$

The computed value of the energy is less negative and is therefore *higher* than the exact value of $\frac{1}{2}E_{\mathrm{h}}$ by about 15%.

17.3.2 A GAUSSIAN© HF Calculation of E_{atom}: Computer Files

The gen keyword in Gaussian permits you to input a value for α as the first entry in the penultimate line of the input file, 0.280000 in File 17.1. We have done this for $\alpha = 0.18,\ 0.28$, and 0.38. The energy results extracted from much larger Gaussian output files are shown as File 17.2.

Two things are clear even from this tiny extract from the output. The value for $\alpha = 0.28$ chosen from among the three α values leads to the lowest energy, but the optimized $\alpha = 0.28$ isn't much better than the other two choices. Evidently the energy minimum is at the bottom of a rather shallow energy well because none of the three energy values is obviously wrong. Changing α from .28 to .38 gives an increase in the energy value of only 0.0107 E_{h}. Changing the input value in the other direction by 0.1 results in a slightly larger energy change. This small result of 0.0107 E_{h} is a little deceptive, however, because the hartree is such a large unit relative to chemical problems. The conversion factor from E_{h} to kcal mol^{-1} being 627.51 kcal mol^{-1}, the calculated value in the hydrogen atom problem is a substantial $0.0107 E_{\mathrm{h}}\,(627.51) = 6.7$ kcal mol^{-1}.

```
0.180000          HF=-0.4070275
0.280000          HF=-0.4244132
0.380000          HF=-0.4136982
```

FILE 17.2 Energies drawn from the Gaussian gen output file for the hydrogen atom.

17.4 HELIUM

The helium atom is similar to the hydrogen atom, with the critical difference that there are two electrons moving in the potential field of the nucleus. The nuclear charge is +2. The Hamiltonian for the helium atom is

$$\hat{H} = -\tfrac{1}{2}\nabla_1^2 - \tfrac{1}{2}\nabla_2^2 - \frac{2}{r_1} - \frac{2}{r_2} + \frac{1}{r_{12}}$$

Regrouping, we obtain

$$\hat{H} = \left(-\tfrac{1}{2}\nabla_1^2 - \frac{2}{r_1}\right) + \left(-\tfrac{1}{2}\nabla_2^2 - \frac{2}{r_2}\right) + \frac{1}{r_{12}}$$

The first two terms on the right replicate the hydrogen case, except for a different nuclear charge. The third term on the right, $1/r_{12}$, is due to electrostatic repulsion of the two electrons acting over the interelectronic distance r_{12}. This term does not exist in the hydrogen Hamiltonian. The sum of two nuclear and one repulsion Hamiltonians is

$$\hat{H}_{\mathrm{He}} = \hat{H}_1 + \hat{H}_2 + \frac{1}{r_{12}}$$

If this Hamiltonian were to operate on an exact, normalized wave function for helium, the exact energy of the system would be obtained:

$$E_{\mathrm{He}} = \int_0^\infty \Psi(r_1, r_2)\hat{H}_{\mathrm{He}}\Psi(r_1, r_2)\, d\tau$$

The helium atom, however, is a three-particle system for which we cannot obtain an exact solution. The orbital and the total energy must, of necessity, be approximate.

As a naive or *zero-order* approximation, we can simply ignore the "r_{12} term" and allow the simplified Hamiltonian to operate on the $1s$ orbital of the H atom. The result is

$$E_{\mathrm{He}} = -\frac{2^2}{2} - \frac{2^2}{2} = -4.00\ E_{\mathrm{h}}$$

which is 8 times the exact energy of the hydrogen atom ($-\tfrac{1}{2}E_{\mathrm{h}}$). The 2 in the numerators are the nuclear charge $Z = 2$. In general, the energy of any hydrogen-like atom or ion is $-Z^2/2$ hartrees per electron, provided that we ignore interelectronic electrostatic repulsion.

We can compare this result with the first and second *ionization potentials* (IP) for helium, which are energies that must go into the system to bring about ionization:

$$\mathrm{He} \rightarrow \mathrm{He}^{+} + \mathrm{e}^{-} \rightarrow \mathrm{He}^{2+} + \mathrm{e}^{-}$$

These energies can be measured experimentally with considerable accuracy. Since they are the energy necessary to pull an electron away from the helium atom, they are equal and opposite in sign to the binding energy of the ionized electron. In this way we have a measure of both (a) the first or "outside" electron attracted to the He^+ system and (b) the inside electron attracted to the He^{2+} nucleus. This second energy IP_2 can be calculated exactly because He^+ is a one-electron (two-particle) system. The calculated second ionization potential IP_2 is exactly 2.0 E_h.

If we compare the calculated total ionization potential, IP $= 4.00$ E_h, with the experimental value, IP $= 2.904$ hartrees, the result is very bad. The magnitude of the disaster is even more obvious if we subtract the known second ionization potential, $IP_2 = 2.0$, from the total IP to find the *first* ionization potential, IP_1:

$$\mathrm{IP}_1 = \mathrm{IP}_{\mathrm{total}} - \mathrm{IP}_2 = 2.904 - 2.000 = 0.904\ E_h$$

Under the approximations we have made, IP_1 (calculated) $= 2.000$ E_h is about 110% in error. Clearly, we cannot ignore interelectronic repulsion.

17.4.1 An SCF Variational Ionization Potential for Helium

One approach to the problem of the r_{12} term is a variational *self-consistent field* approximation. We shall start from Hartree's *orbital approximation*, assuming that the orbital of helium is separable into two one-electron orbitals $\Phi(1,2) = \phi_1(1)\phi_2(2)$. It is reasonable to use the H atom as our starting point with the same kinetic energy operator for each starting orbital $\phi_1(1)$ and $\phi_2(2)$and $-2/r$ as the potential energy part for attraction of each electron to a nuclear charge of 2:

$$\hat{H} = -\frac{1}{2r^2}\frac{d}{dr}r^2\frac{d}{dr} - \frac{2}{r}$$

Although we are solving for one-electron orbitals, ϕ_1 and ϕ_2, we do not want to fall into the trap of the last calculation. This time we shall include a potential energy term V_1 to account for the repulsion between the negative charge on the electron arbitrarily designated electron 1, exerted by the other electron which we shall call electron 2.

We don't know where electron 2 is, so we must integrate over all possible locations $d\tau$:

$$V_1 = \int_0^\infty \phi_2 \frac{1}{r_{12}}\ \phi_2\, d\tau$$

The entire Hamiltonian for electron 1 is

$$\hat{H}_1 = -\frac{1}{2r_1^2}\frac{d}{dr_1}r_1^2\frac{d}{dr_1} - \frac{2}{r_1} + \int_0^\infty \phi_2 \frac{1}{r_{12}}\ \phi_2\, d\tau$$

The same treatment produces a similar operator for electron 2:

$$\hat{H}_2 = -\frac{1}{2r_2^2}\frac{d}{dr_2}r_2^2\frac{d}{dr_2} - \frac{2}{r_2} + \int_0^\infty \phi_1 \frac{1}{r_{12}}\phi_1 \, d\tau$$

We do not know the *orbitals* of the electrons either. We can reasonably assume that the ground state orbitals of electrons 1 and 2 are similar but not identical to the 1s orbital of hydrogen:

$$\phi_1 = \sqrt{\frac{a^3}{\pi}}\, e^{-\alpha r_1}$$

and

$$\phi_2 = \sqrt{\frac{b^3}{\pi}}\, e^{-\alpha r_2}$$

The integral in $\hat{H}_1$, representing the Coulombic interaction between electron 1 at r_1 and electron 2 somewhere in orbital ϕ_2, has been evaluated for Slater-type orbitals (Rioux, 1987; McQuarrie, 1983) and is

$$V_1 = \int_0^\infty \phi_2 \frac{1}{r_{12}}\phi_2 \, d\tau = \frac{1}{r_1}\left[1 - (1 + b\, r_1)e^{-2b\, r_1}\right]$$

Now the approximate Hamiltonian for electron 1 is

$$\hat{h}_1 = -\frac{1}{2r_1^2}\frac{d}{dr_1}r_1^2\frac{d}{dr_1} - \frac{2}{r_1} + \frac{1}{r_1}\left(1 - (1 + br_1)\, e^{-2br_1}\right)$$

with a similar expression for $\hat{h}_2$ involving ar_2 in place of br_1 in the Slater orbital. The orbital is normalized, so the energy of electron 1 is

$$E_1 = \int_0^\infty \phi_1 \hat{h}_1 \, \phi_1 \, d\tau$$

with a similar expression for E_2.

Calculating E_1 requires solution of three integrals:

$$E_1 = \int_0^\infty \phi_1 \left(-\tfrac{1}{2}\nabla_1^2\right) \phi_1 \, d\tau - \int_0^\infty \phi_1 \left(\frac{Z}{r_1}\right) \phi_1 \, d\tau + \int_0^\infty \phi_1 \,(V_1)\, \phi_1 \, d\tau$$

They yield (Rioux, 1987) three terms for the energy of the electron in orbital ϕ_1:

$$E_1 = \frac{a^2}{2} - Za + \frac{ab\left(a^2 + 3ab + b^2\right)}{(a+b)^3}$$

with a similar expression for E_2 except that b replaces a in the first two terms on the right

$$E_2 = \frac{b^2}{2} - Zb + \frac{ab\left(a^2 + 3ab + b^2\right)}{(a+b)^3}$$

The parameters a and b in the Slater-type orbitals for electrons 1 and 2 are minimization parameters representing an effective nuclear charge as "experienced" by each electron, partially shielded by the other electron from the full nuclear charge. The SCF strategy is to minimize E_1 using an arbitrary starting b and to find a at the minimum. This a value is then used to find b at the minimum E_1. This value then replaces the starting b value and a new minimization cycle produces a new a and b, and so on until there is no progressive difference in E_1. The electrical field experienced by the electrons is now self-consistent, hence it is a *self-consistent field* SCF.

In this particular case, the calculations are completely symmetrical. Everything we have said for a we can also say for b. At self-consistency, $a = b$ and we can substitute a for b at any point in the iterative process, knowing that as we approach self-consistency for one, we approach the same self-consistent value for the other.

A reasonable thing to do at the end of each iteration would be to calculate the total energy of the atom as the sum of its two electronic energies $E_{\text{He}} = E_1 + E_2$, but in so doing, we would be calculating the interelectronic repulsion $ab(a^2 + 3ab + b^2)/(a+b)^3$ twice, once as an r_{12} repulsive energy and once as an r_{21} repulsion. The r_{21} repulsion should be dropped to avoid double counting, leaving

$$E_{\text{He}} = E_1 + E_2 = E_1 + \frac{b^2}{2} - Zb$$

as the correct energy of the helium atom.

Although the Hartree procedure for atoms reaches self-consistency and gives a qualitative picture of the electron probability densities, the ionization energy is still in error by a substantial amount (0.014 h $\cong$ 37 kJ mol^{-1} on three iterations). Something is missing. A problem arises with the Hartree product when we exchange one atomic orbital with its adjacent neighbor

$$\psi_i(r_i)\psi_j(r_j) \rightarrow \psi_j(r_j)\psi_i(r_i)$$

The Hartree product is the same after exchange as it was before exchange. This is equivalent to saying that $\psi_i(r_i)$ and $\psi_j(r_j)$ are identical in all respects, even though we recall from elementary chemistry that no two electronic wave functions can be exactly alike and that each electron is represented by a unique set of *four* quantum numbers n, l, m, and s. In three-dimensional Cartesian space, it is not difficult to suppose that there are three quantum numbers $n, l,$ and m in the complete solution for the hydrogen atom, one for each dimension; but by using four quantum numbers, we imply something beyond the solution in 3-space.

If helium is in the excited state, the electrons are in different spatial orbitals; but in the ground state, both electrons have the same $1s$ orbital description. A distinguishing fourth quantum number called the spin quantum number $s = \pm\frac{1}{2}$ must be introduced to give two distinct $1s$ orbitals in helium, one designated $1s\alpha$ and the other designated $1s\beta$.

17.5 SPIN

If we ionize the electrons and examine their spins, we may find that the first electron ionized (the outer electron) has spin α implying that the inner electron (2) has spin β and that the atomic orbital Ψ_{1s} is

$$\Psi_{1s} = 1s(1)\alpha + 1s(2)\beta$$

On the other hand, we may find that the first ionized electron has spin β leading to

$$\Psi_{1s} = 1s(1)\beta + 1s(2)\alpha$$

These two results have *exactly* the same probability. We are in a logical dilemma that results from the Hartree independent orbital hypothesis. It cannot be said that an electron is in either the α spin orbital or the β spin orbital, only that they are both in an orbital that is a *linear combination* of equally weighted space-spin basis functions. The two plausible basis functions

$$1s(1)\alpha 1s(2)\beta \qquad \text{and} \qquad 1s(1)\beta 1s(2)\alpha$$

can be combined in two possible ways:

$$1s(1)\alpha 1s(2)\beta + 1s(1)\beta 1s(2)\alpha$$

and

$$1s(1)\alpha 1s(2)\beta - 1s(1)\beta 1s(2)\alpha$$

Only the last *antisymmetrical* linear combination is acceptable for the $1s$ electrons in helium.

17.6 BOSONS AND FERMIONS

There are only two kinds of elementary particles in the universe, *bosons* and *fermions*. Bosons are symmetrical under exchange and fermions are antisymmetrical under exchange. Electrons are fermions. If both electrons in the $1s$ orbital had the same spin, the negative (antisymmetric) combination above would integrate to zero and there would be no probability of finding the electron. That is why electrons must have opposite spins if they are to reside in the same space orbital.

17.7 SLATER DETERMINANTS

The linear combination selected above for ground state helium is the same as the expansion of a 2×2 determinant, which is the simplest example of a *Slater determinant* (Section 15.3):

$$1s(1)\alpha 1s(2)\beta - 1s(1)\beta 1s(2)\alpha = \begin{vmatrix} 1s(1)\alpha & 1s(1)\beta \\ 1s(2)\alpha & 1s(2)\beta \end{vmatrix}$$

Larger Slater determinants also coincide with uniquely acceptable linear combinations for fermions. In the general $n \times n$ case with some simplification of notation, the Slater determinant is

$$\psi(1, 2, \ldots, n) = \frac{1}{\sqrt{n!}} \begin{vmatrix} \phi_1\alpha_1 & \phi_1\beta_1 & & \ldots & \phi_n\alpha_1 & \phi_n\beta_1 \\ \phi_1\alpha_2 & \phi_1\beta_2 & \ldots & & & \\ \ldots & \ldots & & & & \\ \phi_1\alpha_n & \phi_1\beta_n & \ldots & & \phi_n\alpha_n & \phi_n\beta_n \end{vmatrix}$$

where the premultiplier $1/\sqrt{n!}$ is a normalization constant. The many-electron wave function for orthonormal basis functions ψ_i, in more concise notation, is now written (Pople, Nobel Prize, 1998)

$$\psi = (n!)^{-\frac{1}{2}} \det [(\psi_1\alpha)(\psi_1\beta)(\psi_2\alpha)\ldots]$$

The Slater determinant always produces a linear combination of one-electron orbitals that allows for electron indistinguishability by giving equal weight to electrons of opposing spins in each of n orbitals. The Slater determinantal molecular orbital and *only* the Slater determinant satisfies the two great generalizations of quantum chemistry, the Heisenberg principle of uncertainty (which is why you can't tell in advance whether the first electron ionized from He will be α or β) and the Fermi–Dirac principle of antisymmetric fermion exchange.

In addition to his work on determinantal wave functions, Slater also fitted functions to numerical compilations of SCF data and obtained orbitals in analytic form. These *Slater-type orbitals* (STO) resemble hydrogen wave functions but have adjustable parameters to account for field differences between hydrogen and many-electron *hydrogen-like* atoms. A few STOs are

$$\begin{aligned}\psi_{1s} &= N_{1s}e^{-\alpha r}\\ \psi_{2s} &= N_{2s}re^{-\alpha r/2}\\ \psi_{2p_x} &= N_{2p_x}r^2e^{-\alpha r/2}\\ &\text{etc.}\end{aligned}$$

The normalization constants N are $N_{1s} = (\alpha^3/\pi)^{1/2}$, $N_{2s} = (\alpha^3/96\pi)^{1/2}$, and so on. Clearly, the entire system depends upon α, which is an empirically fitted parameter. The parameter is written as

$$\alpha = Z - S$$

where Z is the atomic number (charge on the nucleus) and S is a *shielding constant* which accounts for the diminution of nuclear charge experienced by an outer electron owing to shielding by inner electrons.

This method can, with computational difficulty, be extended to atoms larger than helium and to a few small molecules. In a molecular problem as simple as methane, however, the dimension of the Slater determinant is 16×16. Clearly, molecular problems are daunting to anyone attempting hand calculations. Practical applications awaited widespread availability of powerful digital computers. Although we shall soon consider more challenging problems of correlated wave functions, Slater orbitals and the Hartree–Fock (HF) equations contain the essence of atomic and molecular orbital theory. Much important structural and thermochemical information can be gleaned from them (Hehre, 2006).

17.8 THE AUFBAU PRINCIPLE

Given the orbital structure of the elements in the first three rows of the periodic table, one can predict, with reasonable certainty, their chemical properties. Hydrogen ionizes to form the H^+ ion (hydrated in aqueous media), and it can be persuaded to take on one electron to form the H^- hydride ion but helium does neither because its 1s orbital is "full" (Pauli). The electronic configuration of helium is He $1s^2$.

The first full row of the periodic table starts with lithium, which, like hydrogen, can lose an electron, this time from the $2s$ orbital as its characteristic reaction. The $2s$ probability density antinode is considerably more distant from the nucleus than the $1s$ orbital of hydrogen, making this loss of an electron from Li even more facile than the equivalent loss from H. Beryllium has two electrons in the $2s$ orbital. Electron loss from Be takes place easily, but not as easily as Li because its nuclear charge is one more than in Li. Be is a metal but is not as metallic as Li. Here the "metallic nature" of an element is defined by its tendency to lose an electron. The next six elements—B, C, N, O, F, and Ne—correspond to the filling of the $2p_x$, $2p_y$, and $2p_z$ orbitals with two electrons each. The 3*s* and 3*p* orbitals are filled in a similar way presenting elements that are roughly similar to their first row counterparts.

Atomic energy levels are not energetically very far apart, and one should not expect complete regularity in the table. Small energetic perturbations cause them to shift around a bit. Sometimes the order of orbital filling, *s*, *p*, *d*, ... is disrupted by external fields brought in by orbital interactions, ligands, solvents, or intentionally imposed magnetic fields.

17.9 THE SCF ENERGIES OF FIRST-ROW ATOMS AND IONS

Starting from any reasonable value of *a*, one can calculate a value for *b* (Section 17.4.1), substitute it as a parameter at the top of a ***Mathcad©*** program, and repeat the calculation to self-consistency for other small atoms. (Ignore negligibly small imaginary roots.) Usually only three or four iterations are necessary.

The same calculation can be carried out for the single positive ions. The energy difference between the atom and its single + ion is the energy necessary to drive off one electron, that is, the first ionization potential IP_1. A collection of IP_1 values calculated in this way is depicted in Fig. 17.1. Notice the sharp drop at Li ($Z = 3$), B($Z = 5$), and O($Z = 8$). The drop at Li results from the greater distance of the probability antinode for 2*s* relative to 1*s*. The drop at B results from shielding of the $2p_x$ orbital by the $1s^2$ electrons, and the drop at O results from the increased shielding of the $2p_x^2 2p_y 2p_z$ configuration relative to the $2p_x 2p_y 2p_z$ of N ($Z = 7$).

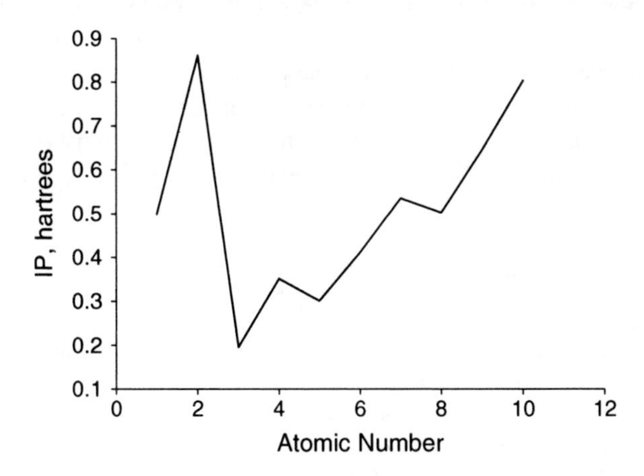

FIGURE 17.1 Calculated IP_1 for elements 1–10. Experimental results follow this pattern, but they are not identical. A more complete graph of this kind is presented in most general chemistry texts.

17.10 SLATER-TYPE ORBITALS (STO)

In the absence of an exact solution of the Schrödinger equation for atoms beyond hydrogen in the periodic table, John Slater devised a set of empirical rules for writing down approximate wave functions. Every entry in Table 17.1 is a function involving the following: the radial distance in units of bohrs, r/a_0; a negative exponential e^{-Zr/na_0}, where n is the principal quantum number, 1, 2, and 3 for H, He, and the first two full rows of the periodic table; and a_0 is the Bohr radius. Z is the nuclear charge. Slater wrote an approximate wave function involving only the radial part for the first two full rows in the periodic table, ignoring the spherical harmonics. The function takes the form

$$\phi(r) = re^{-\frac{-(Z-s)}{na_0}}$$

An adjustable parameter s is called the *screening constant*, and Z is also an adjustable parameter called the *effective quantum number*. For the first-, second-, and third-row elements, Z simply is 1, 2, and 3, though it becomes nonintegral later in the table. What will concern us most is the screening constant, which can be arrived at by fitting experimental data.

Slater's rules for determining what are now called *Slater-type orbitals* (STO) are simple for the first three rows of the table. They become more complicated and less reliable lower in the table. We have already seen that the screening constant for the second electron in the helium atom is about 0.3. For $1s$ electrons in higher atoms, Slater modified this slightly to 0.35. When a $1s$ electron screens a p electron, screening is more effective than simple electron–electron screening in helium because the probability density lobes of the $2p$ orbitals lie outside of the $1s$ orbital. Slater chose a screening constant of 0.85. Notice that the polynomial parts of the wave function (Table 17.1) are gone. Only the "tail" of the wave function is represented because that is the part involved in the first ionization potential and, for the most part, in chemical bonding. For more detail on Slater's rules, see Problem 15.79 in Levine (2000).

TABLE 17.1 Slater's Rules

1. Array orbitals 1s, 2s, 2p, 3s, 3p, 3d, ...
2. Consider only the orbital containing the electron in question and the one below it.
3. Electrons lower than (2) are *interior orbitals.*
4. Orbitals higher than (2) are *exterior orbitals.*
5. Electrons designated s (1s, 2s, etc) have a screening constant of 0.30.
6. Electrons in the same orbital have a screening constant of 0.85.

Electrons are in the orbital $\phi(r) = re^{\frac{-(Z-s)}{na_0}}$. For example, in bohrs and hartrees, the orbital electron in helium has $\phi_{He}(r) = re^{\frac{-(Z-s)}{n}} = re^{\frac{-(Z-s)}{n}} = re^{\frac{-(1-.30)}{(1)}} = 1.7r$ due to shielding.

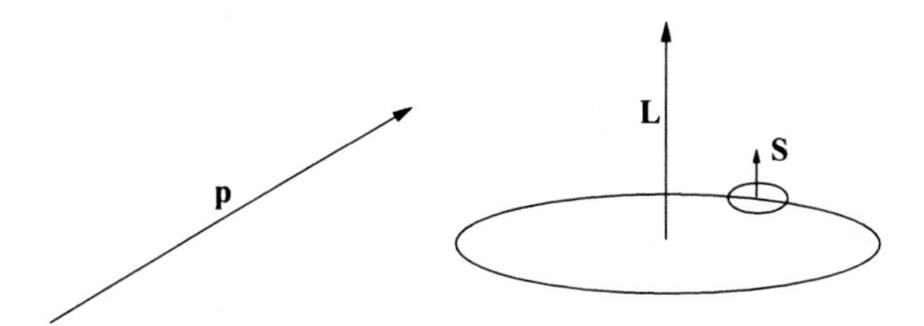

FIGURE 17.2 Linear and angular momentum vectors.

17.11 SPIN–ORBIT COUPLING

An object moving in a straight line tends to conserve its linear momentum **p** by continuing to move in a straight line. A spinning wheel tends to conserve its angular momentum **L** by continuing to spin. Although electrons cannot be correctly described by a deterministic circular path, they have an *orbital angular momentum* **L** and, as we have seen, electrons have a property analogous to spin, so we anticipate (correctly) a *spin angular momentum* **S** (Fig. 17.2). These momenta are vectors.

An atom has an angular momentum that is the sum of its electronic orbital and spin angular momenta. Orbital and spin angular momenta *couple* either by vector addition or vector subtraction according to whether they are in the same direction or are opposed. This results in anomalous spectral splitting. The appearance of a pair of sodium D lines where one line was expected is a result of *spin–orbit coupling*. The hydrogen spectrum also shows spectral splitting. For example, the 656.2 nm (6562 Å) "line" of hydrogen is not really a line, but a *doublet* at 656.272 and 656.285 nm under high resolution.

In many-electron atoms, many vector combinations at different angles produce complicated vector combinations of **L** and **S** which result in complicated spectral splittings. These patterns are not completely understood. Partial explanations valid for lower atomic mass elements include *Russell–Saunders* or LS coupling patterns and spectral splittings.

PROBLEMS AND EXAMPLES

Example 17.1 A Mathcad© SCF Calculation

A calculation of the first three approximations to the SCF energy $\varepsilon\,(a, b)$ of the helium atom is shown in File 17.3.

The first iteration of the SCF procedure in File 17.3 produces an approximation to the first ionization potential of He $-\varepsilon(a, b) = \mathrm{IP_1(calc)} = -(-0.812)$ hartrees—that is, 10.2% too small. This is not very good, but it is a great improvement over the >100% error we found when the r_{12} term was ignored. Continuing the calculation and substituting for the initial value of b, we minimize to find a new value of $\mathrm{IP_1(calc)} = 0.925$, 2.4% in error, followed by $\mathrm{IP_1(calc)} = 0.889$, 1.5% in error on

$Z := 2.000 \qquad a := 2.000 \qquad b := 2.000$

$$\varepsilon(a, b) := \left[\frac{a^2}{2} - Z \cdot a + \frac{a \cdot b(a^2 + 3 \cdot a \cdot b + b^2)}{(a+b)^3}\right]$$

$$a := \text{root}\left(\frac{d}{da}\varepsilon(a, b), a\right)$$

$a = 1.601 \qquad \varepsilon(a, b) = -0.812$

$b := 1.601$

$$a := \text{root}\left(\frac{d}{da}\varepsilon(a, b), a\right)$$

$a = 1.712 \qquad \varepsilon(a, b) = -0.925$

$b := 1.712$

$$a := \text{root}\left(\frac{d}{da}\varepsilon(a, b), a\right)$$

$a = 1.681 \qquad \varepsilon(a, b) = -0.889$

FILE 17.3 Mathcad© calculation of the ionization potential of helium. An approximate screening constant of 0.3 gives $a = (z - s) \cong 1.7$.

the third iteration, approaching the experimental value. Notice that the calculated IP_1 on the second iteration is too large. Sometimes the solution of the iterative procedure is approached asymptotically, and sometimes the approach is oscillatory.

Example 17.2 The Slater Orbital of Oxygen

Find the Slater orbital of oxygen.

Solution 17.2 First observe that the nuclear charge is $Z = 8$. The $1s^2$ part of the screening constant on the valence electron is $2(0.85) = 1.70$ and the part of the screening constant for the electrons in the $2s^2$ and $2p^3$ shell is $s = 5(.35) = 1.75$. The interior electrons are more effective (0.85) in screening the valence electron than the electrons that share the second valence shell (0.35). We do not count the valence electron because it cannot screen itself. The screening electrons are always one less than the number of electrons in the neutral atom, in this case $2 + 2 + 3 = 7$. The total screening constant is $1.70 + 1.75 = 3.45$. This gives an STO for oxygen:

$$s = 2(0.85) + 5(0.35) = 1.7 + 1.75 = 3.45$$

$$Z(\text{shielded}) = \frac{8 - 3.45}{2}$$

$$\phi(r) = re^{-(z-s)/na} = re^{-(8-3.45)/2a} = re^{-2.28/a}$$

Other atomic STOs can be found by applying the same routine.

Problem 17.1

Evaluate the following determinants:

$$\begin{vmatrix} 1 & 0 \\ 0 & 1 \end{vmatrix}, \qquad \begin{vmatrix} x & 1 \\ 1 & x \end{vmatrix}, \qquad \begin{vmatrix} \sin\theta & \cos\theta \\ -\cos\theta & \sin\theta \end{vmatrix}$$

Problem 17.2

What is the energy increase relative to the ground state when one of the quantum numbers, say n_z, for the particle in a cubic box is raised from 1 to 2? What is the degeneracy of the resulting wave function and probability distribution?

Problem 17.3

1,3-Pentadiene shows a strong absorption peak at 45,000 cm^{-1} (Ege, 1994). What is the wavelength of this radiation in (a) centimeters, (b) meters, (c) nanometers, (d) picometers, and (e) angstroms? What is its frequency in hertz? What is its energy in joules?

Problem 17.4

What is the actual energy increase for the single excitation $n_z = 1 \rightarrow n_z = 2$ in Problem 17.2 if the trapped particle is an electron and the dimension of the box is approximately a bond length, 1.5 Å? Give your answer in joules. What wavelength of light will promote an electron from the ground state to one of the degenerate excited states? Give your answer in nanometers.

Problem 17.5

The length of an ethene molecule is about 153 pm according to MM3. Using Hückel theory and the particle in a one-dimensional box as a model, in what region of the electromagnetic spectrum (X-ray, UV, vis, IR, etc.) is the radiation necessary to promote an electron from the highest occupied molecular orbital (HOMO) to the lowest unoccupied molecular orbital (LUMO)?

Problem 17.6

Find the Slater-type orbital (STO) of nitrogen.

Problem 17.7

Write down the unnormalized Slater determinant for He in the ground state. The He atom is a three-particle, two-electron system. Set the Slater determinant, including a normalization constant, equal to the wave function of He. Expand the determinant. What is the normalization constant?

Problem 17.8

We have seen that the linear combination that is the *difference* between terms is equivalent to the *Slater determinantal* wave function for helium.

$$\psi_{\mathrm{He}}(1,2) = \frac{1}{\sqrt{n}}(1s(1)\alpha(1)1s(2)\beta(2) - 1s(2)\alpha(2)1s(1)\beta(1))$$

$$\equiv \frac{1}{\sqrt{n}}\begin{vmatrix} 1s(1)\alpha(1) & 1s(1)\beta(1) \\ 1s(2)\alpha(2) & 1s(2)\beta(2) \end{vmatrix}$$

What is wrong with the positive combination?

$$\psi_{\mathrm{He}}(1,2)\frac{1}{\sqrt{n}}(1s(1)\alpha(1)1s(2)\beta(2) + 1s(2)\alpha(2)1s(1)\beta(1))$$

Hint: Try switching electrons $\psi_{\mathrm{He}}(1,2) \to \psi_{\mathrm{He}}(2,1)$.

Problem 17.9

A variational treatment of atomic helium gave

$$E_{\mathrm{He}} = -(-2Z_{\mathrm{eff}}^2 + \frac{5}{4}Z_{\mathrm{eff}} + 4Z_{\mathrm{eff}}(Z_{\mathrm{eff}} - 2))E_{\mathrm{H}}$$

Find Z_{eff} and E_{He} given that $E_{\mathrm{H}} = -13.6$ eV. Give your answer also in electron volts. The experimental result is $E_{\mathrm{He}} = -79.0$ eV. What is the % error of this variational treatment?

18

EXPERIMENTAL DETERMINATION OF MOLECULAR STRUCTURE

Acceptance, at least provisional acceptance, of atomic theory goes back two and a half millennia, but it inevitably raises the question of what atoms look like. The nature of the world around us—air, earth, fire, and water—is governed less by atomic properties than by the physical and chemical properties of molecules. But acceptance of molecular theory, a child of the nineteenth century, raises the same question: What do molecules look like? Experimental evidence for the existence and physical appearance of molecules comes from interactions of confined pure samples with their environment, largely through energy transfer by the very broad energy spectrum of electromagnetic radiation extending all the way from gamma and X rays to low-energy radio waves.

18.1 THE HARMONIC OSCILLATOR

A mass attached to a fixed beam by a spring can be set into *oscillation* by pulling it down and letting it go (Fig. 18.1). An ordinary spring obeys Hooke's law, at least approximately:

$$f = -k_f z(t)$$

where f is the force exerted by the spring on the moving mass when it is displaced by a distance z from the equilibrium distance z_0. The proportionality constant k_f is

Concise Physical Chemistry, by Donald W. Rogers

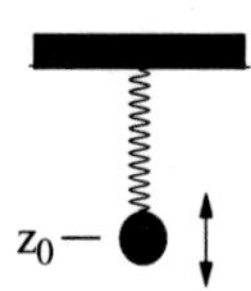

FIGURE 18.1 A classical harmonic oscillator. The equilibrium position on the vertical axis is z_0.

called the *force constant*, and z is a function of time $z(t)$. The sign is negative because the force is a restoring force acting in opposition to the excursion of the mass away from z_0.

By Newton's second law, $f = ma$ where a is the acceleration, $d^2z(t)/dt^2$. These two expressions for the force can be set equal to one another:

$$m\frac{d^2z(t)}{dt^2} = -k_f z(t)$$

$$\frac{d^2z(t)}{dt^2} = -\frac{k_f}{m}z(t)$$

This is a wave equation of the kind described in Section 16.2:

$$\frac{d^2\phi(x)}{dx^2} = -\frac{4\pi^2}{\lambda^2}\phi(x)$$

In the analogous harmonic oscillator case, we have

$$\frac{d^2z(t)}{dt^2} = -\frac{4\pi^2}{\lambda^2}z(t)$$

Comparing the two expressions for acceleration,

$$\frac{k_f}{m} = \frac{4\pi^2}{\lambda^2}$$

leads to

$$\frac{1}{\lambda} = \frac{1}{2\pi}\sqrt{\frac{k_f}{m}}$$

The speed of propagation of electromagnetic radiation is $c = 2.998 \times 10^8$ m s^{-1}, which is the number of waves per second (frequency ν) times the distance covered by each wave (wavelength λ) $c = \nu\,\lambda$. When electromagnetic radiation of many frequencies falls upon an ideal *quantum* harmonic oscillator, most of it bounces off but a selected frequency is absorbed, the one that promotes the oscillator from one quantum state to the next higher quantum state. One of the most common laboratory instruments is the spectrophotometer, which is calibrated to record selected absorbed frequencies called *resonance frequencies*. These are often reported as *wave numbers* $\bar{\nu}$, which are frequencies divided by the speed of radiation. This is just the experimental measurement we want in order to calculate force constants. In units of kg m^{-2}, we have

$$\bar{\nu} = \frac{\nu}{c} = \frac{1}{\lambda} = \frac{1}{2\pi}\sqrt{\frac{k_f}{m}}$$

$$k_f = 4\pi^2 m\,\bar{\nu}^2$$

The force constant is also expressed in newtons per meter (N m^{-1}) because a newton is a kg m^{-1}.

18.2 THE HOOKE'S LAW POTENTIAL WELL

In the harmonic oscillator, potential energy is increased by the work done on the mass by moving it dz against the opposing force $-f$ of the constraining spring. This work, $dw = -f\,dz$, is not lost. It is stored in the spring as *potential energy* V:

$$dV = dw = -f\,dz = k_f z\,dz$$

Integrating, we obtain

$$\int_{V_0}^{V} dV = k_f \int_{z_0}^{z} z\,dz = \frac{k_f z^2}{2}$$

It is convenient to define the potential energy as zero, $V_0 = 0$, at the equilibrium position of the oscillating mass, $z_0 = 0$ (Fig. 18.1). The oscillating mass passes through $z_0 = 0$ on each oscillation but continues along the z axis to nonzero values in the opposite direction, whereupon $\int_{V_0}^{V} dV = V - V_0 = V$and

$$V = \frac{k_f z^2}{2}$$

The potential energy defined in this way is always positive. It is symmetrical about the equilibrium position of a perfect Hooke's law spring and it is steep for a stiff spring (larger k_f) but open for a weak spring (relatively smaller k_f). The excursions z away from z_0 are positive and negative, but the energy always goes up with z because the square of z is always positive (Fig. 18.2).

Upon solving the Schrödinger equation for this system, one finds that the simple harmonic oscillator has evenly spaced energy levels and that there is a half quantum of energy at the bottom of the well (called the *zero point energy*). Because the energy of transition from one level to an adjacent neighbor level is the same no matter where

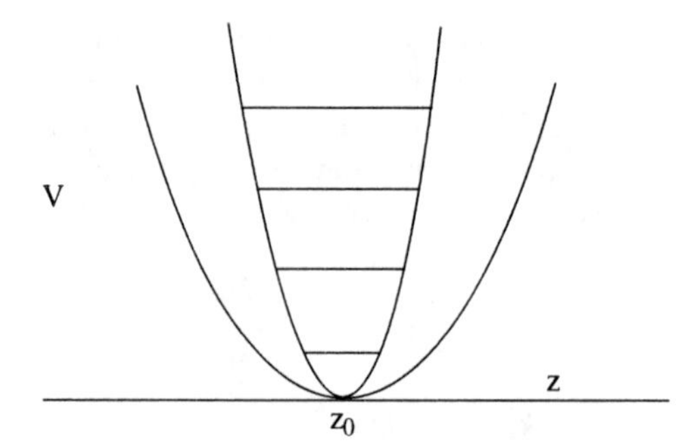

FIGURE 18.2 Parabolic potential wells for the harmonic oscillator. The narrow well has the larger force constant. Only one set of quantum levels is shown.

you start, the absorption spectrum for the harmonic oscillator consists of only one line with frequency $\nu = (\Delta E/h)\ \mathrm{s}^{-1}$ because $\Delta E = h\nu$.

The simple harmonic oscillator is a fairly good model for a hydrogen atom suspended from a heavy carbon atom framework (Fig. 18.1). From C–H resonance frequencies of about 2900–3000 cm^{-1}, one finds the force constant to be about 500 N m^{-1}. Typical experimental values for molecular force constants vary considerably, from about 100 N m^{-1} to about 800 N m^{-1}.

The reader with practical knowledge of infrared spectra will find a discrepancy between the complicated structure of a real IR band spectrum and the single line predicted by Hooke's law. This is the result of many factors, including the failure of Hooke's law, energy coupling among chemical bonds, and many other motions (bending, torsion, etc.) that are possible in a real polyatomic molecule. Nevertheless, the presence of a strong peak near 2900–3000 cm^{-1} is a good indicator of a C–H stretch lurking somewhere in the molecule. Other characteristic frequencies are used in qualitative IR "fingerprint" analysis.

18.3 DIATOMIC MOLECULES

For diatomic molecules attached by a chemical bond, the picture is very similar to the harmonic oscillator of one mass. The atoms vibrate harmonically relative to one another with a natural frequency determined by their mass and the strength of the electronic spring connecting them. One replaces the mass of the simple harmonic oscillator with the *reduced mass* of the diatomic molecule $\mu = m_1 m_2/m_1 + m_2$, where m_1 and m_2 are the atomic masses, and proceeds with the calculation. The problem has indeed been *reduced* from one of two masses vibrating relative to one another to one of a single fictitious mass μ vibrating relative to a central point.

18.4 THE QUANTUM RIGID ROTOR

A small mass on a circular orbit in a fixed plane shows quantum phenomena. The energy level spacing shows a pattern that is similar to the energy levels of the particle in a box (Chapter 16) except that the quantum number n is replaced by J and m times the length of the constraining geometry l^2 is replaced by the moment of inertia of the rotating mass $I = mr^2$

$$E = \frac{J^2\hbar^2}{2I}$$

Because the mass can rotate in either direction (or stand still at $J = 0$), the allowed quantum numbers are $J = 0, \pm 1, \pm 2, \ldots$.

When diatomic molecules rotating about their center of mass (balance point) are considered, the reduced mass of the diatomic molecule is used in the same way as it was for the vibrating two mass problem. Diatomic molecules are not constrained to move in a fixed plane, rather, their plane of rotation can tilt at angles from 0 to π. This added degree of freedom changes the energy level spacing to

$$E = J(J+1)\frac{\hbar^2}{2I}$$

The energy levels diverge with increase in the quantum number $J = 0, \pm 1, \pm 2, \ldots$, giving the values $J(J+1) = 0, 2, 6, 12, \ldots$. The problem is again reduced to one of a fictitious mass, this time with a moment of inertia $I = \mu r^2$, rotating anywhere on the surface of a sphere. For absorption of a *resonance frequency* to occur, the rotational state must be changed by an increase in energy from one level to the next $J \rightarrow J+1$ with

$$E_{J+1} - E_J = 2, 4, 6, 8, \ldots \text{joules}$$

The spacing between resonance frequencies is $4 - 2 = 2, 6 - 4 = 2, 8 - 6 = 2, \ldots$ joules, by which we predict an absorption spectrum consisting of a series of lines of different frequencies separated by an energy of $2(\hbar^2/2I)$ (Fig. 18.3). Measuring $\bar{\nu}$ leads to I, which gives the bond length r through $I = \mu r^2$.

Vibrational and rotational spectroscopic transitions can occur simultaneously, leading to increased complexity of the experimental spectrum. Conversely, some

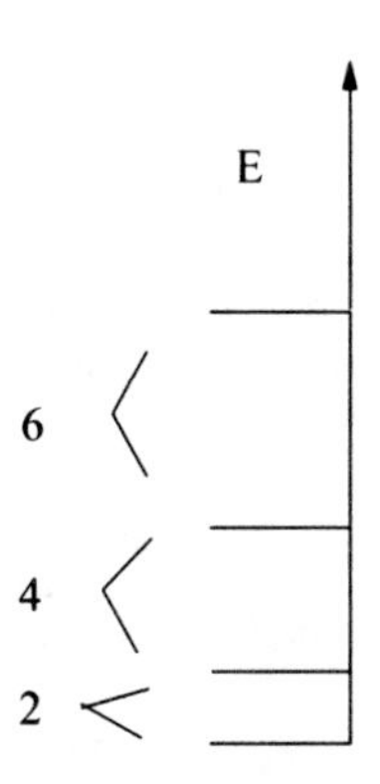

FIGURE 18.3 Energy levels within a simple rotor.

expected transitions are conspicuously absent from experimental spectra owing to selection rules which do not permit them. Despite these deviations from the simple models described here, much invaluable bond distance information has been obtained from microwave rotational spectra. For example, CO has a rotational spacing of $\bar{\nu} = 3.9 \text{ cm}^{-1}$ (in the far-infrared region), which leads to a bond length of 113 pm.

By the way, your microwave oven activates a rocking frequency in H_2O molecules, thereby transmitting energy (heat) to your morning coffee.

18.5 MICROWAVE SPECTROSCOPY: BOND STRENGTH AND BOND LENGTH

Given the input experimental data of the fundamental vibrational frequency ν_0 and the line separation $2\left(\frac{\hbar^2}{2I}\right)$ usually written 2B, in a vibration–rotation band of a diatomic molecule, the bond strength in terms of the Hooke's law force constant k_f and the bond length in picometers (pm) can be calculated.

18.6 ELECTRONIC SPECTRA

Electronic transitions require more energy than vibrational or rotational transitions. They produce spectral peaks that frequently fall in the UV or visible part of the electromagnetic spectrum. The $\pi \rightarrow \pi^*$ transition in ethene is an electronic transition from the highest occupied molecular orbital (HOMO) to the lowest unoccupied molecular orbital (LUMO). A simple model of electronic transition spectroscopy of this kind is that of a *free electron* trapped in a one-dimensional potential well, which is the length of the π system in unsaturated molecules. An MM calculation (Chapter 19) of the distances between the terminal carbon atoms in ethene and 1,3-butadiene gives the values 134 and 359 nm. The two electrons in ethene are in the lowest energy level with $E = h^2/8m_e l^2$. In the $\pi \rightarrow \pi^*$ transition, one of them progresses to the next higher energy state with $E^* = 4h^2/8m_e l^2$. The energy difference is

$$\Delta E = \frac{3h^2}{8m_e l^2} = \frac{3h^2}{8m_e(134)^2} = 5.56 \times 10^{-5}\left(\frac{3h^2}{8m_e}\right)$$

The 1,3-butadiene case has 4 electrons in two π bonds (Fig. 18.4); hence the lowest two levels are occupied and the transition is from the $n = 2$ to the $n = 3$ or $n^2 = 4$ to the $n^2 = 9$ level:

$$\Delta E = \left(\frac{9h^2}{8m_e l^2} - \frac{4h^2}{8m_e l^2}\right) = \frac{5h^2}{8m_e l^2} = \frac{5h^2}{8m_e(359)^2} = 3.88 \times 10^{-5}\left(\frac{3h^2}{8m_e}\right)$$

Even though the quantum number change in the second case is larger (5 vs. 3), the transition for 1,3-butadiene is the smaller of the two energy differences because of

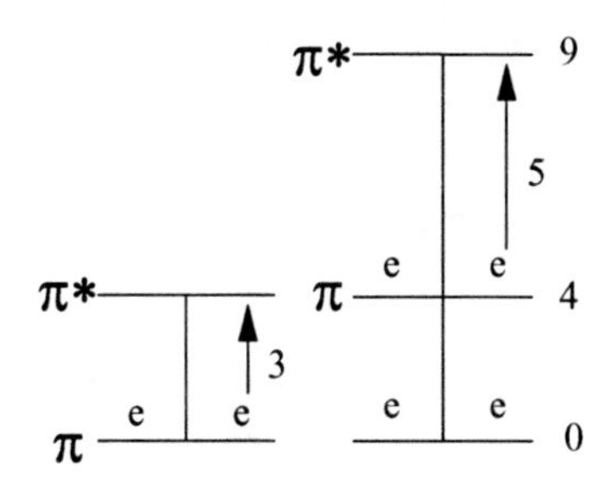

FIGURE 18.4 Electronic promotion in alkenes.

the square of the length in the denominator. The extended double-bond system in 1,3-butadiene decreases the energy and frequency and, because of the inverse relationship between ν and λ, it increases the wavelength. Experimentally, this predicted shift to longer λ is verified.

It is true in general that extended conjugated double-bond systems absorb at wavelengths that tend toward the longer wavelengths in the UV, and even into the visible region of the electromagnetic spectrum. This can be seen in the orange color of carotene from carrots and the red of lycopene, the principal natural coloring agent in tomatoes. Both of these molecules have long conjugated π electron chains and follow the qualitative trend predicted by the free electron model.

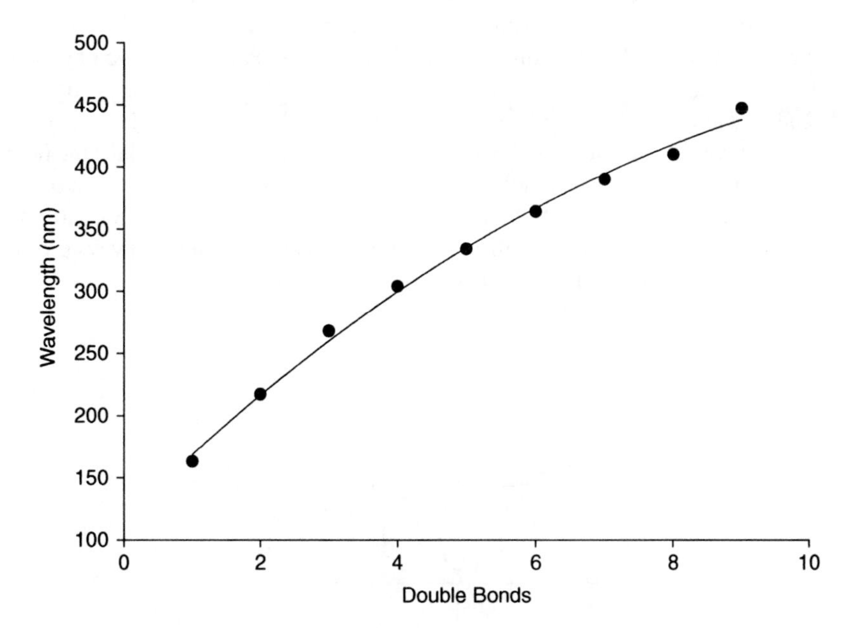

FIGURE 18.5 Absorption wavelengths of conjugated polyalkenes. The wavelengths range from the far ultraviolet well into the visible region. (Taken in part from Streitwieser, 1961.)

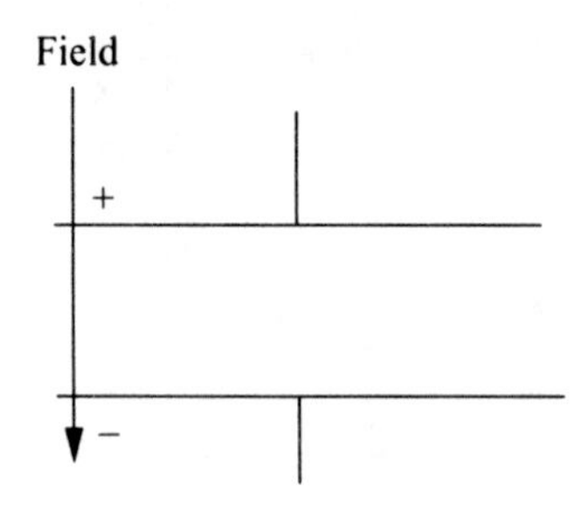

FIGURE 18.6 A charged parallel plate capacitor.

18.7 DIPOLE MOMENTS

If a parallel plate capacitor with a vacuum between its plates is brought up to a charge of Q coulombs by imposing a potential difference V volts between the plates, its *capacitance* is C_0 (Fig. 18.6):

$$C_0 = \frac{Q}{V}$$

where C_0, as the name implies, is analogous to the capacity of a container. It tells how much charge the capacitor can hold per volt of potential difference.

There are two ways of increasing the capacitance of a capacitor: (i) by making the plates larger or (ii) by allowing the space between the plates to be filled by some substance called a *dielectric* (Fig. 18.7). All substances have two poles, a positive pole and a negative electrical pole, when they are in an electrical field. These *dipoles* align themselves in opposition to the field between the plates, so they reduce the field. The potential difference between the plates is decreased, so more charge can be put into the capacitor to reestablish the original potential. Since the capacitor can hold more charge, its capacitance C_x with substance x between the plates is always greater than the capacitance C_0 with the apparatus pumped out to a vacuum.

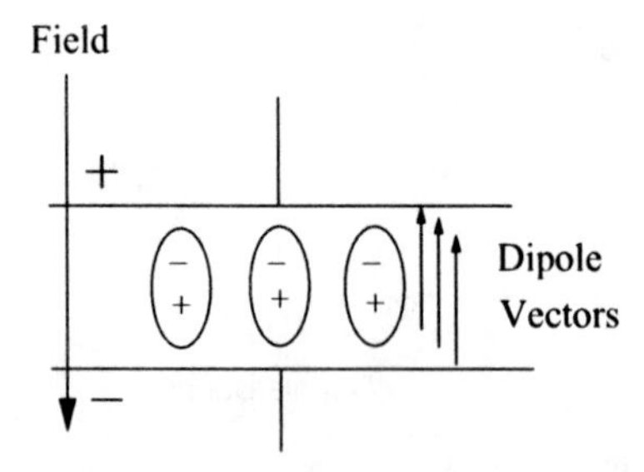

FIGURE 18.7 A charged capacitor with a dielectric.

The *dielectric constant* of substance x is defined as the ratio

$$\varepsilon_x = \frac{C_x}{C_0}$$

which is always greater than one. Dielectric constants are very different for different molecules; for example, the dielectric constants are 1.000272 for $H_2(g)$, 2.283 for benzene, and 78.0 for water at 298 K. We describe benzene as a *nonpolar* solvent and we describe water as a *polar* solvent.

The molar polarization of a substance P is related to the dielectric constant by

$$P = \frac{\varepsilon - 1}{\varepsilon + 2}\frac{\mathrm{M}}{\rho}$$

where M is the molar mass. This is sometimes called the *total molar polarization* because it is made up of two parts: the *distortion polarization* and the *orientation polarization*, $P_T = P_d + P_o$.

Distortion polarization exists for all substances, which is the reason that ε is never less than one. Consider a collection of atoms of a monatomic gas such as neon or argon. The electronic charge distribution within each atom is spherically symmetrical in the absence of a field; but in the presence of a field, the atomic charge distribution is distorted into something like the ellipses shown in Fig. 18.7, and a dipolar nature is *induced* in it.

Orientation polarization results from the permanent dipole brought about by an unsymmetrical charge distribution in the unperturbed molecule. A familiar example is HCl, which has a negative end (Cl) and a positive end (H).

The mathematical form of these two distinct types of polarization enables us to determine both. Distortion polarization is not a function of temperature,

$$P_d = \frac{4}{3}\pi N_A \alpha$$

but orientation polarization is inversely dependent upon T:

$$P_o = \frac{4\pi N_A}{9k_B T}\mu^2$$

where N_A is the Avogadro number and k_B is the Boltzmann constant. One can rearrange the equation for the total polarization as a function of $1/T$ to emphasize its linear nature:

$$P_T = \frac{4\pi N_A \mu^2}{9k_B}\frac{1}{T} + \frac{4}{3}\pi N_A \alpha = \frac{4\pi N_A}{3}\left[\left(\frac{\mu^2}{3k_B}\right)\frac{1}{T} + \alpha\right]$$

Plotting P_T, which we obtain from measured values of ε, against $1/T$ gives a linear function with a slope of $(4\pi N_A/3)(\mu^2/3k_B)$. This gives us μ and enables us to subtract the temperature-dependent term from P_T at any specific temperature to determine $\frac{4}{3}\pi N_A\alpha$ and thus α. (Note that μ is not the reduced mass in this context.)

The value μ is the twisting *moment* on a dipole in a field. The moment is the sum of each charge times the lever arm separating it from the center of rotation of the molecule $\sum_i q_i r_1$. In the simplest case of a diatomic molecule, the charges are equal in magnitude and μ is equal to the moment of one charge times the entire bond length:

$$\mu = |qr_1| + |qr_2| = q(r_1 + r_2) = qr$$

The *polarizability* α is a proportionality constant telling how much of a dipole can be induced in a given molecule by a given field. It is inversely related to the tightness with which a system of nuclei hangs onto its electrons and is widely used in organic chemistry to predict the ease with which electrons can be shifted to or away from a reaction site in the course of a reaction.

18.7.1 Bond Moments

It is useful to break molecular moments into contributions from each bond. The dipole moment of the molecule is, to a good approximation, the vector sum of individual moments of its bonds. We expect that chlorine will draw electrons toward it and constitute the negative end of CH_3Cl. By the principle of bond moments, we predict that *cis*-1,2-dichloroethene has a permanent dipole moment while the *trans* isomer has none. The values of μ given below the isomeric structures in Fig. 18.8 are experimentally determined.

The definition of dipole moment gives us both a convenient unit to express μ and a way of characterizing the polarization of bonds like the H–Cl bond. Since $\mu = qr$ and we know that the electric charge is $q = 4.80 \times 10^{-10}$ esu, a typical bond that is

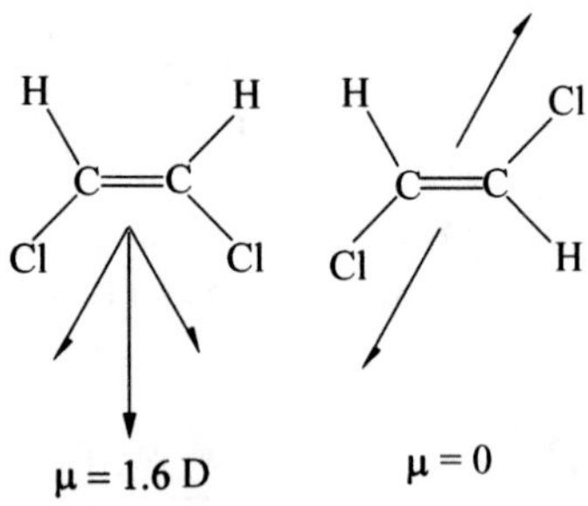

FIGURE 18.8 The total dipoles of two dichloroethene isomers.

1 Å in length should have

$$\mu = qr = \left(4.80 \times 10^{-10}\right) \times \left(1.00 \times 10^{-8}\right) = 4.80 \times 10^{-18} \text{ esu cm}$$

This is called a debye, D. In SI units of coulomb meters, 1 D = 3.338×10^{-30} C m.

We have just assumed that the negative charge is at one end of the molecule and the positive charge is at the other, that is, that the bond is completely ionic. Molecules are not completely ionic and many of them are not completely covalent either. The intermediate case is a compromise called a *polar molecule*, which has some charge separation but which is not totally ionic. One can obtain a numerical parameter telling where this compromise lies by computing the dipole moment expected for a perfectly ionic bond and comparing it with the measured moment. The ratio of the measured moment to the moment calculated assuming complete charge separation leads to the *% ionic character* of the actual molecule.

18.8 NUCLEAR MAGNETIC RESONANCE (NMR)

Some atoms, including hydrogen H, have the property of nuclear spin, which produces a small magnetic field. In the absence of an external field, the spin magnetic fields of nuclei are randomly oriented; but in a magnetic field, they orient themselves with or against the external field according to their spin quantum numbers, $\pm\frac{1}{2}$ in the case of the proton H. One orientation is energetically favorable, whereas the opposite orientation is unfavorable. Twofold energy splitting occurs:

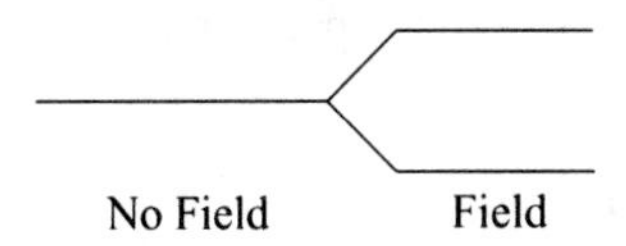

As usual, the lower energy level is more populous but the interaction between the spin field and the external field is weak, so energy splitting is very small (of the order of a few thousandths of a calorie per mole). A Boltzmann calculation for this minute energy separation shows that the lower level has a population that is only a few nuclei per hundred thousand greater than that of the upper level.

In principle, NMR is the same as other kinds of spectroscopy, but important technical differences exist. In absorption spectroscopy, incident electromagnetic radiation is varied until a resonant frequency coincides with the energy-level separation necessary to promote the system from a lower quantum state to a higher state. Emission spectroscopy, which is preferred for some purposes, involves the reverse process, emission of radiation occasioned by the fall of particles from a higher state to a lower state. If both upper and lower energy states are appreciably populated, electromagnetic radiation can induce both absorption and emission of energy.

In NMR, resonance is found by varying the magnetic field until the energy separation matches an input frequency. At a resonance frequency, protons are both absorbing and emitting radiation. In addition to a powerful electromagnet and radiation source, an NMR spectrograph is equipped with a receiver to detect and record the resonance emission frequencies from the sample. A record of the various resonances in a molecule (there will normally be more than one) is its NMR spectrum.

Examination of the details of molecular structure by NMR is possible because of *chemical shifts*. The field under which a given proton acts is primarily the external field, but it is slightly *shielded* by its surroundings. The electron density around H (or other nuclei like ^{13}C) is a function of its chemical environment, particularly the electronegativity of neighboring atoms. Thus a CH_2 group absorbs and emits radiation at a different field strength than a CH_3 group. Chemical shielding is normally recorded in relative terms to minimize differences from one experimental setup to another. One compound, tetramethylsilane TMS, is usually selected as a reference point and other resonances are reported relative to it in units of parts per million difference between the resonance frequency brought about by hydrogens in the sample and those in TMS. The chemical shift is denoted δ (ppm). Tables of the approximate chemical shift of various groups are available.

Methyl hydrogens have a chemical shift of about 1 (ppm), CH_2 hydrogens have $\delta \cong 2$ (ppm), and COOH hydrogens have $\delta \cong 10$ (ppm). NMR spectra, like IR spectra, are unique and serve as "fingerprints" of compounds. If a pure unknown has an NMR spectrum that is identical to an authentic sample, the substances are identical. NMR is, however, more than an expensive way of carrying out qualitative analysis. NMR spectra provide information on the internal details of the molecules examined.

Ethanol CH_3CH_2OH shows three peaks at low resolution because of the three distinct kinds of protons CH_3, CH_2, and OH. The peaks are about equally spaced, but they can be identified because the area under each peak is directly proportional to the number of protons producing the peak, in the ratio 3:2:1.

18.8.1 Spin–Spin Coupling

Spin–spin coupling refers to the minute interaction between the spin field of a proton and the spin fields of adjacent protons. Normally, a proton with one immediate neighbor can couple with the adjacent spin in two ways: $\uparrow\downarrow$ and $\downarrow\uparrow$.[1] A proton with two neighbors can couple in three ways: $\uparrow\uparrow$, $\uparrow\downarrow$ $\downarrow\uparrow$, and $\downarrow\downarrow$. A proton with three neighbors can couple in four ways. We arrive at a peak multiplicity of $n + 1$, where n is the number of neighboring protons. Arrow diagrams show that the intensity of the split peaks is in the ratio 1:1, 1:2:1, 1:3:3:1, and so on (Fig. 18.9).

Extrapolating these simple examples to more complicated molecules gives an idea of the use of NMR in experimental studies of molecular structure. Very powerful magnetic fields are necessary to separate proton and other nuclear peaks in complicated molecular structures. Combination of NMR with the times necessary for various

[1] The case of ethanol is an exception because of fast proton transfer at the OH site, yielding only one peak.

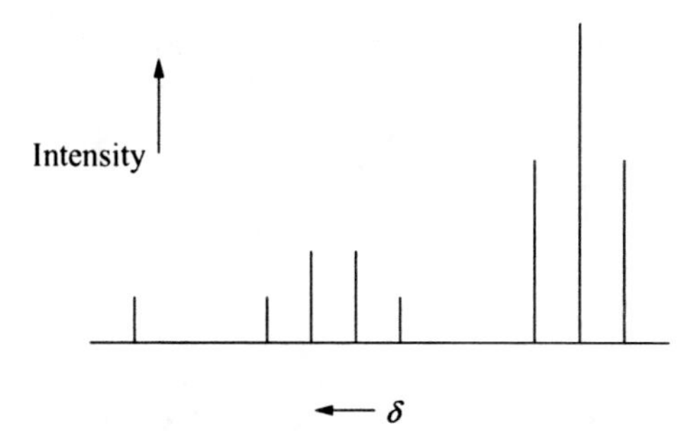

FIGURE 18.9 Schematic NMR spectrum of ethanol, CH_3CH_2OH. The CH_3 peak on the right is split into a triplet because of the two neighboring protons on $—CH_2—$.

nuclei to relax to their equilibrium states after being pulsed by frequency excitation leads to powerful diagnostic methods known as *magnetic resonance imaging* (MRI).

18.9 ELECTRON SPIN RESONANCE

Electrons also have the property of spin and, if spins are unpaired, as in free radicals, electron spin gives rise to an electron spin resonance ESR spectrum, which is analogous to NMR. ESR spectra are useful in the characterization of free radicals, in their detection as reactive intermediates, and in specifying the electron probability distribution within the radical. Free radicals have been implicated in carcinogenesis and in aging, but other free radicals appear to act as sweepers reacting with destructive species before they can harm a host organism.

PROBLEMS AND EXAMPLES

Example 18.1 The Bond Strength and Bond Length of Carbon Monoxide

Experimental results are that a low-resolution peak can be found for $^{12}C^{16}O$ centered around 2142 cm^{-1}, which separates at high resolution to show a peak separation of 3.8 cm^{-1} . Find the bond length and the force constant. The superscripted 12 and 16 indicate that these data are for the predominant isotopes of C and O. Other, far smaller peaks can be found for the other isotopes. Because this is the experimental information for $^{12}C^{16}O$, the isotopic weights can be treated as integers.

Solution 18.1

1. One needs the reduced mass for both calculations. For $^{12}C^{16}O$ the reduced mass is

$$\mu = \frac{12.00\,(16.00)}{12.00 + 16.00} 1.661 \times 10^{-27} = 1.139 \times 10^{-26}\ \text{kg}$$

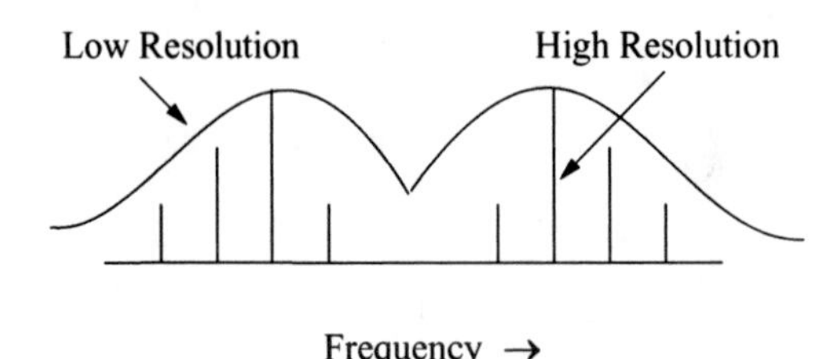

FIGURE 18.10 Schematic diagram of a vibration–rotation band. The low-resolution vibrational band contains many rotational lines that may be discernable at high resolution. The separation between rotational lines is given the term 2B.

2. The force constant is extracted from the classical expression for simple harmonic motion:

$$\nu = \frac{1}{2\pi}\sqrt{\frac{k_f}{m}}$$

Given the experimental measurement of $\bar{\nu} = 2142\ \text{cm}^{-1} = 2.142 \times 10^5\ \text{m}^{-1}$, the frequency is

$$\nu = c\bar{\nu} = \left(2.998 \times 10^8\ \text{m s}^{-1}\right), \qquad \left(2.142 \times 10^5\ \text{m}^{-1}\right) = 6.422 \times 10^{13}\ \text{s}^{-1}$$

From the harmonic oscillator approximation

$$\nu = \frac{1}{2\pi}\sqrt{\frac{k_f}{m}}$$

we get

$$\frac{k_f}{m} = 4\pi^2\nu^2 = 1.628 \times 10^{29}$$

Substituting the reduced mass of $^{12}C^{16}O$ for m, we obtain

$$k_f = \mu \times \left(4\pi^2\,\nu_0^2\right) = \left(1.138 \times 10^{-26}\right) \times \left(1.628 \times 10^{29}\right) = 1853\ \text{N m}^{-1}$$

3. The bond length is found from the moment of inertia of the molecule modeled as a rigid rotor. The energy spacing is

$$2B = \frac{2\hbar^2}{2I} = \frac{h^2}{4\pi^2 I}$$

The spectral input to this calculation is the line separation

$$2B = 3.8604\ \text{cm}^{-1} = 386.04\ \text{m}^{-1}$$

so $B = 193.02\ \text{m}^{-1}$. This is multiplied by c to obtain the frequency in s^{-1} and bond length r in pm:

$$I = \frac{h}{4\pi^2(2Bc)} = \frac{2.799 \times 10^{-44}}{B} = \frac{2.799 \times 10^{-44}}{193.02} = 1.450 \times 10^{-46}$$

$$r^2 = \frac{1.450 \times 10^{-46}}{1.138 \times 10^{-26}} = 1.27 \times 10^{-20}$$

$$r = 1.129 \times 10^{-10}\ \text{m} = 112.9\ \text{pm}$$

Example 18.2 The Dipole Moment of Sulfur Dioxide

Sulfur dioxide, SO_2, has a total molar polarization of 68.2 $\text{cm}^3\ \text{mol}^{-1}$ at 298 K and 56.0 $\text{cm}^3\ \text{mol}^{-1}$ at 398 K. What is the dipole moment of SO_2?

Solution 18.2 The inverse of the lower temperature is 0.00336 and the inverse of the upper temperature is 0.00251. The molar polarizations are 68.2 and 56.0 $\text{cm}^3\ \text{mol}^{-1}$. The slope of the curve of molar polarization vs. the inverse of T is $1.44 \times 10^4\ \text{cm}^3\ \text{K mol}^{-1}$. (Probably for historical reasons, the nonstandard unit system of erg cm is often used.) The experimental slope is related to the dipole moment as

$$\text{slope} = \frac{4\pi N_A \mu^2}{9k_B} = 1.44 \times 10^4$$

Solving for μ^2, we get

$$\mu^2 = \frac{9k_B}{4\pi N_A}\text{slope} = 1.641 \times 10^{-47}\left(1.44 \times 10^4\right)$$

It is convenient at this point to change the units of k_B from J K^{-1} to erg K^{-1}. This requires multiplication by 10^7 to give

$$\mu^2 = 1.641 \times 10^{-40}\left(1.44 \times 10^4\right) = 2.363 \times 10^{-36}$$

$$\mu = \sqrt{2.363 \times 10^{-36}} = 1.54 \times 10^{-18} = 1.54\ \text{D}$$

Problem 18.1

The molecule $^1H^{35}Cl$ absorbs light at $\lambda = 2.991 \times 10^{-4}\ \text{cm} = 2.991 \times 10^{-6}\ \text{m}$. Taking $^1H^{35}Cl$ to be a harmonic oscillator, what are the differences in energy between

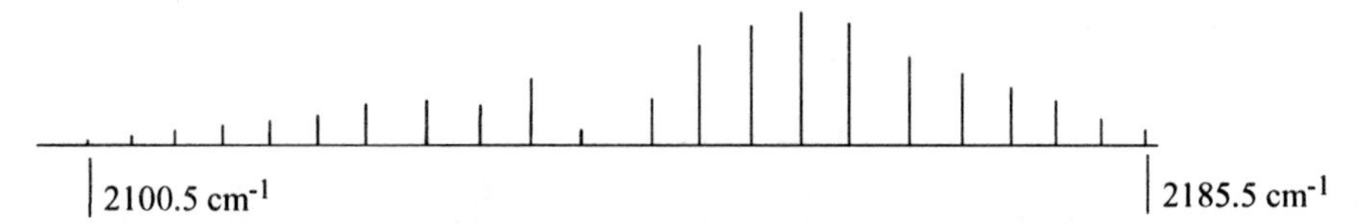

FIGURE 18.11 The vibration–rotation spectrum of CO. Only one vibrational absorption and the accompanying rotational absorptions are depicted.

the levels in the parabolic Hooke's law energy well? (The superscripted 1 and 35 are the isotopic numbers. The mass of each isotope can be taken as an integer.)

Problem 18.2

What is the force constant for the diatomic molecule $^{1}H^{35}Cl$? Take advantage of the fact that it absorbs light at $\lambda = 2.991 \times 10^{-4}$ cm $= 2.991 \times 10^{-6}$ m. Assume that the molecule is a harmonic oscillator.

Problem 18.3

It is reasonable to suppose that the force constant governing a stretch has units of force (newtons) per unit displacement (how long the stretch is). Show that the units of k_f are Nm^{-1}.

Problem 18.4

In early experiments on HCl, line separation in the rotational part of its spectrum was found to be $2\tilde{B} = 20.794\ \text{cm}^{-1}$. What is the length of the H—Cl bond?

Problem 18.5

A somewhat idealized microwave spectral band of carbon monoxide CO is shown in Fig. 18.11. The actual band would show some anharmonic distortion. Find the vibrational frequency. Locate the transition from the 0 to the 1 rotational states. Calculate the force constant of the CO bond.

Comment: Notice that the speed of light (electromagnetic radiation) is used in units of cm s^{-1} because the frequency is given in cm^{-1}. The reduced mass is in kg because the force constant is in Nm^{-1}.

Problem 18.6

From the information is Problem 18.5, estimate the CO bond length.

Problem 18.7

The rotational partition function at 298 can be written as an integral:

$$q_{\text{rot}}(T) = 2(J+1)\,e^{-\Theta_{\text{rot}} J(J+1)/T}\,dJ$$

where Θ_{rot} is a parameter called the rotational temperature, Show that $q_{\text{rot}} = T/\Theta_{\text{rot}}$.

Problem 18.8

The fraction of molecules in a rotational state is given by the ratio of the partition function for that state relative to the total partition function. But we know that $q_{\text{rot}} = T/\Theta_{\text{rot}}$ from the previous problem, so

$$\frac{Q_J}{Q_{\text{total}}} = \frac{2(J+1)\,e^{-\Theta_{\text{rot}} J(J+1)/T} dJ}{T/\Theta_{\text{rot}}} = \frac{2(J+1)\,\Theta_{\text{rot}} e^{-\Theta_{\text{rot}} J(J+1)/T} dJ}{T}$$

Plot Q_J/Q_{total} as a function of J for NO which has $\Theta_{\text{rot}} = 2.34$ at $T = 298$ K.

Problem 18.9

The dipole moment μ is a twisting moment in an electrical field between particles of charge q separated by a distance r:

$$\mu = qr$$

The unit of $q\,r$ is coulomb meters. The unit charge in atomic problems is the charge on a proton, 1.6019×10^{-19} C. The Br–F bond length is 176 pm. What would the dipole moment for the molecule BrF be if it were completely ionic $Br^{+}F^{-}$?

Problem 18.10

If the actual dipole moment of Br–F is only 1.42 D, what is the percent ionic character of the bond? Is the bond predominantly ionic or covalent?

Problem 18.11

The molar polarization of bromoethane was measured at five different temperatures with the following results:

T	205	225	245	265	285
P_T	104.2	99.1	94.5	90.5	80.8

What are the polarizability and dipole moment of bromoethane?

Problem 18.12

Some polyenes, ... C=C–C=C–C=C–C ..., and so on, are colored (lycopene in tomatoes, for example). Use the particle in a one-dimensional box to predict how long the conjugated chain must be for its absorption frequency to be in the visible region.

Problem 18.13

A reaction product is thought to be either methyl acetate or ethyl formate. NMR analysis produced the spectrum shown below. Which is it?

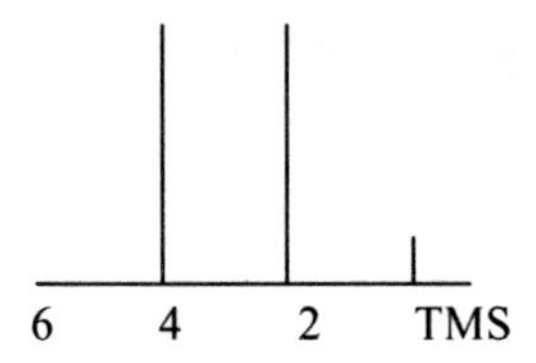

19

CLASSICAL MOLECULAR MODELING

Although the chemical bond is a result of quantum mechanical interactions of electrons with nuclei and with each other, the atom itself is large enough to be treated classically. One way of understanding and predicting chemical properties is by treating the molecule as an aggregation of atoms bound by classical bonds and interacting with each other in a classical way. Such a study is called *molecular mechanics*, MM.

19.1 ENTHALPY: ADDITIVE METHODS

It has long been known that extension of an alkane chain by one CH_2 group

$$CH_3(CH_2)_nCH_3 \rightarrow CH_3(CH_2)_{n+1}CH_3$$

brings about a *decrease* in the $\Delta_f H^{298}$ of about 5 kcal mol^{-1} = 21 kJ mol^{-1} (Fig. 19.1). For example, if we subtract 21 kJ mol^{-1} from $\Delta_f H^{298}$(ethane) = -84 kJ mol^{-1}, we get (correctly) $\Delta_f H^{298}$(propane) = -105 kJ mol^{-1}. Thus we can call –21 kJ mol^{-1} a CH_2 *group enthalpy* for the extension of ethane to propane. The group enthalpy is *transferable* to give the enthalpy change for extension of other alkanes to an alkane larger by one CH_2 group.

If this works for CH_2, why not try it with other alkyl groups? The enthalpy of the CH_3 group must be one-half the enthalpy of formation of the ethane molecule CH_3CH_3, $\Delta_f H^{298}$(ethane) = –84 kJ mol^{-1}. Hence we take the CH_3 group enthalpy

Concise Physical Chemistry, by Donald W. Rogers

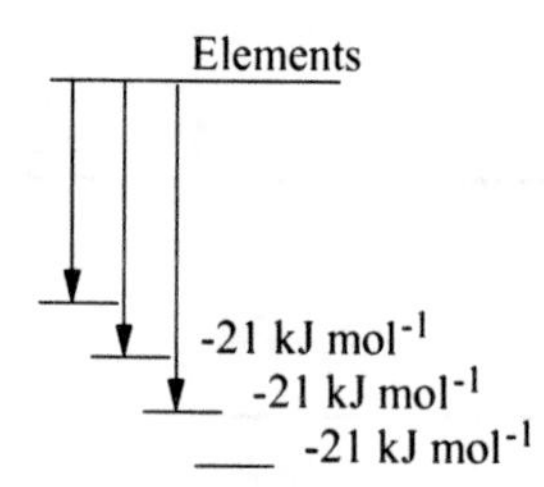

FIGURE 19.1 Enthalpies of formation of "adjacent" *n*-alkanes.

as –42 kJ mol^{-1}. Provided that these enthalpies are transferable, the enthalpy of formation of *n*-pentane $CH_3(CH_2)_3CH_3$, for example, is estimated as $2(-42) + 3(-21) = 147$ kJ mol^{-1}. The calculation agrees with the experimental value:

$$\Delta_f H^{298}\,(CH_3(CH_2)_nCH_3) = 147\,\text{kJ mol}^{-1}$$

to a precision of ± 1 kJ mol^{-1}. Although the agreement is not always this nice, the procedure is very encouraging.

What of groups other than alkyl groups? Can this *group additivity* strategy be extended to include the CH group and the C atom in branched alkanes like 2-methylbutane (isopentane) and 2,2-dimethylpropane (neopentane)? The answer is yes, and this strategy has been extended to cover many functional groups in organic molecules including oxygenated, nitrogenous, and halogenated hydrocarbons (Cohen and Benson, 1993).

19.2 BOND ENTHALPIES

Molecular enthalpy can be segregated in other ways. One familiar way involves associating an enthalpy with each bond in the molecule. For example, inserting a CH_2 group into an *n*-alkane requires breaking the backbone of the alkane molecule at a cost of one C—C bond followed by forming two C—C bonds and increasing the number of C—H bonds in the molecule by two:

$$R{-}CH_2{-}CH_2{-}R \rightarrow R{-}CH_2{-}CH_2{-}CH_2{-}R$$

The sum total of the process is a gain of one C—C bond and two C—H bonds. If we associate –348 kJ mol^{-1} with the C—C bond and –413 kJ mol^{-1} with the C—H bond relative to the isolated atoms C and H, the insertion entails a change in total bond enthalpies of

$$\text{C—C} + 2(\text{C—H}) = -348 + 2(-413) = -1174\,\text{kJ mol}^{-1}$$

Why such a big number? This big number arises because theoretical folks find it convenient to use a reference state that is different from the thermodynamicist's

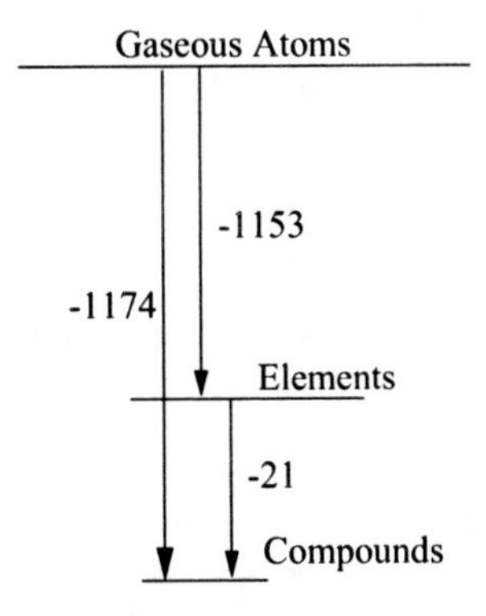

FIGURE 19.2 Bond enthalpies calculated in CH_2, from the reference state of gaseous atoms (top), and relative to elements in their standard state (H_2(g) and C(graphite)). Not to scale.

standard state of the elements in their most stable form at 298 K. To translate from one reference state to the other, we need to know their enthalpy differences. These values have been measured with care (and some difficulty). They are $\Delta_{\text{atomization}} H^{298}(H_2) = 436\,\text{kJ mol}^{-1}$ and $\Delta_{\text{atomization}} H^{298}(C_{gr}) = 717\,\text{kJ mol}^{-1}$. These larger enthalpies must be subtracted from the bond enthalpies relative to the atoms to find $-1174-(-1153) = -21\ \text{kJ mol}^{-1}$, the sum of bond enthalpies in CH_2 relative to the thermodynamic standard state of H_2(g) and C(graphite) (Fig. 19.2).

This result, relative to a new standard state, is the same as the one used with success in Section 19.1. The rationale for what seems to be a rather exotic standard state will be clearer in the next chapter. It is also important to remember that without correction factors for molecular deformations of the real molecule relative to the additive model we have postulated here, either summing appropriate bond enthalpies or summing simple group enthalpies gives a *strainless* molecule.

19.3 STRUCTURE

Given that the hydrogen atoms in methane repel one another, the only reasonable structure we can assign to methane is that of a tetrahedron in which four bound H atoms achieve maximum separation in 3-space about the central C atom:

Using this simple structural symmetry and an arbitrary bond length (the length of a wooden peg, perhaps), we can construct a unique "ball-and-stick" model of methane and, by extension, very many models of alkanes (Fig. 19.3). By the nature

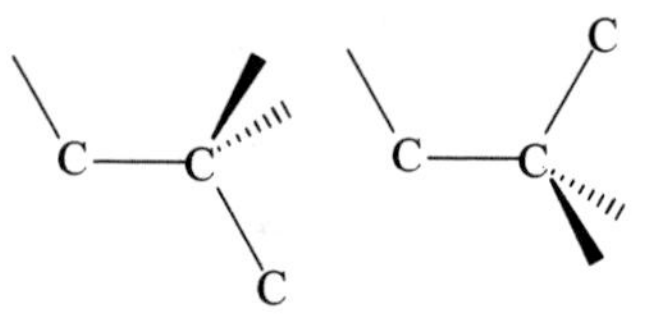

FIGURE 19.3 Structurally distinct alkane conformers resulting from the tetrahedral symmetry of carbon.

of tetrahedral symmetry about all C atoms, we soon find that our models include numerous *conformers* of the higher members of the alkane series.

We have the force constants k_f of the C—H bond from spectroscopic studies (Sections 18.1 and 18.2). Using these force constants, we can relieve the arbitrary nature of the C—H bond lengths in CH_4 by starting at some reasonable value for the bond length r and calculating the energy of the CH_4 molecule. The result will be, no doubt, very high. Making systematic small changes in r (Problem 18.4) and repeating the process, perhaps many times, the calculated energy can be brought to a minimum value. At the optimum bond length, each H atom is as close as it can get to the bottom of its Hooke's law potential well while still respecting its neighbors. In optimizing E, we have also optimized r. This is the best value of the C—H bond length in methane we can get from the Hooke's law force parameter k_f we have chosen.

This raises the question of the best force parameter we can choose. Will it be exactly the spectroscopic constant? What if there is no spectroscopic constant? Experience has shown that all things considered, the best MM force parameters express knowledge derived by fitting bond lengths and stretching energies to spectroscopy, thermodynamics, X-ray crystallography, and that old favorite "chemical intuition."[1] MM *parameters* have evolved away from spectroscopic force *constants*; the two are not the same. MM parameters can be proposed, and then they can be revised in the light of new experimental results. Spectroscopic force constants are determined by a specific category of experiments. They can be made more accurate, but they do not change.

We now have the tetrahedral bond angles and the bond lengths of CH_4. From this we can express the structure of CH_4 in Cartesian coordinate locations of all of the atoms in the molecule. There is nothing more we can know about the geometry of CH_4.

The method can be elaborated to determine the complete MM structure of any molecule, provided that we have the force parameters for all of the structural forces in the molecule.

There is the rub. It turns out that there are many subtle forces operative in molecules larger than CH_4. It is the task of anyone who wishes to find the structure of a complex molecule to seek out all of the forces within it and to determine the force parameters of the stretch, bend, torsional deformations, van der Waals, or other interactions. One is not advised to attempt this daunting task. Fortunately, N. L. Allinger and coworkers

[1] "Chemical intuition" is not as silly as it sounds. Like the scriptures, it is the accumulated knowledge of very many years of observation and experience.

(Nevins et al. 1996 a, 1996 b) and other groups have devoted four decades to the job, resulting in rather large MM computer programs that have evolved with time. Among them is the Allinger series MM1, MM2, ..., MM4, with some variants in between.

19.4 GEOMETRY AND ENTHALPY: MOLECULAR MECHANICS

Suppose that we have suitable force parameters for the C—H and the C—C bonds in alkanes and r_0 at the minima of the Hooke's law potential energy wells (Fig. 18.2). We should be able to construct models of ethane and higher alkanes and find the structure and energy that best satisfy the k_f values.

But not quite. Like the members of a large family, the bond–atom combinations in a molecule do not always agree. The final structure of the molecule is a compromise in which no bond is exactly at its ideal length r_0 but no bond is too far removed from it. The distance between the idealized bond length r_0 and the real one r is assumed to follow Hooke's law, $F = -k_f(r - r_0)$. The restoring force F results in a parabolic potential energy well with the potential energy defined as $V = 0$ at its minimum:

$$V(r - r_0) = k_f \tfrac{1}{2}(r - r_0)^2$$

Like a large family, the final compromise engenders some strain. The energy of the real molecule is above the energy of a hypothetical strainless molecule by the summation of the strain energy experienced by all its members at the compromise position. The energy of the real molecule is at least

$$E = E_{\text{strain-free}} + \sum_{\text{all bonds}} V(r - r_0)$$

Bond stretching is not the only energy that enters into the final compromise. Chemical bonds can also be bent through an angle θ. It is reasonable to assign a Hooke's law potential energy for bending. If the compromise structure involves some bond bending, the potential energy of bending must be included in the sum

$$E = E_{\text{strain-free}} + \sum_{\text{all bonds}} V_{\text{stretch}}(r - r_0) + \sum_{\text{all simple angles}} V_{\text{bend}}(\theta - \theta_0)$$

Intramolecular interactions result in forces called van der Waals forces, and molecules may suffer dihedral angular distortion called *torsional forces*. These give rise to potential energies V_{vdW} and V_{tor}, respectively. The sum of all potential energies over a hypothetical strainless enthalpy is

$$E = E_{\text{strain-free}} + \sum V_{\text{stretch}} + \sum V_{\text{bend}} + \sum V_{\text{vdW}} + \sum V_{\text{tors}}$$

Intramolecular interactions are not independent of one another. For example, bending of a stretched bond is not the same as bending of a bond at its equilibrium length r_0. Conversely, stretching of a bent bond is not the same as stretching of a bond at $\theta = \theta_0$. A stretch-bend term $V_{\text{s-b}}$, called a *cross term*, is added. Other terms, especially electrostatic terms, have been added (somewhat reluctantly) as necessary to achieve closer and closer agreement with known experimental data. The collection of terms and the parameters that produce them is called a *force field*. The MM force field has become larger as new experimental evidence has been added to the basis set and as a larger set of molecular problems has been attempted. Allinger et al. (1996, 2010) have augmented the strainless energy by an 11-term equation in his programs MM3 and MM4.

19.5 MOLECULAR MODELING

The term *molecular model* includes ball-and-stick physical models used as visual aids to facilitate thinking about the invisible world of the molecule. In contemporary usage, however, the term usually denotes a mathematical model, possibly generating a computer output set of Cartesian triples, each of which specifies the position of an atom in the model.

The most fundamental molecular property obtained from a molecular model is its geometry. Geometry is found by optimization of an estimated input geometry. One of many possible geometry optimization procedures is shown in Example 19.1.

19.6 THE GUI

Input files for more complicated molecules can be constructed using a point-and-click *graphical user interface* (*GUI*)—for example, PCModel©. Having the optimized model in mathematical form permits use of the coordinate set to draw two- or

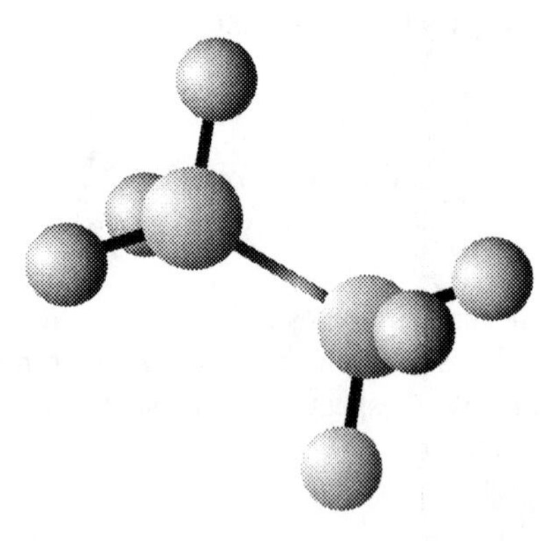

FIGURE 19.4 Visualization of the output for the ethane molecule (PCModel 8.0©).

three-dimensional diagrams (*visualization*) on the computer screen to appeal to the pictorial thinking favored by many chemists. Indeed, the drawing may be made to mimic the familiar wooden ball-and-stick models, with the difference that we expect it to be more accurate.

19.7 FINDING THERMODYNAMIC PROPERTIES

Suppose that, with some labor, we have arrived at a collection of force parameters sufficient to describe a real molecule at its optimized geometry. Suppose further that we have arrived at a set of bond enthalpies that enables us to write idealized "heats" of formation as (a) a sum of bond energies SBE derived from a set of molecules taken to be *strainless* and (b) another sum of bond energies NBE derived from a set of molecules taken to be *normal*. Some of the resulting information available from a computational run using MM4 for ethane is shown in File 19.1. The difference

```
HEAT OF FORMATION AND STRAIN ENERGY CALCULATIONS

                         (UNIT = KCAL/MOLE)
                         (  #  = TRIPLE BOND)

         NORMAL (BE) AND STRAINLESS (SBE) ENTHALPY OF INCREMENTS
                  (CONSTANTS AND SUMS OF INCREMENTS)

     BOND OR STRUCTURE     NO      ----NORMAL----          --STRAINLESS--
     C-C SP3-SP3            1    -89.2005    -89.2005    -84.0718     -84.0718
     C-H ALIPHATIC          6   -105.7262   -634.3572    -98.4375    -590.6250
     C-METHYL (ALKANE)      2     -0.0964     -0.1928     -0.5379      -1.0758
     -----------------------------------------------------------------------
                                      NBE =  -723.7505      SBE =   -675.7726

     HEAT OF FORMATION OF ELEMENTS (HATOM) =        -655.9401 KCAL/MOLE
     IN STANDARD STATE FROM ATOMS

     MOLAR HEAT CONTENT OF COMPOUND (MH)   =          48.6328 KCAL/MOLE
     (STERIC + ZPE + THERMAL ENERGIES)


     PARTITION FUNCTION CONTRIBUTION (PFC)
          CONFORMATIONAL POPULATION INCREMENT (POP)       0.00 (ASSUMED VALUE)
          TORSIONAL CONTRIBUTION (TOR)                    0.00 (ASSUMED VALUE)
                                                  ----------------
                                                PFC =     0.00 (ASSUMED VALUE)

     HEAT OF FORMATION (HFN) AT  298.2 K         =      -19.18 KCAL/MOLE
     (HFN = MH + NBE + PFC - HATOM)


     STRAINLESS HEAT OF FORMATION FOR SIGMA SYSTEM (HFS) =          -19.83
     (HFS = SBE - HATOM)

     INHERENT SIGMA STRAIN (SI) = HFN - HFS                           0.65

     SIGMA STRAIN ENERGY (S) = POP + TOR + SI                         0.65
```

FILE 19.1 Partial MM4 enthalpy output for ethane. (Units are kcal/mol.)

between SBE and HATOM [the enthalpy input to drive hydrogen and graphite into the atomic state (Section 19.2) for the strainless molecule] is

$$\Delta_f H^{298}(\text{strainless}) = -675.77 - (-655.94) = -19.83\,\text{kcal/mol}$$

A similar sum of normal bond enthalpies gives a different result for the normal molecule:

$$\text{NBE} + \text{HATOM} = \Delta_f H^{298}(\text{normal}) + \text{MH} = -67.81 + 48.63 = -19.18\,\text{kcal/mol}.$$

The new term MH (molar heat content) arises from a slightly different way of defining the normal bond energies; they should be represented by an arrow going all the way to the bottom of the energy wells in the real molecule. Positive enthalpies—consisting primarily of the zero point energies of vibration (Section 18.2), with a contribution from the *steric energy*—and small statistical factors add up to a positive correction of 48.63 kcal/mol. The difference between HFS and HFN is an accumulated strain energy:

```
INHERENT SIGMA STRAIN (SI) = HFN - HFS                    0.65
```

The partial MM4 output for ethane (File 19.1) illustrates some other features of the MM procedure and the full file provides more. See `STATISTICAL THERMODYNAMICS ANALYSIS` from the MM4 output and Nevins et al. (1996 b, pp. 703–707) for more detail.

19.8 THE OUTSIDE WORLD

In any molecular modeling enterprise, we seek to compare physical properties predicted by the model with the same properties measured by experiment. One obvious choice is bond lengths as measured by X-ray crystallography. Agreement is usually good but not perfect. Even assuming that the impossible task of removing all defects from the model has been achieved, the X-ray crystallographer and the MM modeler are not really looking at the same thing. The idea of a model as developed so far implies an isolated gas-phase species, but molecules in crystals are subjected to powerful forces holding them in the crystalline state due to the sum of forces exerted by their neighbors.

Somewhat surprisingly, these forces do not influence bond lengths very much. Agreement between X-ray bond lengths and MM4 lengths is usually within about 0.002 Å. Simple bond angles, being weaker, are distorted by a few degrees, and torsional angles, being weaker still, are distorted more. X-rays are scattered by electrons, so crystallographic structures represent the centers of electron densities which tend to be the same as the location of the nuclei, but which may be displaced in some chemical bonds. For example, the difference between the C—H bond length determined by X-ray scattering and that determined by neutron scattering (which yields the nuclear locations) is more than 0.1 Å in some cases (Allinger, 2010). Geometric parameters from different sources should not be used indiscriminately. Motion of the entire molecule in the crystalline lattice is called *rigid-body motion*. The amplitude of rigid-body motion is not spherically symmetrical, so the position of the molecule is not represented as a sphere but as an ellipsoid.

As molecular spectroscopy has contributed force constants to molecular mechanics, so MM contributes vibrational frequencies to spectroscopy. Because force parameters are not the same as force constants, MM values of bond stretching frequencies are not the same as known experimental values but they generally agree to within $\pm 25\,\mathrm{cm}^{-1}$. Knowing the geometry and atomic masses, angular momenta can also be calculated along with spectral intensities, heat capacities, entropies, Gibbs free energy functions, and temperature variations of the thermodynamic functions.

19.9 TRANSITION STATES

Existence of several potential energy minima, one for each of the atoms in a more or less complicated molecule, implies a potential surface analogous to a mountainous terrain with peaks, valleys, and mountain passes. Each pass represents a potential path from one minimum to another. In chemistry, these paths are said to be along a possible *reaction coordinate*. A transition from one minimum potential energy to another goes over a relative maximum called the transition state. A transition state is a maximum relative to the two minima it connects but, analogous to a mountain pass in high country, it is a least energy *path* between the minima, Mathematically, all second derivatives of the energy are positive except one, leading to vibrational frequencies all of which are real except for one which is an imaginary number.[2]

An example is conversion of the "chair" form of cyclohexane to the "boat" form:

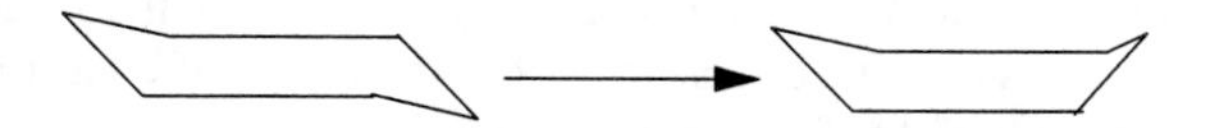

For this change to take place, cyclohexane must go through a planar conformation that is higher in energy than either the chair or the boat. The planar conformation is the transition state. One way (not necessarily the best way) of finding the transition state energy is by starting with a planar input geometry along with the chair and boat.

The MM4 difference between the chair and boat forms is $5.7\ \mathrm{kcal\,mol^{-1}} = 23.8\ \mathrm{kJ\,mol^{-1}}$ at 226 K, and the experimental value is $5.5\ \mathrm{kcal\,mol^{-1}} = 23.0\ \mathrm{kJ\,mol^{-1}}$. The transition enthalpy from chair to boat is a little less than twice this amount and involves somewhat more complicated geometries like the twist boat conformation (Allinger et al., 1996, p. 650 ff).

[2]Please do not think that there is something wrong with it: The term "imaginary" is just mathematical jargon meaning that if you square it, you get a negative number.

PROBLEMS AND EXAMPLES

Example 19.1 A Geometric Optimization of Water

Construct an input file for the MM4 geometry optimization of water.

Solution 19.1 First and most obvious, we name the molecule and enter the number of atoms. In MM4, the name goes on line 1 and can occupy up to 60 columns (spaces). In this case, we enter `water` in columns 1 to 5 of the first line in File 19.1 and specify the number of atoms as 3 in column 65. On the second line or "card," the number 0 in column 5 shows that there are no *connected atoms* (see Example 19.3). The 2 in column 30 gives the number of *attached atoms*, namely the two hydrogens. Line 3 identifies he attached atoms. Atom 2 (oxygen) is attached to atom 1 (one of the hydrogens), and atom 2 (oxygen) is also attached to atom 3 (the other hydrogen)

Next, we input a starting geometry in Cartesian coordinates for all the atoms in the molecule. The geometry chosen is a simple angle > with oxygen at the origin. The position vectors of the two hydrogens in an arbitrary Cartesian coordinate space are (−0.5, 0.5, 0.0) and (−0.5, −0.5, 0.0). The input geometry has been constrained to the x, y plane by setting the z components to zero. The Cartesian coordinates are given in the order x, y, and z from left to right. The O atom has an identifying number 6, and the two hydrogens are given the number 21 in this MM4 convention. Numerical identifiers are used to represent atoms, which may be of different kinds—for example, sp, sp^2, or sp^3 carbon atoms. These two parts, the descriptive cards and the Cartesian geometry, constitute the input file. Save your input files as `h20.mm4` or some similar name with the suffix `.mm4`. It is to be emphasized that this is only one of very many possible input files.

Example 19.2 A Typical Small Molecule Run

Run the MM4 input file, File 19.2. The run is initiated by typing `mm4` followed by the input filename `h20.mm4`. Enter 0 for `parameters`, 1 for the `line number`, and 2 for the `program choice`. Several alternatives are offered. Running them later should make most of the differences self-evident.

Solution 19.2 A successful run should give a small steric energy of <1 kcal mol^{-1}. Hit enter three times to exit MM4 and to save TAPE4.MM4 and TAPE9.MM4. Enter

```
water                                                            3
   0                        2   0   0   0   0   0   0   0
   2    1   2    3
 -0.50000   0.50000  0.00000    21
  0.00000   0.00000  0.00000     6
 -0.50000  -0.50000  0.00000    21
```

FILE 19.2 An input file for water. The geometry is an arbitrarily chosen estimate. The zeros in line 2 and in the Cartesian coordinates are only place indicators to make counting the columns easier in more complicated files (see File 19.4).

```
water                                                    0 3 0 0 0 0 10.0
0   0       0.0000000          2    0    0  0  0  0  0          1    0
    2    1    2    3
  -0.76769    0.52221  0.00000 H 21( 1)
   0.00000   -0.06580  0.00000 O  6( 2)
   0.76769    0.52221  0.00000 H 21( 3)
```

FILE 19.3 The MM4 geometry output TAPE9.MM4 for water.

`cat` TAPE9.MM4 to display the optimized geometry of your input file. File 19.3 shows an MM4 output for water. The output File 19.3 has the same format as the input File 19.2, but it contains some more information beyond column 65 which we do not need at this point. These differences notwithstanding, the geometry is clear. The simple angle > has been up-ended to a V and moved down 0.06580 Å during the optimization. By a Pythagorean calculation, we get an O—H bond length of 0.967 Å. This length is confirmed by a calculation within the program. We find, along with much other information in TAPE4.MM4,

```
H(    1)- O(    2)    0.9670
O(    2)- H(    3)    0.9670
```

A search toward the end of the output file will reveal a computed dipole moment (Section 18.7):

$$\text{DIPOLE MOMENT} = 2.0188\,\text{D}$$

The experimental value is 1.85 D.

Example 19.3 An MM4 Calculation of the Geometry and $\Delta_f H^{298}$ of Methane

Convert the H_2O input file to a methane input file and run it in the MM4 protocol. The atomic identifiers are 1 for an sp^3 carbon and 5 for an alkane hydrogen. A 1 in column 65 of line 2 requests $\Delta_f H^{298}$.

Solution 19.3 Assuming considerable ignorance about the structure of methane, let us place the carbon atom at the origin and suppose that the projection of each atom upon the Cartesian coordinates centered on the carbon is 1.0 Å. For convenience, it is useful to orient the molecule with two C—H bonds in the x, y plane. Examination of the TAPE4.MM4 file circumvents awkward trigonometric hand calculations within the tetrahedral output geometry and gives the C—H bond length as `C(1)- H(2) 1.1070` expressed in Angstroms. The experimental value is 109.3 pm = 1.093 Å. The tetrahedral H—C—H bond angles are `H(2)- C(1)- H(3) 109.471` expressed in degrees. The experimental value is 109.5°. A very important line in TAPE4.MM4 is the $\Delta_f H^{298}$ "heat" (enthalpy) of formation that was requested in entry 2, 65 of the input file:

```
HEAT OF FORMATION (HFN) AT 298.2 K = -17.89 KCAL/MOLE
```

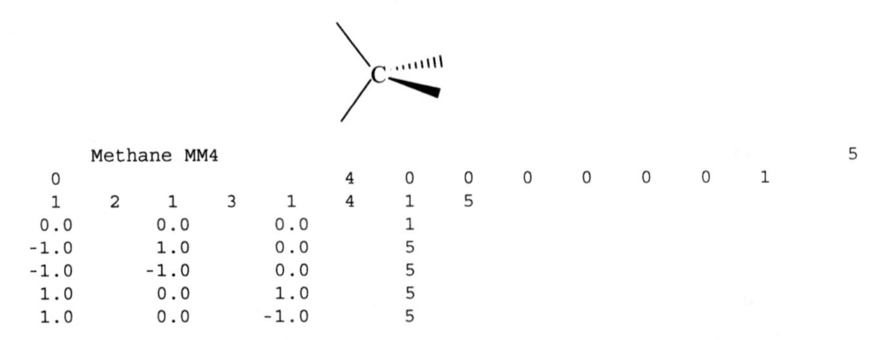


```
          Methane MM4                                                                5
    0                             4    0    0    0    0    0    0    1
    1    2    1    3    1         4    1    5
  0.0       0.0       0.0              1
 -1.0       1.0       0.0              5
 -1.0      -1.0       0.0              5
  1.0       0.0       1.0              5
  1.0       0.0      -1.0              5
```

FILE 19.4 MM4 input geometry for methane. The 1 in column 65 requests calculation of $\Delta_f H^{298}$.

The experimental value is $\Delta_f H^{298} = -74.8 \pm 0.3\ \text{kJ mol}^{-1} = -17.9 \pm 0.1\ \text{kcal mol}^{-1}$. Molecular properties beyond geometry and heat capacity, such as dipole moments, spectra, partition functions, and entropy, are calculated in MM4.

Problem 19.1

(a) What is the enthalpy of formation $\Delta_f H^{298}$ of neopentane (dimethylpropane) by the simple group additivity method in Section 19.1?

(b) The enthalpy of isomerization of isopentane (methylbutane) to neopentane is $-14\ \text{kJ mol}^{-1}$. What is the enthalpy contribution of the C—H group in alkanes?

Problem 19.2

Determine the O—H bond length from the Cartesian coordinates given in File 19.3 by hand calculation using Pythagoras's theorem. The experimental value is 97.0 pm = 0.970 Å.

Problem 19.3

Find the H—O—H bond angle from the optimized geometry in File 19.3. The experimental value is 104.5°.

```
Methane MM4                                                        0   5 0   0 0   0 10.0
0  0        0.0000000           4     0          0    0    0    0    1          1    0
   1    2    1    3    1    4    1    5
   0.00000   0.00000   0.00000 C   1(  1)
  -0.60766   0.92497  -0.02506 H   5(  2)
  -0.66877  -0.88172  -0.02769 H   5(  3)
   0.60089  -0.02167   0.92947 H   5(  4)
   0.67554  -0.02158  -0.87672 H   5(  5)
```

FILE 19.5 MM4 output geometry for methane.

Problem 19.4

What is the dipole moment of water according to an MM4 calculation? Compare this value with the value given in a current textbook.

Problem 19.5

Expand the input file for water, File 19.2, which has the geometry > to an input file for ethene, which has the geometry >=<. The atom designator for an sp^2 carbon is 2 and the hydrogens are designated 112. Run the resulting program. The steric energy for a successful run should be approximately 2 kcal mol^{-1}.

Problem 19.6

Modify either the input or the output files resulting from Problem 19.5 after the manner of Example 19.3 to obtain an input file for ethane. Run the file and obtain the full output file TAPE9.MM4 and TAPE4.MM4, part of which is given as File 19.1.

Problem 19.7

Be sure that there is a 1 in column 65 of both the ethene and ethane programs obtained from Problems 19.5 amd 19.6 to produce values for the enthalpy of formation of each. Determine the enthalpy of hydrogenation of ethene to ethane, each in the standard state. The result from a successful geometry optimization should be about –32 kcal mol^{-1}. The experimental value is -32.60 ± 0.05 kcal mol^{-1} $= -136.4 \pm 0.2$ kJ mol^{-1}.

Problem 19.8

In the previous problem, why don't you need to compute the standard state enthalpy of formation of H_2?

$$C_2H_4(g) + H_2(g) \rightarrow C_6H_6(g)$$

Problem 19.9

If you have access to a GUI, find the enthalpies of formation of mono- di- tri- and tetra(*t*-butyl)methane. Explain the curious pattern of these results. Print the structure of tetra(*t*-butyl)methane.

Problem 19.10

There are three vibrational frequencies given for water in the output file TAPE4.MM4. Two are for the symmetric and antisymmetric O—H stretch. Diagram these two modes of motion. What molecular motion corresponds to the third frequency? Why it this frequency so different from the other two?

20

QUANTUM MOLECULAR MODELING

With Slater determinants and the Hartree–Fock equations, one had everything necessary to solve problems in molecular structure, energy, and dynamics. But the equations could not be solved. Quantum chemistry awaited two spectacular advances. First, Roothaan derived a way to express its apparently insoluble integrodifferential equations as equations in linear algebra. This was accompanied by an exponential rise in power of the digital computer, which now routinely carries out trillions of simple mathematical operations per second.

The total energy of a molecule is almost entirely that of its constituent atoms, leaving only a fringe energy, which might otherwise be considered trivial, to hold the molecule together. But it is the *chemical bond* that dominates the world we see around us with all its color, life, and diversity, and so it is the chemical bond that we seek to calculate. We will be able to extend the atomic orbital concept to our molecular calculations but, because chemical energies and energy differences are so small, we must achieve a very high level of accuracy. Accurate programs now exist that can be extended to all molecules (in principle at least). Finite computer speed and memory place strict limitations on these grandiose plans, but barriers fall almost every day in this active research field.

20.1 THE MOLECULAR VARIATIONAL METHOD

The variational method applies to molecules as well as to atoms. By it, we can approach an optimized molecular energy E:

$$E = \frac{\langle \Psi | \hat{H} | \Psi \rangle}{\langle \Psi \mid \Psi \rangle}$$

Concise Physical Chemistry, by Donald W. Rogers

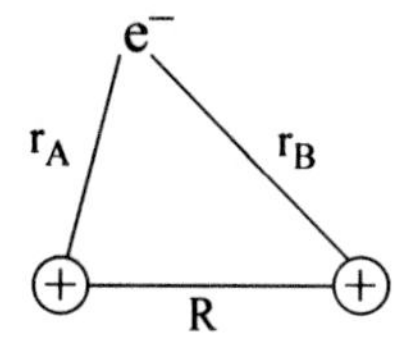

FIGURE 20.1 The hydrogen molecule ion, H_2^+.

where $|\Psi\rangle$ is the state vector for the *entire molecule*. The inner product $\langle\Psi \mid \Psi\rangle$ is 1, so

$$E = \langle\Psi|\, \hat{H}\, |\Psi\rangle$$

In chemical applications the state vector is usually written as the equivalent wave function:

$$E = \int \Psi \left|\hat{H}\right| \Psi \, d\tau$$

If the state vector or wave function is exact, the energy will be exact.

20.2 THE HYDROGEN MOLECULE ION

A stepping stone toward full-scale molecular structure and energy calculations is the hydrogen molecule ion H_2^+. Conversion of the problem to elliptic coordinates and subsequent solution by numerical methods has been carried out, so we know the answer before we start. The energy of the bound state of H_2^+ is about 268 kJ mol^{-1}. Unfortunately, this numerical method cannot be extended to larger molecules or ions, so we shall use the known result to help us to develop of an approximate method, which can then be applied to molecular systems large enough to be important in chemistry.

The geometry of the H_2^+ problem is given in Fig. 20.1. Assume that the nuclei are stationary[1] at a distance R from one another. This gives us a problem of one electron in the field of positive nuclei A and B over distances r_A and r_B. The Schrödinger equation is similar to that of the hydrogen atom, except that there are two centers of positive charge rather than just one. For any selected value of R,

$$\left[-\frac{1}{2}\nabla^2 - \frac{1}{r_A} - \frac{1}{r_B}\right]\Psi = E_{\text{electronic}}\ \Psi$$

[1]This is the Born–Oppenheimer approximation.

There are many possible values of R, each of which leads to a unique value of the electronic energy $E_{\text{electronic}}$. Although R is constant for any single calculation, the total energy of the system is a function of R:

$$E_{\text{total}} = E_{\text{electronic}} + \frac{1}{R}$$

This means that a curve of E_{total} vs. R can be drawn. From this point on, we shall drop the subscripts on E, taking the nature of E, electronic or total, to be clear from context.

By the *LCAO* approximation, a molecular orbital can be expressed as a *L*inear *C*ombination of two hydrogenic *A*tomic *O*rbitals:

$$\psi = N_1 e^{-r_A} \pm N_2 e^{-r_B}$$

The atomic orbitals are *basis functions* which define a vector space that includes the molecular orbital. The LCAO basis set is not complete, so the molecular orbital we obtain will not be correct. If functions $N_1 e^{-r_A}$ and $N_2 e^{-r_B}$ are normalized hydrogen $1s$ orbitals, we call them $1s_A$ and $1s_B$ and the approximate molecular orbital for H_2^+ is $\psi_{H_2^+} = 1s_A \pm 1s_B$. Physically, one basis function represents hydrogen atom A at some distance from proton B, while the other basis function represents hydrogen atom B at some distance from proton A. Neither basis function alone recognizes the simultaneous interaction of the electron with both nuclei. It is this interaction we seek because it brings about the chemical bond.

The energy of the positive combination $\psi_{H_2^+} = 1s_A + 1s_B$ is

$$E = \int \psi \hat{H} \psi \, d\tau = N^2 \int (1s_A + 1s_B) \left[-\tfrac{1}{2}\nabla^2 - \frac{1}{r_A} - \frac{1}{r_B} \right] (1s_A + 1s_B) \, d\tau$$

where N^2 is the product of the two normalization constants and the integration is taken over the entire space τ. If we expand and simplify according to the symmetry of the problem, the energy gives three functions often denoted J, K, and S:

$$J = \left(1 + \frac{1}{R}\right) e^{-2R}$$

$$K = \left(\frac{S}{R} - 1 + R\right) e^{-2R}$$

and

$$S = \left(1 + R + \frac{R^2}{3}\right) e^{-R}$$

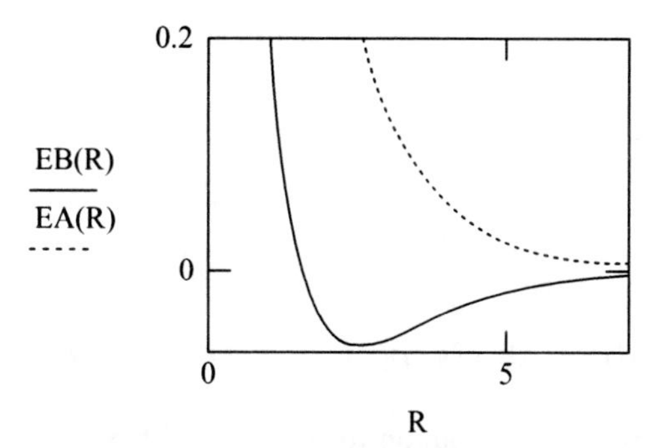

FIGURE 20.2 Bonding and antibonding orbitals for H_2^+. The units of E are hartrees and the units of R are bohr = 52.9 pm.

which are combined to give

$$E = \frac{J + K}{1 + S}$$

The functions J, K, and S are all dependent upon the internuclear separation R because orbital interaction is greater when the nuclei are close together and smaller when they are far apart. The function $E = f(R)$ for the positive combination of atomic orbitals is shown by the solid curve in Fig. 20.2.

Positive nuclei attracted to an electron between them are bound to each other. The nuclei do not crash into each other because at some R, internuclear repulsion becomes dominant over bonding and the energy begins to rise sharply. The compromise between attractive and repulsive forces results in a minimum in $E = f(R)$. The bond has a definite and fixed length of 2.5 bohr = 132 pm in this first approximation. To find the energy of the H_2^+ bond at this distance, we must solve equations for S, J, and K to give

$$S = 0.461, \qquad J = 0.00963, \qquad K = -0.1044, \qquad E = -0.065 \ E_h$$

The hartree of energy is $1 \ E_h = 2625 \text{ kJ mol}^{-1}$. The minimum is at $E = -0.065 \ E_h = -171 \text{ kJ mol}^{-1}$. This is about 64% of the experimental value. The quantitative result is not very good, but the qualitative result shows that the energy is negative at 132 pm, which indicates bonding. It is noteworthy that the chemical bond arises as a natural consequence of quantum mechanics without further assumptions.

There is a second solution for E in Fig. 20.2 which arises from the negative combination $\psi_{H_2^+} = 1s_A - 1s_B$. The energy does not go through a minimum with this *antibonding* wave function but rises monotonically with decreasing R. In H_2^+, bonding and antibonding orbitals are above and below the energy of the separated system H^+ and H. The higher and lower bonding and antibonding molecular orbitals at the hond distance are often shown simply as Fig. 20.3.

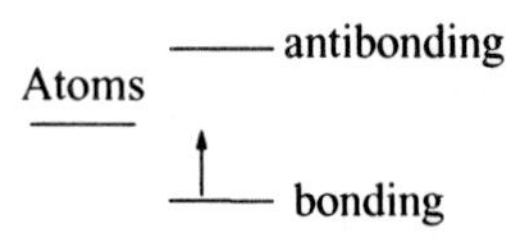

FIGURE 20.3 Bonding and antibonding solutions for the H_2^+. One electron in the lower (bonding) orbital of H_2^+ is indicated by an arrow.

20.3 HIGHER MOLECULAR ORBITAL CALCULATIONS

In obtaining an approximate solution to the molecular Schrödinger equation, the many-electron wave function $\Psi(\mathbf{r}_i)$, which is a function of all radial distance vectors $\mathbf{r}_i$, is broken up into orbitals ψ_i:

$$\Psi(\mathbf{r}_i) = (n!)^{-1/2} \det[(\psi_1\alpha)\,(\psi_1\beta)(\psi_2\alpha)\,\ldots]$$

The orbitals $\psi_1\alpha, \psi_1\beta, \psi_2\alpha, \ldots$ accommodate single electrons. The symbols α and β designate opposite spins and "det" indicates a Slater determinant. Because of spin pairing, each orbital can contain two electrons; hence the minimum number of molecular orbitals $\psi_1, \psi_2, \psi_3, \ldots$ is one-half the number of electrons.

In 1951 Roothaan further divided single-electron molecular orbitals ψ_i into *linear combinations* of basis functions χ_μ:

$$\psi_i = \sum_{\mu=1}^{N} c_{\mu i}\chi_\mu$$

($\mu = 1, 2, 3, \ldots N$), where N > n. Having selected a basis set χ_μ, one wishes to find the coefficients $c_{\mu i}$. A large $c_{\mu i}$ means that the corresponding basis vector makes an important contribution to the total molecular orbital, while a small $c_{\mu i}$ means that the corresponding basis vector makes a small contribution. This gives a set of *algebraic* equations in place of the set of coupled differential equations in the original problem. Roothaan's equations can be written in matrix form as

$$\mathbf{FC} = \mathbf{SCE}$$

where **C** is the column vector of coefficients, **E** is the diagonal matrix of energies, with elements $E_{ij} = \varepsilon_i\delta_{ij}$, the elements of **S** are $S_{\mu\nu} = \int \chi_\mu\chi_\nu d\tau$, and **F** is the Fock matrix.[2]

Elements in the **F** matrix are

$$F_{\mu\nu} = H_{\mu\nu} + \sum_{\lambda\sigma} P_{\lambda\sigma}[\,(\mu\nu|\lambda\sigma) - (\mu\lambda|\nu\sigma)\,/2]$$

[2]It is unfortunate that **F** is used to represent both the force field matrix in molecular mechanics and the Fock matrix in quantum mechanics. Be careful not to confuse the two.

defined by

$$H_{\mu\nu} = \int \chi_\mu \hat{H} \chi_\nu d\tau, \qquad P_{\mu\nu} = 2\sum_{1}^{n} c_{\mu i} c_{\nu i}$$

and

$$(\mu\nu|\lambda\sigma) = \iint \chi_\mu(1)\chi_\nu(1) \times \frac{1}{r_{12}} \chi_\lambda\,(2)\chi_\sigma(2)\; d\tau_1 d\tau_2$$

$$(\mu\lambda|\nu\sigma) = \iint \chi_\mu(1)\chi_\lambda(1) \times \frac{1}{r_{12}} \chi_\nu\,(2)\chi_\sigma(2)\; d\tau_1 d\tau_2$$

The matrix elements $H_{\mu\nu}$ are elements of the core Hamiltonian that would be imposed by the nuclei on each electron in the absence of all other electrons, and elements ε_i of the diagonal matrix E are one-electron energies. Many computer routines exist for multiplication, inversion, and diagonalization of matrices.

The integrals $(\mu\nu|\lambda\sigma)$ and $(\mu\lambda|\nu\sigma)$ are difficult to evaluate, which caused a bifurcation of the field of molecular orbital studies into subdisciplines called *semiempirical* and *ab initio*. Research groups led by Dewar and by Stewart were devoted to obtaining solutions by substituting empirical constants into $F_{\mu\nu}$. A second approach was followed by groups led by Pople, Gordon, and others, who used very efficient computer codes and relied upon the increasing power of contemporary computing machines to solve the integrals in $F_{\mu\nu}$. In general, the rule of speed vs. accuracy applies. Semiempirical substitution is faster, hence applicable to larger molecules. *Ab initio* methods are more accurate but they are very expensive in computer resources.

20.4 SEMIEMPIRICAL METHODS

Solution of the Schrödinger equation requires evaluation of many integrals. A large proportion of these integrals make a very small contribution to molecular energy and enthalpy. When they are dropped, the calculation is simplified in the hope that the sacrifice in accuracy will be small. Dropping the integrals in the equation set leaves only $H_{\mu\nu}$ in place of the Fock matrix elements $F_{\mu\nu}$. Dropping *some* integrals and replacing others with empirical parameters gives legitimate Hamiltonian elements but elements that are approximate because of the use of empirical parameters. They are elements $H_{\mu\nu}$ of a *semiempirical* Hamiltonian operator. The general rule is that if you are modifying the **F** matrix to obtain an approximate Hamiltonian, the method is semiempirical. If you are working with the full **F** matrix and attempting to approach a complete basis set, the method is *ab initio*. Neither method is exact.

The most important steps in development of a computer-based semiempirical method were in deciding which of the many integrals $(\mu\nu|\lambda\sigma)$ and $(\mu\lambda|\nu\sigma)$ in a polyatomic molecule can be dropped, which of the integrals must be retained and parameterized, and how they should be parameterized. *Neglect of differential overlap*

(NDO) and *neglect of diatomic differential overlap* (NDDO) approximations led to a series of semiempirical programs denoted CNDO (complete neglect of differential overlap), INDO, MINDO, and so on. Initially, NDDO programs were parameterized to reproduce *ab initio* values for simple molecules but later semiempirical programs such as the AM1 programs of Dewar et al. and the PM3 method of Stewart are parameterized against experimental thermochemical results so as to calculate energies and enthalpies. Presently, they are both in wide use for this purpose. Along with thermochemical data, dipole moments, geometries, and isomerization potentials are also calculated by modern semiempirical programs.

20.5 AB INITIO METHODS

An exact solution of the Schrödinger equation would employ no empirical parameters beyond mass and charge of the subatomic particles, and it would be an absolute solution to the problem that could never be changed or revised. In practice, the absolute properties of a molecule are never found, though they may be approached by ever-improving approximate methods. Today, what we call "*ab initio* procedures" contain small empirical "corrections," but a very serious effort has been put forth to minimize these aspects of the procedure. One condition we wish for the set of functions $\{\chi_\mu\}$ we choose as a basis set is that they be as nearly complete as possible, that is, we hope that the set will *span* the vector space.

20.6 THE GAUSSIAN BASIS SET

We saw in Chapter 17 that the single Gaussian function $\phi(r) = Ce^{-\alpha r^2}$, where $C = 1.0$ and $\alpha = 0.2829$, does not give a very good approximation to the energy of the hydrogen atom. The result, HF $= 0.4244\ E_\mathrm{h}$, is 84.9% of the true (defined) energy of the hydrogen atom. The Gaussian, with r^2 in the exponent, drops off faster than the $1s$ orbital, which has r in the exponent (Fig. 20.4). The Gaussian is too "thin" at larger distances from the nucleus.

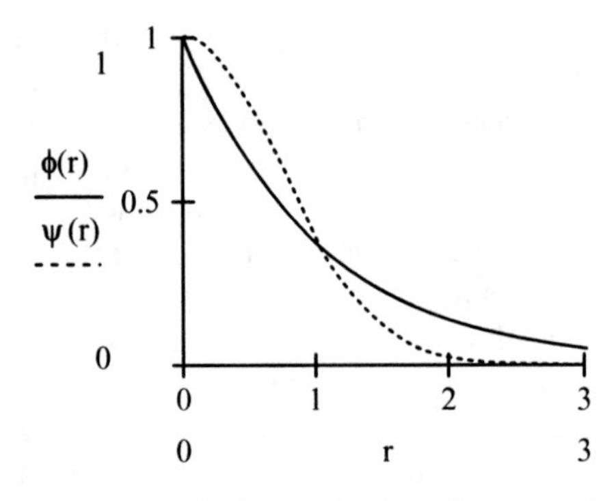

FIGURE 20.4 The $1s$ STO (solid line) and a Gaussian approximation (dotted line).

The next step is to take two Gaussian functions, called *primitives*, parameterized so that one fits the STO close to the nucleus and the other contributes to the part away from the nucleus. We seek a function

$$\text{STO-2G} = C_1 e^{-\alpha_1 r^2} + C_2 e^{-\alpha_2 r^2}$$

But how shall we apportion this linear combination so that we have one tall basis function contributing to the orbital near the nucleus and one for a "fat" tail? Let us take $\alpha_1 = 1.0$ and $\alpha_2 = 0.25$. Both $C_1 e^{-r^2}$ and $C_2 e^{-0.25\, r^2}$ contribute to the sum. The larger negative exponent is tall near the nucleus but drops off faster than the small negative exponent. Now $\phi(r)$ will extend to larger values of r and give us the fat tail we seek:

$$\phi(r) = C_1 e^{-r^2} + C_2 e^{-0.25\, r^2}$$

But we still have two parameters to worry about C_1 and C_2. They control the relative contribution of each primitive to the final wave function. Let us take a 60/40 split and favor the fat tail:

$$\phi(r) = 0.40 e^{-1.0\, r^2} + 0.60 e^{-0.25\, r^2}$$

We now have a four-parameter basis set for use with the gen keyword (File 20.1). They are entered in the format

$$\begin{matrix} \alpha_1 & C_1 \\ \alpha_2 & C_2 \end{matrix}$$

The GAUSSIAN© input file becomes File 20.1. The STO curve fit is shown in Fig. 20.5.

$$\text{STO}(r) := e^{-r}$$
$$\phi(r) := 0.40 e^{-1.0\, r^2} + 0.60 e^{-0.25\, r^2}.$$

```
# gen
hatom gen
0 2
h

1
S          2
1.0  0.40
0.25 0.60
****
```

FILE 20.1 (Input) A four-parameter Gaussian File for the hydrogen atom. Line 8 designates the first center (the only one in this case) and line 9 identifies it as an s orbital with 2 basis functions.

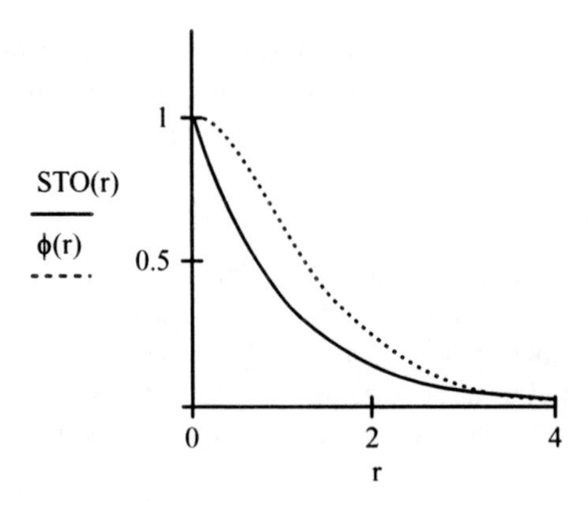

FIGURE 20.5 Comparison of the 1*s* STO of hydrogen with an arbitrarily parameterized two-Gaussian function $\phi(r) := 0.40e^{-1.0r^2} + 0.60e^{-0.25r^2}$.

The fit is certainly not perfect, but it is better than what we got with a single basis function. The energy obtained from this basis set is `HF=-0.4572106` which is 91.4% of the defined value of 0.5 E_h. The error has been reduced from about 15% to about 8.5%. Variational optimization in the C parameters takes place during the program run as we see by adding `GFInput` to the `# gen` input line in File 20.1. This change produces the output shown in File 20.2.

```
AO basis set in the form of general basis input:
        1 0
S     2 1.00         0.000000000000
        0.1000000000D+01   0.4304660143D+00
        0.2500000000D+00   0.6456990214D+00
****
```

FILE 20.2 (Output) The STO-2G basis set written as a 1s orbital consisting of functions with arbitrarily selected exponents 1.00 and 0.25.

20.7 STORED PARAMETERS

We can also run an STO-2G *ab initio* calculation on the hydrogen atom using the GAUSSIAN stored parameters rather than supplying our own. The input file is shown in file 20.3. We find that there are two Gaussian primitives and one unpaired electron from the output:

```
1 basis functions          2 primitive gaussians
1 alpha electrons          0 beta electrons
```

which agrees with the picture of the STO-2G basis set that we are trying to build. Of course we want to know what the parameters are for the two Gaussians. The keyword

```
# sto-2g

hatom

0 2
h
```

FILE 20.3 (Input) An STO-2G input file using a stored basis set.

`GFinput` inserted after `# sto-2g` in the *route section* of the input file produces the added information. The parameterized STO-2G basis function is

$$\text{STO-2G} = 0.4301e^{-1.309\,r^2} + 0.6789e^{-0.233\,r^2}$$

which is not too far from the arbitrary function we guessed for input file, File 20.1. The stored function is graphed in Fig. 20.5. The coefficients `0.4301` and `0.6789` give the intercepts of the two Gaussians at $r = 0$. As before, the two α parameters determine how extended the Gaussian is in the r direction (how fat the tail of the function is), and two C parameters determine how much of a contribution each Gaussian makes to the final STO approximation. Stored parameters have been optimized for general application to molecular problems more complex than the hydrogen atom, which accounts for the "overshoot" of $\alpha_1 = 1.31$ arrived at in the final compromise.

$$\begin{aligned} \text{STO-2G} &= C_1 e^{-\alpha_1 r^2} + C_2 e^{-\alpha_2 r^2} \\ \alpha_1 &= 1.31, \qquad \alpha_2 = 0.233 \\ C_1 &= 0.430, \qquad C_2 = 0.679 \end{aligned}$$

We now have two ways of inserting parameters into the STO-2G calculation. We can write them out in a `gen` file like input File 20.1 if we know the parameters we want, or we can use the stored parameters as in input File 20.3 if we don't. And we can use the `GFInput` keyword to find out what stored parameters were used. This process can be carried further for the STO-3G, STO-4G, and many other atomic orbital approximations. The stored parameters for some orbitals are quite cumbersome, and one would not want to enter them by hand.

```
AO basis set in the form of general basis input:
      1 0
 S   2 1.00        0.000000000000
      0.1309756377D+01  0.4301284983D+00
      0.2331359749D+00  0.6789135305D+00
 ****
```

FILE 20.4 (Output) GAUSSIAN© stored parameters for the STO-2G basis set.

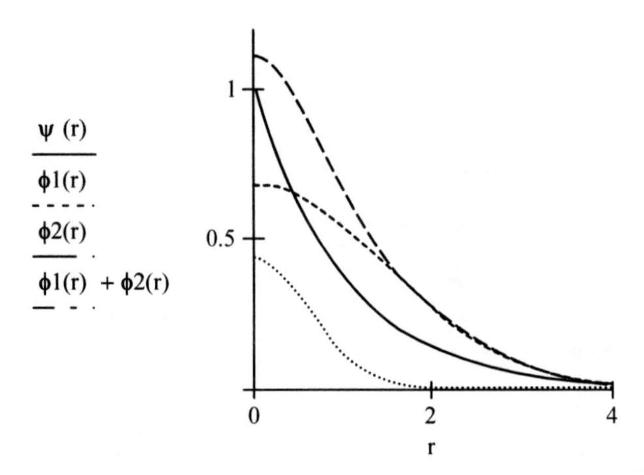

FIGURE 20.6 Approximation to the 1s orbital of hydrogen by 2 Gaussians. The upper curve is the sum of the lower two curves. The solid curve is the hydrogen 1*s* atomic orbital.

Upon extending the basis set along the series STO-2G, STO-3G, ..., STO-6G, and more complicated functions, one reaches a point of diminishing return, beyond which further elaboration produces little gain. On the positive side, the basis functions fall into natural groups, which are replicated from one calculation to the next, hence a way of using computer resources more efficiently is by treating each group as though it were a single function. Each natural group of basis functions, treated as a unit, is called a *contracted Gaussian-type orbital* (CGTO). For example, if we start out with 36 primitives (by no means a large number in this context) and segment them into groups of 6, we have reduced the problem six-fold.

20.8 MOLECULAR ORBITALS

Extension of the method we have developed so far to molecules is quite straightforward. One simply puts a basis set into the input file for each atom in the molecule. In a simple illustration of the transition from atomic to molecular input files, we can convert the input file for a single hydrogen atom using our arbitrarily chosen STO-2G atomic basis set into an input file for the H_2 molecule. The result is File 20.5.

The second atom is added to the first, in *z-matrix format*. This is done by selecting the first atom as a reference point and placing the second hydrogen atom a distance r from it. Line 7 in File 20.5 reads `h 1 r` meaning "the distance between the second hydrogen h and atom number 1 is r." The approximate distance chosen is specified in angstrom units immediately below in line 9 as `r = 0.7`.

In the previous problem of the uncharged atom, the spin multiplicity $M = 2S + 1$, where $S = \pm 1/2$ is the total spin of the system, was 2 (line 5 of input File 20.3), but the multiplicity for the paired σ electrons in molecular H_2 is 1. In molecular hydrogen File 20.5, lines 11–15 are (almost exactly) repeated in lines 16–20. The first set of

```
#gen

 H2mol gen

 0 1
 h
 h 1 r

 r=0.7

 1
  S 2
      1.00  0.40
      0.25  0.60
  ****
 2
  S 2
      1.00  0.40
      0.25  0.60
  ****
```

FILE 20.5 (Input) A molecular orbital input file for H_2. The file is in *z*-matrix GAUSSIAN format.

parameters refers to H atom 1, and the second set of parameters refers to H atom 2. The only difference is in the atom identifiers: 1 in line 11 and 2 in line 16.

The energy output is designated a *restricted* Hartree Fock energy `E(RHF)`,

`E(RHF) = -1.07637168255` E_h

because it is the energy of the 1σ orbital calculated on the assumption that both spin paired electrons will be in it. If the spins were unpaired and the electrons were in different orbitals, as in an excited state, the *unrestricted Hartree Fock* energy `E(UHF)` would be used. Unrestricted HF calculations are somewhat more demanding on computer resources because calculations are done on each set of electrons for α and β spin rather than on the pair. In this simple case, the difference is negligible.

The experimental dissociation energy of H_2 is 0.174 E_h. The STO-2G calculated energy is only about 44% of the experimental value in this crude approximation. The input file for the same task using *z*-matrix input and stored parameters is shown in File 20.6.

Now the energy output `E(RHF) = -1.0934083` E_h is 53% of the experimental value of 0.174 E_h. Neither the arbitrary parameters in File 20.5 nor the STO-2G stored basis set gives a very convincing calculation of the total bond energy of H_2 (44 and 54%, respectively). However, when compared to the energy of two H atoms calculated *using the same basis set*, the result $-1.0764 - 2(-0.4572) = -0.162$ E_h is 93% of the experimental bond energy. This is due to cancellation of error between two

```
# sto-2g

H2 molecule

0 1
H
H 1 r
variables
r=0.7
```

FILE 20.6 (Input) A GAUSSIAN input file for H_2. The file is in *z*-matrix format.

systems: atom and molecule. Cancellation of error is often helpful but cannot be relied upon. (Bear in mind that there is some uncertainty in the experimental value as well.)

20.8.1 GAMESS

GAMESS (General Atomic and Molecular Energy Structure System) is a molecular orbital program intended for general and academic use as open source software. The authors do not charge for this program, but the user may not sell it to anyone else. GAMESS gives the STO-2G energy calculated from input File 20.7. The geometry of the molecule is given in Cartesian coordinates. Only one nonzero coordinate is necessary to designate the geometry of a diatomic linear molecule. The bond length is 0.7 Å (estimated).

Translating File 20.7, each statement begins with $ and ends with $END. The statement may occupy more than one line. The $CONTROL statement specifies the SCFTYPe as RHF and the COORDinates as CARTesian $END. The $BASIS set is a Gaussian BASIS, which is STO and contains a Number of GAUSSians equal to 2 $END. The $DATA set follows with an identifier `Hydrogen` for the human reader, the symmetry Dn1, and the coordinates in Cartesian format $END.

The single entry 0.7 Å for the bond length is an element in the 2×3 matrix of locations in Cartesian space using three components for each atom. One atom is placed at the origin and the other is placed on the x axis at a distance of 0.7.

```
  $CONTRL SCFTYP=RHF COORD=CART $END
 $BASIS GBASIS=STO NGAUSS=2 $END
 $DATA
Hydrogen
 Dn 1
 H    1.0  0.0       0.0      0.0
 H    1.0  0.7       0.0      0.0
  $END
```

FILE 20.7 (Input) GAMESS file for hydrogen molecule. The SCF calculation is RHF, the coordinate system is Cartesian and the stored basis set is STO-2G. The symmetry space group is Dn 1, the charge on each hydrogen atom is 1.0, and the bond length is 0.7 Å.

The energy, taken from a much more detailed output file, is

```
TOTAL ENERGY= -1.0934083240
```

The value calculated here used two Gaussian basis functions STO-2G and arrived at an energy of $-1.09340\ E_h$ which replicates the output from File 20.6. The bond energy BE is the difference between the molecular energy and the level of the atoms, which have an energy of 2(.5000):

$$BE = -1.0934 - 2(-0.5000) = -0.0934\ E_h = -245\ \text{kJ mol}^{-1}$$

This is about half of the experimental value of 435 kJ mol^{-1}, slightly better than the one calculated using the estimated coefficients in File 20.5. Using six Gaussian functions improves the result to 0.0991 E_h = 260 kJ mol^{-1}, which is about 57% of the experimental value:

```
TOTAL ENERGY= -1.0991105267
```

20.9 METHANE

The carbon atom can be combined with four hydrogen atoms to form input File 20.8 for methane. The geometry for methane is given in approximate coordinates, which will be changed during the program run. Once carbon and hydrogen can be combined, the whole vista of alkanes, alkenes, and alkynes (many thousands of molecules) is open. Extension to heteroatomic molecules other than hydrocarbons follows by the same method (Example 20.1).

An optimized geometry is part of the output file. Neither the basis set nor the optimized coordinate set is unique. Many (strictly, infinitely many) sets are possible.

Center Number	Atomic Number	Atomic Type	Coordinates (Angstroms) X	Y	Z
1	6	0	0.000000	0.000000	0.000162
2	1	0	0.000000	0.000000	1.089051
3	1	0	0.000000	1.026565	-0.363340
4	1	0	-0.889032	-0.513283	-0.363340
5	1	0	0.889032	-0.513283	-0.363340

FILE 20.8 One of many possible STO-2G optimized coordinate sets for methane. The carbon atom has been arbitrarily placed at the origin of a Cartesian coordinate system.

20.10 SPLIT VALENCE BASIS SETS

A hundred years of theory and experiment persuades us that the electrons in atoms reside in one or more "shells" surrounding the nucleus, each at a more or less fixed radius. One way to improve our basis set is recognize the shell structure for atoms Li and higher by splitting it into a part for the inner shell *core* electrons and a part for the outer shell *valence* electrons. An example is the 3–21G split valence basis set, which consists of core electrons expressed by 3 basis functions followed by 2 and 1 separate orbitals for the valence electrons. The single valence basis function 1 has a larger excursion from the nucleus than the double basis function 2. The total orbital is the sum of all the basis functions. Addition of new basis functions leads to larger basis sets, which we hope will be more complete and will be a better representation of the actual orbital. Larger split valence basis sets become complicated as can be seen from the response to `GFInput` in the route section of a relatively simple 6-31G calculation for methane:

```
17 basis functions  38 primitive Gaussians
```

20.11 POLARIZED BASIS FUNCTIONS

The electron probability density of an atom participating in a chemical bond is distorted somewhat into the region between the atoms participating in that bond. It is not spherically symmetrical. The new probability density can be represented by adding some kind of a directional basis function to the spherical orbitals already used. The new orbital is said to be *polarized*. A convenient change in the basis set is a *p* orbital added to the spherical *s* orbital of a bonded H atom. This would be denoted, for example, 6–31G(*p*). It turns out that adding *d* orbitals to the *p* orbitals of carbon in a C—H bond has a greater effect on the energy than does the the *p*-hydrogen addition. Thus we see the more common notation 6–31G(*d*). A step further is to add basis functions for both distortions, resulting in the widely used 6–31G(*d*,*p*) basis set. The notation 6–31G** is also used.

By the time split basis sets were being used, the development of *ab initio* molecular orbital theory had shifted from an effort to reproduce Slater-type orbitals to an effort to reproduce the results of experimental measurements of the chemical and physical properties of the molecules themselves. Basis sets were further improved by adding new functions, each representing some refined aspect of the physics of the molecular orbital. Electrons do not have a very high probability density far from the nuclei in a molecule, but the little probability that they do have is important in chemical bonding, so *diffuse functions*, denoted + as in 6-311 + *G*(*d*,*p*), were added in some very high level basis sets. While these basis set extensions were being made, the power of computing hardware was growing to accommodate them.

20.12 HETEROATOMS: OXYGEN

We shall extend our calculations to the STO-3G methanol input file given as File 20.9. The stored basis functions for H, C, and O are used. For this calculation, STO-3G is requested in the # route section of the input file. The keyword

```
# STO-3G opt=z-matrix

methanol

  0   1
 C
 O,1,R1
 H,1,R2,2,A
 H,1,R2,2,A,3,D1
 H,1,R2,2,A,3,D2
 H,2,R2,1,A,3,D3
      Variables:
 R1=1.5
 R2=1.1
 A=110.
 D1=-120.
 D2=120.
 D3=60.
```

FILE 20.9 Input file for a GAUSSIAN© STO-3G calculation on methanol. The dihedral angle D3 accounts for the refusal of O to conform to the H—C—C plane.

`opt=z-matrix` requests optimization in internal coordinates and output in *z*-matrix format. Construction of the z-matrix input geometry goes as follows:

Take carbon as the reference atom. Call it atom 1. The oxygen atom is attached to the carbon atom at a distance R1. Three hydrogen atoms are connected to C1, each at a distance R2. Distance R1 is specified as 1.5 Å in the `Variables:` section of the input file and distance R2 is assigned the value 1.1 Å. Bond distances are guesses taken from prior experiments, possibly spectroscopic studies.[3] The simple O—C—H angles are guessed to be pretty close to the tetrahedral angle of methane, $A \cong 110°$, also entered into the `Variables`: section of the input file. Along with the simple angles, there are two dihedral angles $D1 = -120°$ and $D2 = 120°$ relative to atom 2. Finally, there is a hydrogen (atom 5) attached to O (atom 2) at a bond length of 1.1 Å and a simple angle $A \cong 110°$. This is the OH hydrogen that is out of the H—C—O plane at a dihedral angle of $D3 = 60°$.

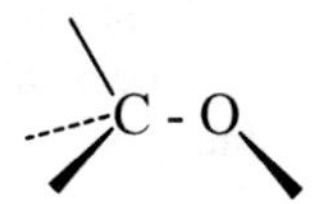

There is no external coordinate system for the *z*-matrix input file. The position of each atom is given only relative to some other atom or atoms in the molecule. The input geometry is said to be given in *internal coordinates*.

[3]Typical spectroscopic bond lengths are 1 to 2 Å (Chapter 18). Electron diffraction studies yield 1.1 Å for the C—H bond length in 1,2-dichloroethane in the vapor phase.

```
                     Z-MATRIX (ANGSTROMS AND DEGREES)
CD Cent Atom N1    Length/X    N2  Alpha/Y    N3    Beta/Z        J
---------------------------------------------------------------
  1   1  C
  2   2  O     1 1.439156( 1)
  3   3  H     1 1.066156( 2) 2 109.116( 6)
  4   4  H     1 1.066156( 3) 2 109.116( 7)  3 -120.422( 10)  0
  5   5  H     1 1.066156( 4) 2 109.116( 8)  3  119.800( 11)  0
  6   6  H     2 1.066156( 5) 1 109.116( 9)  3   60.253( 12)  0
---------------------------------------------------------------
                      Z-Matrix orientation:
---------------------------------------------------------------
Center     Atomic     Atomic            Coordinates (Angstroms)
Number     Number      Type            X           Y           Z
------------------------------------------------------ ---------
   1          6          0        0.000000    0.000000    0.000000
   2          8          0        0.000000    0.000000    1.439156
   3          1          0        1.007363    0.000000   -0.349152
   4          1          0       -0.510100    0.868665   -0.349152
   5          1          0       -0.500630   -0.874157   -0.349152
   6          1          0        0.499824    0.874618    1.788308
```

FILE 20.10 Optimized Geometry from a GAUSSIAN© STO-3G Calculation on Methanol (internal and Cartesian coordinates).

Even this simple molecule would be a discouraging prospect for hand calculation as seen from the printout of the number of basis functions and primitive Gaussians:

```
14 basis functions,     42 primitive gaussians,     14 cartesian
   basis functions
 9 alpha electrons           9 beta electrons
```

The approximate input geometry is optimized during the program run. Output files can be obtained in either internal or Cartesian (external) coordinates. Bond distances and angles have been corrected in small amounts by the program.

The expected distinction between the C—H bond lengths and the O—H length in the optimized geometry (1.439 vs. 1.066 Å) is found in lines 2 and 3 of the *z*-matrix. Also, all simple angles and dihedral angles have been shifted from their original inputs. The presence of an oxygen atom in the molecule distorts the tetrahedral geometry of the —CH_3 group slightly. The calculated energy is

```
TOTAL ENERGY = -113.5386327
```

20.13 FINDING $\Delta_f H^{298}$ OF METHANOL

Either the GAUSSIAN or GAMESS calculation gives the STO-3G molecular energy as -113.538633 hartrees. This energy is relative to the separated nuclei and electrons.

If we knew the energy released when C, O, and H atoms are formed from their subatomic particles, and the energy released when an appropriate number of C, O, and H atoms combine to form C(graphite), O_2(g), and H_2(g), all in the standard state at 0 K, we could find the energy of formation of CH_3OH(g) at 0 K.

$$\mathrm{C(gr)} + \tfrac{1}{2}\mathrm{O_2(g)} + 2\mathrm{H_2(g)} \rightarrow \mathrm{CH_3OH(g)}$$

$$\Delta_f H^{298}(\text{methanol}) = H^{298}(\text{methanol(g)})$$

$$-\left[H^{298}\,(\mathrm{C(gr)}) + \tfrac{1}{2}H^{298}\,(\mathrm{O_2(g)}) + 2H^{298}\,(\mathrm{H_2(g)})\right]$$

The numbers in square brackets have been established both by computation and by experiment. They are shown at the left of Fig. 20.7.

There are six downward steps shown on the left of Fig. 20.7. Three are large steps given in hartrees. Their sum $-114.78647\ E_h$ represents the addition of a requisite number of electrons to a carbon nucleus, four hydrogen nuclei, and an oxygen nucleus to produce the gaseous atoms C, H, and O. The bottom three steps on the left represent the enthalpy in kcal mol^{-1} obtained when a mole of C atoms, two moles of H_2 molecules, and $\frac{1}{2}$ mole of O_2 molecules, all in their standard states, are formed from the gaseous atoms. The elements in their standard states have an enthalpy $H^{298} = -115.471973\ E_h$ at the short solid line.

The enthalpy change from nuclei and electrons to the methanol molecule in its standard state is given by the long vertical line on the right-hand side of Fig. 20.7. The difference between the calculated enthalpy of the molecule in the standard state and that of the composite elements in their standard states is the enthalpy difference we seek, $\Delta_f H^{298}$. Thus the whole argument depends on the length of the vertical line on the right-hand side of Fig. 20.7. Notice that $\Delta_f H^{298}$ has the opposite sign from

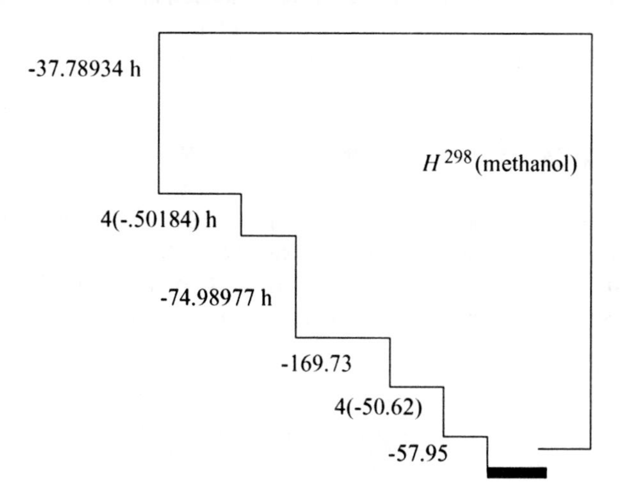

FIGURE 20.7 The G3(MP2) thermochemical cycle for determination of $\Delta_f H^{298}$ of methanol. Step sizes are not to scale. Values not designated h (hartrees) are in units of kcal mol^{-1}.

the difference in calculated enthalpies in Fig 20.7 because it is the enthalpy *input* for the reaction forming the molecule from its constituent elements.

The total drop in enthalpy for all the steps on the left-hand side of Fig. 20.7 is –115.471973 E_h. When we substitute our recently found STO-3G numbers to find $\Delta_f H^{298}$, we get $-115.471973-$`(-113.5386327)` $=-1.93334$ hartrees with a sign change to $+1.93334$ E_h. The result is a disaster, more than 5000 kJ mol^{-1} for a $\Delta_f H^{298}$ that we expect will be no more than 100 kJ mol^{-1}or so on the basis of experimental thermochemistry done on molecules of comparable size. What went wrong?

The procedure as described involves a serious mismatch. The atomic numbers on the left-hand side of Fig. 20.7 are state-of-the-art experimental and computational results, while the molecular number on the right-hand side is an STO-2G approximation. A large approximate number subtracted from a large accurate number gives a very poor result. We could go back and substitute STO-3G energies for the atoms, but a more rewarding approach is to go forward and try to improve the molecular enthalpy.

The experimental value of $\Delta_f H^{298}$(methanol) is about –200 kJ mol^{-1}, that is, $\sim$0.0767 E_h. In order to obtain 1% accuracy in the calculated result, we must achieve a cumulative accuracy of $\pm 0.000767 E_h$ in our calculations. This demanding standard is the reason why computed energies and enthalpies are usually expressed to a precision of five or six digits beyond the decimal point. In other words, we are working in the realm of microhartrees. We must look further into the daunting problem of finding ways to achieve 0.000767% or 767 *parts per million* accuracy in our *ab initio* calculations.

When the STO-2G basis set is expanded to STO-6G, we get a `TOTAL ENERGY = -114.6409388873`. At a higher level of 6-31G (MP2), the output file yields a `TOTAL ENERGY = -115.1928829735`. Using the opt keyword, the output file reads `TOTAL ENERGY = -115.2040086436 OPTIMIZED`. These results lead to $\Delta_f H^{298}$ 0.831034 E_h = 2182 kJ mol^{-1} and $\Delta_f H^{298}$= 0.27909 E_h = 733 kJ mol^{-1} and $\Delta_f H^{298}$ = 0.267965 E_h = 704 kJ mol^{-1}. Evidently we are moving in the right direction but we still have a long way to go.

20.14 FURTHER BASIS SET IMPROVEMENTS

The literature documents a long history of improved basis sets. Details of what may be the culminating effort in this series, the basis set `G3MP2Large`, are given in the original publications (Curtiss et al., 1999). The basis set itself is available on the web (http://chemistry.anl.gov/compmat/g3theory.htm) for use with the `gen` keyword if desired.

20.15 POST-HARTREE–FOCK CALCULATIONS

No matter how good the basis set is made by extension toward an infinite set, one encounters the *Hartree–Fock limit* on the accuracy of molecular energy, because the influence of one electron upon the others has not been fully accounted for in the

SCF averaging procedure. The difference between a Hartree–Fock energy and the experimental energy is called the *correlation energy*. To remedy this fault, *correlated* models are made up which consist of a linear combination of the Hartree–Fock solution plus new basis functions representing singly, doubly, and so on, substituted wave functions:

$$\psi = a\psi_0 + \sum_{ia} a_i^a \psi_i^a + \sum_{ijab} a_{ij}^{ab} \psi_{ij}^{ab} + \cdots$$

In this equation, $ij\ldots$ designate occupied *spin orbitals* (orbitals treated separately according to electron spin α or β), and $ab\ldots$ designate excited orbitals of higher energy than $ij\ldots$ called *virtual orbitals*. Virtual orbitals have small but nonzero occupation numbers. The method is called configuration interaction (CI). The first sum on the right above includes singly substituted orbitals (CIS). Inclusion of the second sum on the right leads to doubly substituted orbitals (CID), while inclusion of both sums is (CISD), and so on. A QCISD(T) method, with a 6-31G(*d*) basis set QCISD(T)/6-31G(*d*), includes exponential terms in the expansion and is generally considered to give a better estimate of the energy than do simple CI terms alone (Pople, 1999).

20.16 PERTURBATION

Another method of progressing beyond the Hartree–Fock limit is by inclusion of *many-body perturbation terms* (Atkins and Friedman, 1997):

$$E = E_0 + \lambda_i E_i \qquad (i = 1,2,\ldots)$$

The terms $\lambda_i E_i$ are small energies due to perturbations of the larger Hartree–Fock calculation of the base energy E_0. The method was described by Moeller and Plesset (1934), long before computers existed that were powerful enough to fully exploit it. Higher values of $\lambda_i E_i$ lead to correlated energies. Perturbations at $i = 2$ or 4 are called *Moeller–Plesset MP2 and MP4* energies. They are used in corrected Gaussian calculated energies, designated, for example, MP2/6-31G.

Atomic spin–orbit coupling energies $\Delta E(\mathrm{SO})$ can be added (C: 0.14 mE_h, H: 0.0 mE_h). A "higher level correction" (HLC) and a zero-point energy *E*(ZPE) are added in the powerful combined methods called G3 or G4 calculations as well. The zero point energy arises because the ground state of a quantum harmonic oscillator is one-half quantum above the bottom of its parabolic potential well (Section 18.2). The summed zero-point energies of all atoms in a molecule, oscillating about their equilibrium positions, is *E*(ZPE). The HLC, 0.009279 E_h per pair of valence electrons for a neutral molecule in the ground state, is a purely empirical factor, parameterized so as to give the minimum discrepancy between a large test set of accurately known experimental energies and calculated energies.

HF/6-31G(d) FOpt	HF geometry
HF/6-31G(d) Freq	ZPE and vib freq
MP2(full)/6-31G(*d*) Opt	MP2(full) geo
QCISD(T)/6-31G(*d*)	QCI energy
MP2/GTMP2Large	GTMP2Large: energy

Scheme 20.1 A computational chemical script. The G3MP2Large set also goes under the names G3MP2large and GTMP2Large. (If one input notation gives you an error message, try one of the others.)

The sum of all these five energy terms with a G3 basis set at the MP2 post-Hartree–Fock level is E_0[G3(MP2)]:

$$E_0[\mathrm{G3(MP2)}] = E(\mathrm{QCISD(T)/6\text{–}31G}(d)) + \Delta E_{\mathrm{MP2}} + \Delta E(\mathrm{SO}) + E(\mathrm{HLC}) + E(\mathrm{ZPE})$$

20.17 COMBINED OR *SCRIPTED* METHODS

A quantum mechanical *script* is a list of procedures that are carried out sequentially and automatically once the program has been started. Combined or scripted programs like the GAUSSIAN© family and the CBS© group by Petersson (1998) are very popular. A sample script is shown as Scheme 20.1.

Using the results from the scripted sequence, an extrapolation can be carried out as shown in Fig. 20.8. The desired energy E[(QCISD(T)/G3MP2Large] cannot be obtained by direct calculation, but the other three energies can. With these three energies, extrapolation to the desired energy is carried out as diagramed in the figure. The line from E[MP2/6-31G(*d*)] to E[MP2/GTMP2Large] represents lowering of the MP2/6-31G(*d*) energy when the basis set is expanded to MP2/GTMP2Large. The line from E[MP2/6-31G(*d*)] to E[QCISD(T)/6-31G(*d*)] is the 6-31G(*d*) energy lowering when the QCISD(T) correlation is taken into account.

All of the calculations above are carried out within the G3(MP2) script. The output portion of the G3(MP2) printout of H^{298} for methanol shown in File 20.11 contains the message `G3MP2 Enthalpy = -115.547933`. This value inserted into Fig 20.7 as H^{298} gives $\Delta_f H^{298}$(methanol) $= -199.4$ kJ mol^{-1} as compared to the experimental value of -201.5 ± 0.3 kJ mol^{-1}. The difference is 1%. We have reached our accuracy goal for methanol and, by implication, for any molecule we choose, subject only to limitations on computer power.

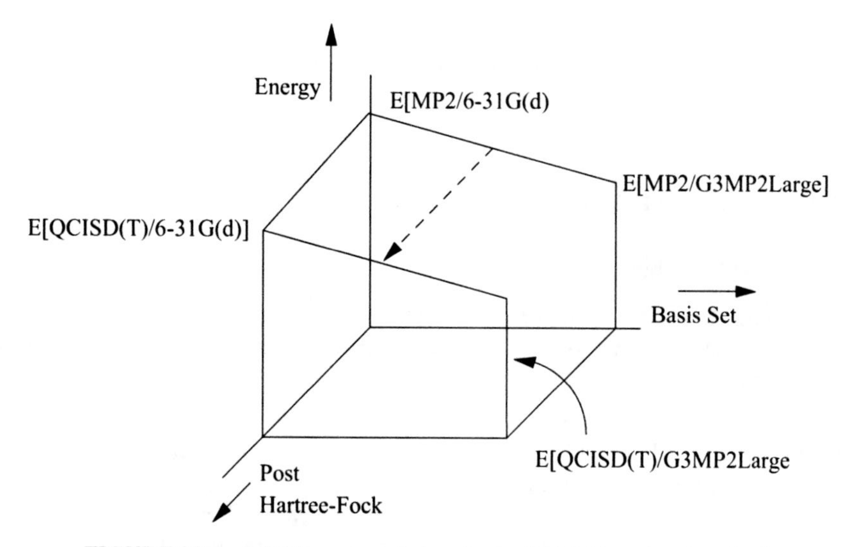

FIGURE 20.8 Additive extrapolations in the G3(MP2) scripted method.

```
Temperature=          298.150000     Pressure=              1.000000
 E(ZPE)=              0.049406       E(Thermal)=            0.052749
 E(QCISD(T))=         -115.374855    E(Empiric)=           -0.064953
 DE(MP2)=             -0.161818
 G3MP2(0 K)=          -115.552220    G3MP2 Energy=          115.548877
 G3MP2 Enthalpy=      -115.547933    G3MP2 Free Energy=   -115.574944
 1\1\GINC-DOUG\Mixed\G3MP2\G3MP2\C1H4O1\DROGERS\08-Jun-2009\0\\# g3mp2\
 \methanol\\0,1\C,0,0.0185989192,0.032214273,0.0391675131\O,0,0.0052594
 495,0.0091096338,1.4617939406\H,0,1.0367987363,0.0123972426,-0.3682155
 629\H,0,-0.5076630411,0.9040926655,-0.3682155629\H,0,-0.5013614095,-0.
 8683834341,-0.2863534272\H,0,0.4651982827,0.8057470613,1.7693787775\\V
 ersion=AM64L-G03RevD.01\State=1-A'\MP2/6-31G(d)=-115.3461339\QCISD(T)/
 6-31G(d)=-115.3748547\MP2/GTMP2Large=-115.5079522\G3MP2=-115.5522202
```

FILE 20.11 Partial GAUSSIAN G3(MP2) output. (G3(MP2) is a slightly simplified version of G3.)

20.18 DENSITY FUNCTIONAL THEORY (DFT)

DFT methods are quite fast relative to G3 and G3 (MP2) scripted molecular orbital model chemistries. For this reason they have enjoyed popularity among computational chemists working on practical problems. Density functional theory is based on the fact that the average energy of an electron in an atom or a molecule is a function of its probability density ρ in the vicinity of the nucleus or nuclei, and ρ is a function of its position in space. A function of a function is a *functional*. DFT methods are largely empirical, containing adjustable parameters, though each rests on a reasonable theoretical model. One of the most accurate functionals, devised by Becke (1932) and by Lee, Yang, and Parr in 1988 is denoted BLYP. The BLYP method using three parameters is denoted B3LYP. At present, B3LYP calculations are quite satisfactory for smaller hydrocarbons but do not compete successfully in studies of larger molecules.

PROBLEMS AND EXAMPLES

Example 20.1 Methane

Extend input File 20.7 to methane by including more basis functions in z-matrix format. Use the `opt` keyword in the route section of File 20.12 to find the optimized bond length.

Solution 20.1 In File 20.12, the hydrogen atoms are added in z-matrix format sequentially below the carbon atom C, designated atom 1, the reference point. For the first H atom, we need to know only the distance relative to C, $r = 1.1$ Å. To locate the second hydrogen, we need r and the simple H—C—H angle, 109°; and for the third and fourth hydrogens we need the distance, the simple angle, and the dihedral angles with the hydrogens already in place, $-120°$ and $120°$.

```
# STO-2G opt

Methane

 0  1
C
H,1,R
H,1,R,2,A
H,1,R,2,A,3,D1
H,1,R,2,A,3,D2
 Variables:
R=1.1
A=109.
D1=-120.
D2=120.0
```

FILE 20.12 A z-matrix input file for methane. The distances and angles are estimates.

Example 20.2 A GAMESS Calculation of Methane

For those wishing to perform the calculation using GAMESS freeware, the input file consists of three command lines, a symmetry specification `cs` followed by the z-matrix as already shown. Be sure to include the blank lines in the `CONTROL` file leave a blank line for the word `Variables:`. GAMESS can be a little more cranky than user friendly GAUSSIAN, but hey, . . . you can't beat the price.

```
$CONTRL SCFTYP=RHF MULT=1 RUNTYP=OPTIMIZE COORD=ZMT $END
 $BASIS GBASIS=STO NGAUSS=2 $END
 $DATA
 Methanol
 cs

 C
 O,1,R1
 etc.

 $END
```

FILE 20.13 Control lines for a GAMESS calculation. NGAUSS can be set to 2,3, . . . , 6 as desired in line 2 of the control file. A z-matrix is selected here but either internal or external coordinates can be used (see File 20.7).

Example 20.3 Symmetry Unique Atomic Coordinates

In some cases, symmetry can be used to reduce the demand on computer resources. Although symmetry is most important in carrying out high-level calculations on large molecules, the concept can be illustrated by a very simple example, the input file for molecular hydrogen in the GAMESS procedure (Section 20.8.1). In this example, we shall go from the GAMESS input file for molecular hydrogen in File 20.7 to one that makes use of the plane of symmetry separating one hydrogen atom from an identical partner at an equal distance from the center of symmetry.

Solution 20.3 The file presented in File 20.7 contains more information and demands more computation than is necessary. Any calculation done on one H atom must be identical to the calculation done on the other. Therefore, why not do one calculation and be done with it?

This is accomplished by replacing the two-line Cartesian coordinate location matrix by one line specifying the location of one of the two hydrogen atoms relative to their center of symmetry. In the CONTRL line, COORD=CART is replaced by COORD=UNIQUE and Dn 1 is replaced by Dnh 4. The distance of the selected H atom from the center of symmetry is one half the bond length, 0.35 Å. If the bond length is not known, RUNTYP=OPTIMIZE can be inserted before COORD=UNIQUE.

```
  $CONTRL SCFTYP=RHF RUNTYP=OPTIMIZE COORD=UNIQUE $END
 $BASIS GBASIS=STO NGAUSS=2 $END
 $DATA
Hydrogen
 Dnh 4

 H    1.0    0.0    0.0    0.35
  $END
```

The energy output of this simplified file is the same as the more complicated File 20.7.

```
TOTAL ENERGY = -1.0934083240
```

Optimizing from a bond length estimate that is 10% in error leads to

```
E = -1.0938179551
```

which is different by about 0.25 kcal mol^{-1} (slightly lower).

Problem 20.1

Find the energy of a particle in a one-dimensional box of length l. by the variational method. Take $\Psi(x) = A \sin n\pi x/l$ as a trial function. Note that the Hamiltonian is $-(\hbar^2/2m)(\partial^2/\partial x^2)$.

Problem 20.2

Electron diffraction studies yield two different Cl—Cl distances (not bonded) in 1,2-dichloroethane $CH_2Cl{-}CH_2Cl$. Explain why this is so. What influence does this fact have on molecular structure and energy calculations like MM and G3(MP2)?

Problem 20.3

The benzene molecule can be modeled as a potential well with a square planar bottom 0.400 nm on each side containing six $2p\pi$ electrons (ignoring all the other electrons in the molecule). What is the degeneracy of the occupied orbitals and what is the degeneracy of the first two virtual orbitals according to this model? In what respect does this model fail to represent the true levels of the benzene molecule?

Problem 20.4

Given the model proposed in Problem 20.3, what is the minimum energy required to promote a ground state electron to its first excited state? What is the wavelength of electromagnetic radiation that will supply this energy? Is this in the visible region? Is benzene colored?

Problem 20.5

(a) Add a basis function of your own to the STO-2G basis set so as to create your personal STO-3G linear combination. Use a graphing program to plot your function, call it $\varphi_{\text{mySTO-3G}} = f(r)$. STO-3G calculations were used at the research level until the 1990s.

(b) What is the energy of the hydrogen atom according to your STO-3G basis set? What is the percent difference between your value and the exact (defined) value of $E_H = 0.5000\ E_h$?

Problem 20.6

Carry out a 3-21G split valence calculation of the energy of atomic helium, He. What is the approximation to the first ionization potential of helium in this model? The experimental value is 0.903 E_h.

Problem 20.7

(a) Carry out a 3-21G calculation of the energy of the molecular ion HeH^+. Do you predict that this molecular ion exists?

(b) Carry out a 3-21G calculation of the energy of the bond energy of the molecular ion ${He_2}^+$. Do you predict that this molecular ion exists?

(c) Carry out a 3-21G calculation of the energy of the bond energy of the molecule He_2. Do you predict that this molecule exists?

(d) Carry out a full 3-21G calculation of the total energy of the lithium hydride molecule.

Problem 20.8

Suppose, on an exam, you are given the following Hartree–Fock 3-21G total enthalpies H^{298}

Methane	HF = 39.9768776	Ethene	HF = −77.6009881
Ethane	HF = 78.793948	1,3-Butadiene	HF = −154.0594565

along with the enthalpies of formation of $\Delta_f H^{298}$ (methane) = −74.4 kJ mol^{-1}, $\Delta_f H^{298}$ (ethene) = 52.5 kJ mol^{-1}, and $\Delta_f H^{298}$ (ethane) = −83.8 kJ mol^{-1}. No other information is given. Find the enthalpy of formation of 1,3-butadiene. See also Chapter 21, Section 21.4.

Problem 20.9

Recalculate the previous result, this time using the 6-31G basis set to calculate the HF values of the four hydrocarbons in the reaction

$$CH_2{=}CH{-}CH{=}CH_2 + 2CH_4 \rightarrow 2CH_2{=}CH_2 + CH_3{-}CH_3$$

Is the agreement with the experimental result improved by using the higher-order basis set? By how much?

21

PHOTOCHEMISTRY AND THE THEORY OF CHEMICAL REACTIONS

So far we have considered chemical reactions in which the energy necessary to surmount the activation barrier is obtained from the environment in the form of heat translated as molecular motion. These are thermal reactions. Light of frequency ν is also able to impart energy $E = h\nu$ to reactant molecules so that they can surmount the barrier. These are *photochemical reactions*.

21.1 EINSTEIN'S LAW

Like any chemical reaction, a photochemical reaction involves excitation of the reactant species up and over an activation energy barrier before transmission to the product state. In a photochemical reaction, the *activation energy* is supplied by incident light instead of ambient heat. The Bohr–Einstein concept of a particle of light (a photon) carrying a *quantum* of energy $E = h\nu$ leads one to suppose that the energy of an incident light particle is concentrated in one reactant molecule only, rather than being distributed evenly over an entire collection of molecules. This mechanism of energy transmission means that if the quantum E is sufficiently large, all struck molecules will acquire the activation energy and the number of molecules reacting will be equal to the number of photons striking and retained by the system. This is Einstein's law.

$$\text{molecules reacting} = \text{photons absorbed}$$

Concise Physical Chemistry, by Donald W. Rogers

Many systems follow Einstein's law, which argues strongly in favor of the photon concept of light and the proposed mechanism for energy transfer. Many systems, however, deviate widely from Einstein's law, so the true nature of photochemical reactions must be more complicated than we have pictured so far.

21.2 QUANTUM YIELDS

There are systems for which the number of reacting molecules is smaller than the number of incident photons and there are those for which it is larger, sometimes much larger. A system can be characterized by its *quantum yield* Φ, defined as

$$\Phi = \frac{\text{number of molecules reacting}}{\text{number of photons absorbed}}$$

A system that obeys Einstein's law has $\Phi = 1$.

When a quantum yield is less than 1, as in the first entry in Table 21.1, we suppose that there are energy dissipating processes going on. In a complicated molecule, the quantum of incident energy may be split up among several degrees of freedom and dissipated eventually as heat. In some molecules, a part of the incoming quantum of energy is reemitted as light. The system might emit part of the excitation energy as light and the remainder as heat.

If the energy emitted is less than the energy taken in, the wavelength of the emitted radiation is the longer of the two. This is the common observation in *fluorescence*; beaming ultraviolet light (short λ, high energy) on a fluorescent material produces visible light (longer λ, lower energy). If there is a time lag between absorption and emission, the phenomenon is called *phosphorescence* (Fig. 21.1). Explanation of a high quantum yield is not quite so simple and involves a new concept, the *chain reaction*. One postulates a reaction mechanism for a *photoinduced* chain reaction in which

1. an energetic species is produced, which
2. brings about a chemical reaction in which it is replicated, and it
3. Passes its energy on to produce another product molecule and another energetic species. By this mechanism, each incident photon can be responsible for many product molecules, so a high quantum yield results. Explosions are possible. The production of HCl from the elements H_2 and Cl_2 is an example of a

TABLE 21.1 Some Experimental Quantum Yields[a].

Reaction	Quantum Yield
$(CH_3)_2C{=}O \rightarrow CH_3C^{\bullet}{=}O + {}^{\bullet}CH_3$	0.17
$CH_3COOH \rightarrow CH_4 + CO_2$	1
$H_2 + Cl_2 \rightarrow 2HCl$	$\sim 10^5$

[a]The symbol • indicates a free radical.

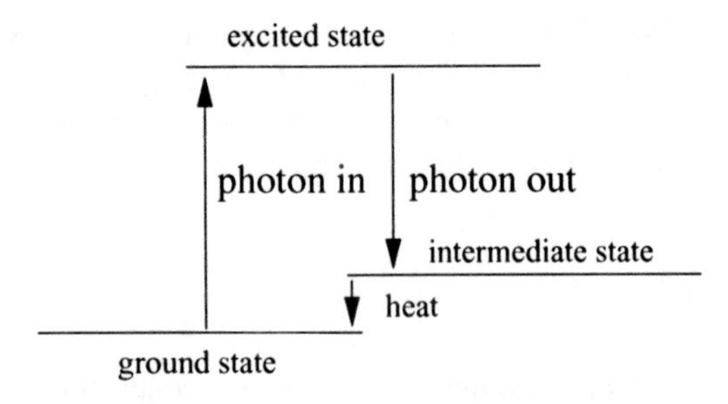

FIGURE 21.1 Mechanism for fluorescent and phosphorescent light emission.

photoinduced chain reaction. The process starts with the production of two free radicals $Cl^{\bullet}$ by absorption of a photon $h\nu$. The first step is called the *chain-initiating step*:

$$Cl_2 + h\nu \rightarrow 2Cl^{\bullet}$$

The chain initiating step is followed by the *chain-propagating steps*:

$$Cl^{\bullet} + H_2 \rightarrow HCl + H^{\bullet}$$

$$H^{\bullet} + Cl_2 \rightarrow HCl + Cl^{\bullet}$$

These steps constitute the heart of the chain reaction and continue indefinitely, which explains the high quantum yield. Ultimately, however, a free radical collides with another free radical

$$Cl^{\bullet} + Cl^{\bullet} \rightarrow Cl_2 + \text{heat}$$

or loses its energy by some other transfer mechanism. These are the *chain-terminating steps*.

21.2.1 Lipid Peroxidation

In *lipid peroxidation* or *autooxidation*, the chain-initiating step is the production of a lipid free radical $R^{\bullet}$ by removal of a hydrogen atom from the ground state molecule by light, heat, or enzymatic initiation. Lipid peroxidation follows a chain sequence. The chain-propagating steps are

$$R^{\bullet} + O_2 \rightarrow ROO^{\bullet}$$

$$ROO^{\bullet} + RH \rightarrow ROOH + R^{\bullet}$$

and the chain-terminating step is

$$ROO^{\bullet} + ROO^{\bullet} \rightarrow \text{molecular products}$$

This chain reaction is damaging to biological membranes, especially cell walls, and has been implicated in both carcinogenesis and the aging process. Within living systems, the unsaturated fatty acids are most vulnerable to autooxidation.

Living biological systems (as contrasted to dead ones) are protected from this autooxidation process through the action of small amounts of exogenous *antioxidants*, including vitamin E. Antioxidants form very reactive free radicals themselves. These radicals interfere with the chain oxygenation cycle in the propagation steps by reacting with the radical $ROO^{\bullet}$. This terminates the chain and limits damage to the organism. Vitamin E is the name given to the group of four *tocopherols*. The tocopherols are aromatic alcohols ArOH, which react with peroxyl radicals.

$$ROO^{\bullet} + ArOH \rightarrow ROOH + ArO^{\bullet}$$

The phenol radical $ArO^{\bullet}$ is stabilized by delocalization of electrons over the aromatic framework. Because of their stability, they persist in the system and eventually react with a peroxyl radical to complete the chain-breaking mechanism

$$ArO^{\bullet} + ROO^{\bullet} \rightarrow \text{products}$$

21.2.2 Ozone Depletion

Neither O nor O_2 absorbs radiation in the 200- to 300-nm (ultraviolet) region, but ozone O_3 absorbs at about 250 nm, shielding the earth's surface and creatures living on it from mutagenic DNA alteration by impact of ultraviolet photons. Among the many photochemical reactions occurring in the earth's stratosphere is a sequence in which the middle two reactions below are in balance, maintaining a low but fairly constant ozone level.

$$O_2 + h\nu\,(< 200\,\text{nm}) \rightarrow 2O$$

$$O_3 + h\nu(UV) \rightarrow O_2 + O$$

$$O + O_2 + M \rightleftarrows O_3 + M$$

$$O + O_3 \rightarrow 2O_2$$

M is a molecule, possibly N_2, that does not enter into the reaction but exchanges energy through collisions with those that do.

Other reactions interfere with this cycle, hence they interfere with the stratospheric ozone level. An example is the chlorine cycle involving Cl_2O_2 derived ultimately from man-made chlorofluorocarbons. Stratospheric chlorine has increased by a factor of about 7 in the past 50 years, and a decrease in stratospheric ozone has been found (primarily for reasons of atmospheric turbulence) over Antarctica. This is the source of the polar "ozone hole." There is no reason to suppose that ozone depletion, though it is most easily monitored over the poles, is limited to the polar stratosphere.

21.3 BOND DISSOCIATION ENERGIES (BDE)

In photoinduced and many other mechanisms, we are concerned with the stability and rate of formation of free radicals, for example,

$$CH_4 \rightarrow CH_3^\bullet + H^\bullet$$

The enthalpy of *homolytic cleavage* is not easy to measure for these fleeting species, but they can be calculated as an ordinary dissociation:

$$AB \rightarrow A^\bullet + B^\bullet$$

The *bond dissociation energy* BDE of ethane

$$CH_3CH_3 \rightarrow CH_3^\bullet + CH_3^\bullet$$

to produce methyl radicals is a simple example of the more general case for hydrocarbons and their radicals:

$$CH_3R \rightarrow CH_3^\bullet + R^\bullet$$
$$\mathrm{BDE}\,[CH_3R] = \Delta_f H_{298}\left[CH_3^\bullet\right] + \Delta_f H_{298}\,[R^\bullet] - \Delta_f H_{298}\,[CH_3R]$$

The radical R may be branched, or it may contain double or triple bonds, or it may contain two or all three of these structural features (Rogers et al., 2006).

21.4 LASERS

Some atoms and molecules can undergo a *population inversion*, such that an upper energy state is more highly populated than the ground state. When this happens, incident radiation is reemitted with its initial energy plus an energy gain obtained as the *gain medium* returns to its ground state. In commercial lasers, atoms or molecules comprising the gain medium are intentionally driven into a population inversion by an optical pump, which emits high-energy radiation or fast electrons.

A *laser* consists of an optical pump that brings about a population inversion in a gain medium which *lases*, bringing about a slight radiative amplification of the light source. The radiation from this laser gain medium is reflected back and forth between two mirrors so that amplification is increased on each pass through the medium. One mirror is intentionally made less reflective than the other so that some of the highly amplified radiation passes through the less reflective mirror and is emitted as a *laser beam*. Einstein developed the theory of lasers long before the first commercial laser was produced.

21.5 ISODESMIC REACTIONS

Ab initio enthalpy calculations of H^{298} for hydrocarbons containing more than 2 or 3 carbon atoms are often converted to thermodynamic values by setting up an *isodesmic* reaction in which an experimental $\Delta_f H^{298}$ is known for all participants but one (Hehre et al., 1970). The single remaining unknown $\Delta_f H^{298}$ is calculated from the known experimental values by difference. Computed $\Delta_f H^{298}$ values for individual molecules on the right and left of the equation may suffer considerable error; but if the bond types are the same on both sides of the reaction, errors tend to cancel because they arise from similar computational defects. Isodesmic reactions are usually set up such that a relatively complicated molecule is compared to several simple molecules like methane and ethane, for which the experimental values are thought to be very accurate.

21.6 THE EYRING THEORY OF REACTION RATES

Henry Eyring (1935) worked out a theory of reaction rates, which explains the *Arrhenius law* of temperature dependence and which leads to a conceptual picture that has been widely used in many branches of chemistry. In Section 20.2, we based our qualitative thinking about complicated molecules on the system of the hydrogen molecule ion because it is the only molecule simple enough to describe completely. So we would like to base our thinking about complicated reaction mechanisms on the simpler system of one hydrogen atom substituting for another[1]:

$$H + H - H \rightarrow H - H + H$$

The attack of a hydrogen atom on a hydrogen molecule encounters an ellipsoidal repulsive field with the minimum repulsion at either end of the molecule.

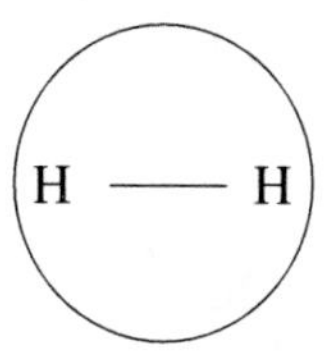

Under the simplifying assumption that the attack of the hydrogen atom on H_2 is an end-on attack, only two distances are necessary to describe the reacting system, the distance from the the attacked hydrogen molecule to the incoming H, and the distance from the opposite end of the molecule to the outgoing H. Call these distances r'_{HH} and

[1]Experimentally, deuterium atoms D were used because they can be distinguished from H by their difference in mass.

r_{HH}. At some instant, the incoming H and the departing H must be equidistant from the central hydrogen atom so that the three atoms form an energetic intermediate:

$$\mathrm{H+H-H} = \left[\mathrm{H} \overbrace{\cdots}^{r'_{HH}} \mathrm{H} \overbrace{\cdots}^{r_{HH}} \mathrm{H} \right] = \mathrm{H-H+H}$$

The *bond-forming* configuration is on the left and the *bond-breaking* configuration is on the right of the central structure, which is the *activated complex*.

21.7 THE POTENTIAL ENERGY SURFACE

In the study of molecular structure, we plotted the potential energy of the molecule as a function of internuclear distance and found that it has a minimum at the equilibrium bond length. Some of the qualities of this potential energy curve carry over to the activated complex, but now the potential energy is a more complicated function of two variables r'_{HH} and r_{HH}. It requires a three-dimensional surface or a two-dimensional contour map for its representation. Eyring has constructed such a surface represented by the equivalent contour map in Fig. 21.2.

This contour map is drawn in the same way that a forest service map of Western Montana would be drawn. Each contour corresponds to a constant potential energy that would occupy the dimension out of the plane of the paper if the three-dimensional representation were used. The hyperbola connecting the two arrows is the locus of potential energy minima of the system as the reaction takes place. Contour 4 bounds a potential energy plateau.

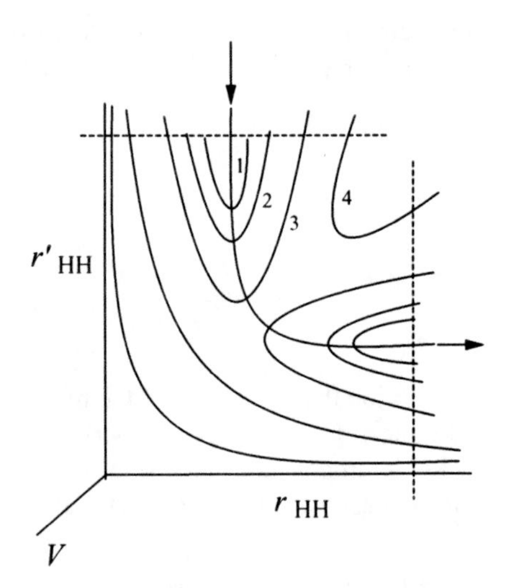

FIGURE 21.2 Eyring potential energy plot for the reaction H + H—H → H—H + H.

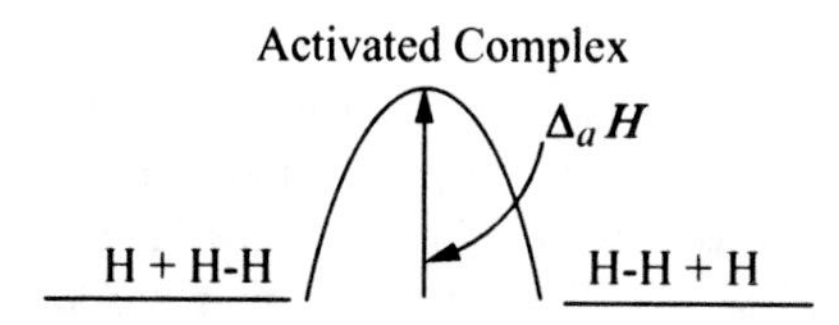

FIGURE 21.3 Activation of the symmetrical reaction H + H—H → H—H + H. The height of the barrier is about 40 kJ mol^{-1}.

If a slice is taken at the horizontal dotted line at the top of Fig. 21.2, its shape is that of a typical two-center molecular potential energy curve with a minimum like those in Fig. 20.2. This is because the distance r'_{HH} is large and the potential energy as a function of r_{HH} is essentially that of an unperturbed H_2 molecule. The same argument holds for a slice taken at the vertical dotted line to the right of the figure; r_{HH} is large and the system is an H_2 molecule with the H atom so far away as to have no effect on it.

The curved line connecting the reactant H—H potential energy basin with the product H—H potential energy basin is the path a system must follow if the transformation from one H—H to the other is to take place. This is the *reaction coordinate*, leading from reactants, over an *activation barrier* to products. During the reaction, as the system moves along the reaction coordinate, there is a rise in potential energy to a saddle point which is a kind of "mountain pass" between the two potential energy basins. At the top of the pass, $r'_{HH} = r_{HH}$ and the system exists as the activated complex. The potential energy as a function of the reaction coordinate for the system appears as shown in Fig. 21.3.

The potential energy of products is equal to the potential energy of the reactants in the case of H + H—H → H—H + H, but they are different in the general case of a more complicated reaction. Figure 21.4 shows the activation energy barrier that a more complicated system must surmount if it is to go from the reactant

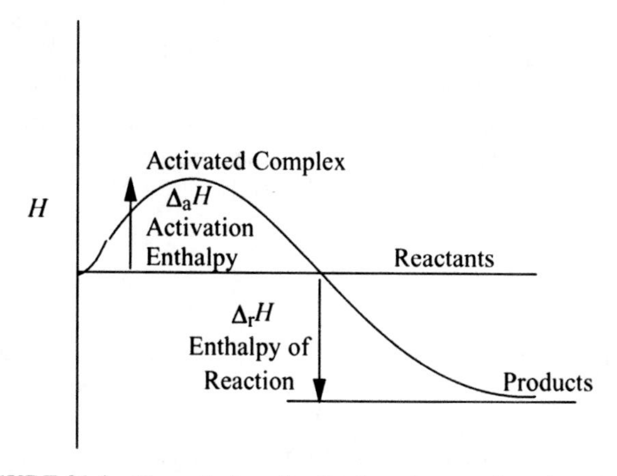

FIGURE 21.4 The enthalpy of activation of an exothermic reaction.

state to products. There are two distinct factors influencing the reaction: (1) the *thermodynamic factor*, which tells where the equilibrium is going and where it will be after infinite time, and (2) the *kinetic factor*, indicating the time scale necessary to reach equilibrium—that is, whether "infinite time" is a matter of microseconds, hours, weeks, eons, and so on. We have seen that the thermodynamic factor is largely determined by the difference between the enthalpy levels of the reactants and products. The kinetic factor is largely determined by the height of the enthalpy barrier between them.

21.7.1 Optical Inversion

An example of the Eyring mechanism is the explanation of certain geometric changes that take place during organic reactions. If a carbon atom is attached to four different groups, it is *optically active* and it rotates the plane of polarized light (Fig. 21.5). The angle of optical rotation can be measured experimentally. Such a substance can be geometrically changed by a chemical reaction involving an attacking ion. If the attacking ion is identical to group D, a substance is produced that is identical to the original compound except that it rotates light in exactly the opposite direction. This *inversion* of optical rotation is readily explained using the concept of the activated complex.

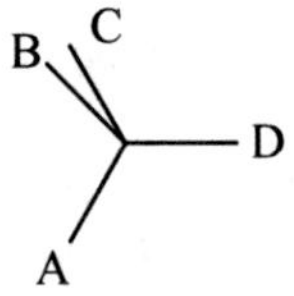

FIGURE 21.5 An optically active species.

The incoming D brings about a geometric inversion of groups A, B, and C attached to the central carbon. The geometric inversion causes a reversal in the angle of optical rotation. The enthalpy barrier to inversion, which is the controlling factor in the *rate* of inversion, is the 5-bonded carbon atom at the center of Fig. 21.6. This is clearly an unstable species, hence it represents an enthalpy maximum.

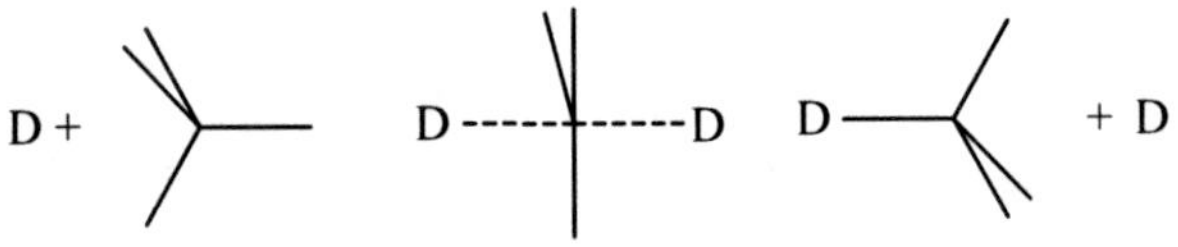

FIGURE 21.6 Inversion of optical activity.

21.8 THE STEADY-STATE PSEUDO-EQUILIBRIUM

The enthalpy difference between the reactant state and the activated complex is not a legitimate equilibrium in the thermodynamic sense; but it is useful to treat it as though it were. It is best called a *pseudo-equilibrium* or *steady-state equilibrium*.

Suppose a second-order reaction proceeds through an activated complex [AB]*:

$$\mathrm{A} + \mathrm{B} = [\mathrm{AB}]^* \rightarrow \text{products}$$

The activated complex is controlled by a pseudo-equilibrium constant:

$$K^* = \frac{[AB]^*}{[A]\,[B]}$$

If the reaction to form products is a first-order breakup of the activated complex, we obtain

$$\text{rate} = -\frac{dA}{dt} = k_1\,[AB]^*$$

The concentration of activated complex through its equilibrium constant is

$$[AB]^* = K^*[A][B]$$

whence

$$-\frac{d[A]}{dt} = k_1 K^*[A][B]$$

This equation explains the observation of second-order overall kinetics for the reaction, even though the rate constant for decomposition of the activated complex is first order. According to this mechanism, $k_1 K*$ is the observed second-order rate constant.

The incoming and outgoing species of the activated complex may be weakly bound. If so, we may regard its breakup as a kind of one-cycle molecular vibration. The complex stretches along its weakest bond, and then it falls apart to give either products or the original reactants. The vibrational energy acquired during this one-dimensional, one-cycle stretch is $E = k_B T$ because there is one degree of vibrational freedom contributing $\frac{1}{2}k_B T$ to the kinetic energy and one contributing $\frac{1}{2}k_B T$ to the potential energy. It is also true for a vibration that $E = h\nu$, hence

$$k_B T = h\nu \qquad \nu = \frac{k_B T}{h} = \frac{RT}{N_A h}$$

where N_A is the Avogadro number.

In the limit of an activated complex that always breaks up to give products, the number of vibrations taking place per second is the number of activated complexes reacting because each one-cycle vibration leads to products. Thus k_1 is ν (notice that both are frequencies) and

$$-\frac{d[A]}{dt} = \frac{RT}{N_A h} K^* [A][B]$$

If the assumption that all vibrations lead to products does not hold, one includes a *transmission coefficient* κ, which is that fraction of activated complexes that break up to give products. Now,

$$-\frac{d[A]}{dt} = \kappa \frac{RT}{N_A h} K^* [A][B]$$

or

$$k_{\text{observed}} = \kappa \frac{RT}{N_A h} K^*$$

21.9 ENTROPIES OF ACTIVATION

In general, the free energy change of a reaction varies as the natural logarithm of its equilibrium constant, so for the steady-state activation equilibrium we write

$$\Delta G^{\circ *} = -RT \ln K^{\circ *}$$

where the superscript designates the standard state of the Gibbs free energy for the activation pseudo-equilibrium. The equivalent form for $K^{\circ *}$ is

$$K^{\circ *} = e^{-\Delta G^{\circ *}/RT}$$

By analogy to a true equilibrium, for which

$$\Delta G^\circ = \Delta H^\circ - T\Delta S^\circ$$

we write

$$K^{\circ *} = e^{\Delta S^*/R} e^{-\Delta H^*/RT}$$

The observed rate constant is

$$k_{\text{observed}} = \kappa \frac{RT}{N_A h} e^{\Delta S^*/R} e^{-\Delta H^*/RT}$$

which can be compared with the Arrhenius equation

$$k_{\text{observed}} = s e^{-\Delta H_a / RT}$$

Noting that ΔH_a in Arrhenius theory is the same thing as ΔH^* in Eyring's theory, the preexponential s must be

$$s = \kappa \frac{RT}{N_A h} e^{\Delta S^* / R}$$

This expression gives a preexponential that is a function of the temperature as contrasted to Arrhenius' conclusion that s is constant. Accurate measurements have indeed shown the temperature dependence of the supposed constant s.

21.10 THE STRUCTURE OF THE ACTIVATED COMPLEX

If the activated complex has a "tight" structure, its formation implies a reduction of freedom in the system, therefore a negative entropy change ΔS. If the activated complex has a loose structure by comparison to the reactants, ΔS is positive hence the preexponential factor is large as well. For some reactions, $e^{-\Delta H_a / RT}$ may be unfavorable but the reaction rate is appreciable because of a large preexponential.

The reactions

$$NO + O_3 = NO_2 + O_2$$

and

$$CH_3I + HI = CH_4 + I_2$$

have been shown to have second-order rate constants $k = 6.3 \times 10^7 \sqrt{T} e^{-2300/RT}$ and $k = 5.2 \times 10^{10} \sqrt{T} e^{-33,000/RT}$, respectively. Analyzing these two results, we conclude that the first reaction has a low activation barrier relative to the second and it has a lower preexponential than the second reaction. The entropy of activation is more negative for the nitric oxide oxidation than the iodine abstraction, suggesting a tightly bound structure for the first activated complex and a loose structure for the second activated complex.

PROBLEMS AND EXAMPLES

Example 21.1 A C—C Bond Dissociation Enthalpy (BDE)

Find the enthalpy of dissociation (BDE) of the C—C bond in ethane in the G3(MP2) model chemistry.

Solution 21.1 The term "G3(MP2) model chemistry" means that G3(MP2) calculations are used for all molecular and radical species. In the case of ethane, homolytic cleavage of the C—C bond

$$CH_3CH_3 \rightarrow 2\,CH_3^{\bullet}$$

has a $\Delta_{\rm dissoc}H^{298}$ and a BDE:

$$\begin{aligned}\text{BDE}\,[CH_3CH_3] &= \Delta_{\rm dissoc}H^{298}\,[CH_3CH_3] = 2\Delta_{\rm f}H^{298}\left[CH_3^{\bullet}\right] - \Delta_{\rm f}H^{298}\,[CH_3CH_3]\\ &= 2H^{298}\left[CH_3^{\bullet}\right] - H^{298}\,[CH_3CH_3]\\ &= 2\,(-39.752873) - (-79.646716) = 0.1410E_{\rm h}\\ &= 88.4\,\text{kcal mol}^{-1} = 370.0\,\text{kJ mol}^{-1}\end{aligned}$$

Calculated enthalpies of formation of the molecule and radicals from the nuclei and electrons H^{298} can be substituted for $\Delta_{\rm f}H^{298}$ of the molecule and radicals in Example 21.1 because the enthalpies of formation of the atoms cancel in the homolytic cleavage reaction (they are the same for 2 $CH_3^{\bullet}$ as for CH_3CH_3). The experimental value for this BDE is 89.7 kcal mol^{-1} = 375.3 kJ mol^{-1}, so the difference between the experimental value and calculated value is 1.4%.

Example 21.2 An Isodesmic Reaction

Find the enthalpy of formation of *tert*-butylmethane (2,2-dimethylpropane) by an isodesmic reaction using known values of $\Delta_{\rm f}H^{298}$(methane) $= -17.8$ kcal mol^{-1} and $\Delta_{\rm f}H^{298}$(ethane) $= -20.1$ kcal mol^{-1}.

Solution 21.2 Cheng and Li (2003) calculated $\Delta_{\rm f}H^{298}$ of *n-tert*-butyl methanes, where n = 1, 2, 3, and 4. Find $\Delta_{\rm f}H^{298}$ of the simplest example *tert*-butylmethane (2,2-dimethylpropane). The isodesmic reaction is

$$CH_3{-}C(CH_3)_2{-}CH_3 + 3CH_4 \longrightarrow 4C_2H_6$$

$C(CH_3)_4$	$3CH_4$	$4C_2H_6$
−197.35354	−40.42210	−79.65120
	−17.8	−20.1

Computed values of $\Delta_{\rm f}H^{298}$ are in the first line below the reaction, and experimental values of $\Delta_{\rm f}H^{298}$ for CH_4 and C_2H_6 are in the second line.

On the reactant side, two alkanes have a total of 24 C—H bonds and 4 C—C bonds. On the product side, one molecule has a total of 4(6) = 24 C—H bonds and 4(1) = 4 C—C bonds. The reaction is isodesmic because the number and types of bonds on the left is the same as the number and types of the bonds on the right. One value is missing from this scheme, that of $\Delta_f H^{298}$ for *t*-butylmethane.

Their calculated enthalpy of reaction $\Delta_r H^{298}$ was

$$\Delta_r H^{298} = 4(-79.64672) - (-197.35354) - 3(-40.41828)$$
$$= 0.02150\,\text{h} = 13.49\,\text{kcal mol}^{-1} = 56.44\,\text{kJ mol}^{-1}$$

This enthalpy change is used to calculate the remaining unknown from the experimental values of $\Delta_f H^{298}$ for methane and ethane:

$$\Delta_r H^{298} = 4\Delta_f H^{298}(\text{ethane}) - \Delta_f H^{298}(t\text{-butylmethane}) - 3\Delta_f H^{298}(\text{methane})$$
$$13.5 = 4(-20.1) - \Delta_f H_{298}(t\text{-butylmethane}) - 3(-17.9)$$

which leads to

$$\Delta_f H^{298}(t\text{-butylmethane}) = -40.2\,\text{kcal mol}^{-1} = -168.2\,\text{kJ mol}^{-1}$$

The experimental value is $-39.9 \pm 0.2\,\text{kcal mol}^{-1} = -166.9\,\text{kJ mol}^{-1}$.

Comments: So far, the isodesmic reaction has merely passed the test of reproducing a known experimental result. Agreement with the experimental value is also good for the second compound in the series, but experimental work is uncertain and under debate for the third compound. An experimental value is nonexistent for the last compound named, which is highly strained and has not yet been synthesized. The authors carry these calculations on in a logical sequence to obtain $\Delta_f H^{298}$ of all four compounds di(*t*-butyl)methane (−59.2), tri(*t*-butyl)methane (−55.3), and tetra(*t*-butyl) methane (−320.7 kcal mol^{-1}). Example 21.2 is a good example of verification of a method for known compounds followed by extension to unknowns that are not amenable to experimental work.

Problem 21.1

Chlorine does not react with toluene in the dark. If light is admitted to a chlorine–toluene mixture, a reaction occurs. Propose a mechanism for this reaction. Propose an experiment to support or contradict your mechanism.

Problem 21.2

The number density of one mole of methane at 1 bar and 298 K is $\rho = n/V = \text{N}_A p/RT = 2.43 \times 10^{25}\,\text{m}^{-3}$ under the ideal gas assumption. The collision cross section is $5.3 \times 10^{-19}\,\text{m}^2$, and the average speed of the methane molecule is $\langle x \rangle = (8RT/\pi M)^{1/2} = 630\,\text{m s}^{-1}$. What is the collision frequency? Carry all units through your calculation and demonstrate that the answer is, indeed, a frequency.

Problem 21.3

Find the BDE for the following reactions within the G3(MP2) model chemistry:

$$CH_2 = CH - CH_3 \rightarrow CH_2 = CH^{\bullet} + CH_3^{\bullet}$$

and

$$CH_2 = CH - CH_2 - CH_3 \rightarrow CH_2 = CH - CH_2^{\bullet} + CH_3^{\bullet}$$

Discuss the implications of this difference for free radical formation in unsaturated fatty acids as contrasted to saturated fatty acids in physiological systems (Section 21.2.1). You will need to use data from Exercise 21.1.

Problem 21.4

Three coffee cups are arranged vertically A, B, C from top cup A to bottom cup C. Cups A and B have holes in the bottom but C does not. A drains into B, and B drains into C.

$$A \rightarrow B \rightarrow C$$

The hole in A is slightly larger than the hole in B. One cup of coffee is poured into A. What happens?

1. Sketch the liquid levels in A, B, and C as a function of time.
2. The hole in A is made a little larger. Sketch A, B, and C as a function of time.
3. The hole in A is made a little larger still. Sketch A, B, and C as a function of time.
4. If the process is continued, what happens to level B?

Problem 21.5

Nitric oxide, NO, is very destructive to ozone in the stratosphere. (Commercial supersonic jet flight has been banned because of the potential increase in stratospheric NO.) Propose a mechanism for this destructive reaction.

Problem 21.6

The helium–neon laser produces light at 1152.3 nm. What is the energy of one mole (an *einstein*) of radiation at this wavelength?

Problem 21.7

Irradiation of a gaseous sample of acetone(g) by light of 313 nm brings about photochemical dissociation to give ethane(g) and CO(g). If, in an experimental run, the total radiant energy was 200 J and 9.00×10^{-5} moles of acetone dissociated, what was the quantum yield?

Problem 21.8

Part of chemical folklore is that heating a reaction by 10 degrees doubles the rate. If this is true, what is the activation energy for the temperature change from 298 to 308 K? Is this rule generally true?

REFERENCES

Allinger, N. L. 2010. *Molecular Structures*, John Wiley & Sons, Hoboken NJ.

Allinger, N. L.; Chen, K.; Lii, J.-H. 1996. *J. Comp. Chem.* **17**, 642–668.

Atkins, P. W. 1998. *Physical Chemistry*, 6th ed., Freeman, New York.

Atkins, P. W.; Friedman, R. S. 1997. *Molecular Quantum Mechanics*, 3rd ed., Oxford, New York.

Barrante, J. R. 1998. *Applied Mathematics for Physical Chemistry*, 2nd ed., Prentice-Hall, Englewood Cliffs, NJ.

Becke, A. D. 1933. *J. Chem. Phys.* **98**, 1372–1377, 5648–5652.

Barrow, G. M. 1996. *Physical Chemistry*, 6th ed., WCB/McGraw-Hill, New York.

Bohr, N. 1913. *Philos. Mag.* **26**, 1–25.

Born, M.; Heisenberg, W.; Jordan, P. 1926. *Z. Phys.* **35**, 557–615.

Cheng, M. F.; Li, W.-K. 2003. *J. Phys. Chem. A* **107**, 5492–5498.

Cohen, N.; Benson, S. W. 1993. *Chem. Rev.* **93**, 2419–2438.

CRC Handbook of Chemistry and Physics, 2008–2009, 89th ed., Lide, D. R.; ed., CRC Press, Boca Raton, FL.

Curtiss, L. A., et al. 1999. *J. Chem. Phys.* **110**, 4703–4709.

De Broglie, L. 1924. Thesis, Paris. 1926. *Ann. Phys.* **3**, 22. See also Pauling, L.; Wilson, E. B. 1935. *Introduction to Quantum Mechanics*, McGraw-Hill, New York. Reprinted 1963, Dover, New York.

Ebbing, D. D.; Gammon. S. D. 1999. *General Chemistry*, 6th ed., Houghton Mifflin, Boston.

Eğe, S. N. 1994. *Organic Chemistry; Structure and Reactivity*, 3rd ed., D. C. Heath, Lexington, MA.

Eyring, H. 1935. *J. Chem. Phys.* **3**, 107–115.

Fock, V. 1930. *Z. Phys.* **61**, 126–148.

Gibson, D. G., et al. 2010. *Science* **329**, 52–56.

Hammes, G. G., 2007. *Physical Chemistry for the Biological Sciences*, John Wiley & Sons, Hoboken, NJ.

Hartree, D. R. 1928. *Proc. Cambridge Philos. Soc.* **24**, 89–111, 111–132.

Hehre, W. 2006. Computational Chemistry, in Engel T., *Quantum Chemistry and Spectroscopy*, Pearson-Benjamin Cummings, New York.

Hehre W. J., et al. 1970. *J. Am. Chem. Soc.* **92**, 4766–4801.

Heisenberg, W. 1925. *Z. Phys.* **33**, 879–893.

Heitler, W.; London, F. 1927. *Z. Phys.* **44**, 455–472.

Henry, W. 1803. *Philsophical Transactions of the Royal Society.*

Houston, P. L. 2006. *Chemical Kinetics and Reaction Dynamics*, Dover Publications, Mineola, New York.

Irikura, K. K. *Essential Statistical Thermodynamics* in *Computational Thermochemistry*, in Irikura, K. K.; Furrip, D. J., eds. 1998. *Computational Thermochemistry*, American Chemical Society, Washington, D.C.

Kistiakowsky, G. B., et al. 1935. *J Am. Chem. Soc.* **57**, 65–75.

Klotz, I. M.; Rosenberg, R. M. 2008. *Chemical Thermodynamics. Basic Theory and Methods*, 7th ed., J. Wiley Interscience, New York.

Kondipudi, D.; Prigogine, I. 1998. *Modern Thermodynamics: From Heat Engines to Dissipative Structures*, John Wiley & Sons, New York.

Laidler, K. J.; Meiser, J. H. 1999. *Physical Chemistry*, 3rd ed., Houghton Mifflin, Boston.

Lee, C. et al. *Phys. Rev.*, **37**, 785–789.

Levine, I. N. 2000. *Quantum Chemistry*, 5th ed., Prentice-Hall, New York.

Levine, I. 2000. *Physical Chemistry* 6th ed. McGraw-Hill Inc., New York.

Lewis, G. N.; Randall, M., revised by Pitzer K. S.; Brewer, L. 1961. *Thermodynamics*, 2nd ed., McGraw-Hill, New York.

Maczek, A. 1998. *Statistical Thermodynamics*, Oxford Science, New York.

McQuarrie, D. A. 1983. *Quantum Chemistry*, University Science Books, Mill Valley, CA.

McQuarrie, D. H.; Simon, J. D. 1997. *Physical Chemistry: A Molecular Approach*, University Science Books, Sausalito, CA.

Metiu, H. 2006. *Physical Chemistry*, Taylor and Francis, New York.

Moeller C.; Plesset, M. S. 1934. *Phys. Rev.* **46**, 618–622. See also Cramer, C. J. 2004. *Computational Chemistry 2nd ed.*, John Wiley & Sons, Hoboken NJ.

Nash, L. K. 2006. *Elements of Statistical Thermodynamics*, 2nd ed. Dover Publications, Mineola, New York.

Nevins, N.; Chen, K.; Allinger, N. L. 1996a. *J. Comp. Chem.* **17**, 669–694.

Nevins, N.; Lii, J.-H.; Allinger, N. L. 1996b. *J. Comp. Chem.* **17**, 695–729.

Pauling, L.; Wilson, E. B. 1935. *Introduction to Quantum Mechanics*, McGraw-Hill, New York. Reprinted 1963, Dover Publications, New York.

Petersson, G. A. 1998. Complete basis set thermochemistry and kinetics, in Irikura, K. K.; Furrip, D. J., eds., *Computational Thermochemistry.* American Chemical Society, Washington, D.C.

Petersson et al. 1991. *J. Chem. Phys.*, **94**, 6081–6090, 6091–6101.

Pitzer K. S.; Brewer, L. Revision of Lewis, G. N.; Randall, M. 1961. *Thermodynamics*, 2nd ed., McGraw-Hill, New York.

Planck, M. 1901. *Ann. Phys.* **4**, 553–563; 717–727. See also Kuhn, T. S. 1978. *Black Body Theory and the Quantum Discontinuity 1894–1912*, The University of Chicago Press, Chicago; and Oxford, New York.

Pople, J. A. 1999. Nobel lectures. *Rev. Mod. Phys.* **71**, 1267–1274.

Rioux, F. 1987. *Eur. J. Phys.* **8**, 297–299.

Rogers, D. W., et al. 2003. *Org. Lett.* **5**, 2373–2375.

Rogers, D. W. 2005. *Einstein's **Other** Theory. The Planck-Bose-Einstein Theory of Heat Capacity*. Princeton University Press, NJ.

Rogers, D. W.; Zavitsas, A. A.; Matsunaga, N. 2005. *J. Phys. Chem. A* **109**, 9169–9173.

Rogers, D. W.; Matsunaga, N.; Zavitsas, A. A. 2006. *J. Org. Chem.* **71**, 2214–2219.

Rogers, D. W.; Zavitsas, A. A.; Matsunaga, N., 2009. *J. Phys. Chem. A*, 113, 12049–12055.

Rosenberg, R. M.; Peticolas, W. L. 2004. *J. Chem. Ed.* **81**, 1647–1652.

Schrödinger, E. 1925. *Z. Phys.* **33**, 879.

Schrödinger, E. 1926. *Ann. Phys.* **79**, 361, *Ann. Phys.* **79**, 734.

Silbey, R. J. et al. 2005. *Physical Chemistry 4th ed.*, John Wiley & Sons, Hoboken NJ.

Slater, J. C. 1930. *Phys. Rev.* **35**, 210–211.

Steiner, E. 1996. *The Chemistry Maths Book*, Oxford Science Publications, NY.

Streitwieser Jr., A. 1961. Molecular Orbital Theory, John Wiley & Sons, Hoboken NJ.

Treptow, R. S. 2010. *J. Chem. Ed.* **87**, 168–171.

Uhlenbeck G. E.; Goudsmit, S. 1925. *Naturwissenschaften* **13**, 953–954.

Webbook.nist.gov

Zavitsas, A. A. 2001. *J. Chem. Ed.* **78**, 417–419.

Zavitsas, A. A.; Rogers, D. W.; Matsunaga, N. 2008. *J. Phys. Chem. A* **112**, 5734–5741.

Zewail, A. H. 1994. *Femptochemistry: Ultrafast Dynamics of the Chemical Bond*, Vols. 1 and 2, World Scientific, Singapore.

ANSWERS TO SELECTED ODD-NUMBERED PROBLEMS

CHAPTER 1

1.1 8.65 m^3; **1.3** 98 g; **1.7** 92.7 H_2; **1.9** 0.0798 m^3, 0.0168 kg; **1.11** 0.032 kg; **1.13** 3.7 kJ mol^{-1}; **1.15** 515 m s^{-1}.

CHAPTER 2

2.1 V = 17.9; **2.3** dm^6 bar mol^{-2}, dm^3 mol^{-2}; **2.5** $b = V_c/3$; **2.7** V = 0.55 dm^3; **2.9** z = 0.944; **2.13** $pV = 24.7881 - 0.0100p + 5.184 \times 10^{-5}p^2 + 1.4977 \times 10^{-7}p^3$.

CHAPTER 3

3.1 0.667; **3.3** V = 3922 J, $v = 19.8$ m s^{-1}; **3.5** 5.46 J K^{-1}; **3.11** 25.08 °C.

CHAPTER 4

4.1 2.39×10^{-4} K; **4.3** −5152 kJ mol^{-1}; **4.5** −1268 kJ mol^{-1}; **4.7** 14.1 kJ mol^{-1}.

CHAPTER 5

5.1 31.2 kJ mol^{-1}; **5.3** 11.5 J; **5.5** negligible; **5.7** 1.24 J K^{-1} mol^{-1}; **5.9a** 22.0 J mol^{-1}; **5.9b** 109.1 J mol^{-1}.

CHAPTER 6

6.3 88 J K^{-1} mol^{-1}; **6.5** 79.2 J K^{-1} mol^{-1}; **6.7** 46.5 J K^{-1} mol^{-1}; **6.9** −893 kJ mol^{-1}.

CHAPTER 7

7.3 $\Delta_{sol} H_{298} > 0$; **7.5** 3.46 J K^{-1} mol^{-1}; **7.7** 179 kJ mol^{-1}, 1.93.

CHAPTER 8

8.1 62.3, 37.7%, >60/40; **8.3** 8.15 pm; **8.5** 3.07 pm; **8.7** 2.61exp33, 1.28exp33.

CHAPTER 9

9.1a 2; **9.1b** 3; **9.5** 279.6 K.

CHAPTER 10

10.1 0.693, 3.46×10^{-2} min^{-1}; **10.3** 384 crabs; **10.5** 4515 ± 86 y; **10.9** 2.1 dm^3 mol^{-1} s^{-1}.

CHAPTER 11

11.1a 4.84 unit unspecified; **11.1b** 6 unit unspecified; **11.1c** sphere: 24.1% smaller; **11.3** 126 nm; **11.5** 6.

CHAPTER 12

12.1a 10%; **12.1b** 0.1711 mol; **12.1c** 1.901 molal; **12.1d** you don't have enough information; **12.1e** 93.34 cm^3; **12.1f** 1.833 molar; **12.5** 100.7; **12.7** 50.

CHAPTER 13

13.1 –54.8 kJ mol^{-1}; **13.3** 59 μg; **13.5a** 45.9 S; **13.5b** 368 S; **13.7** 55.35 mols dm^3, 1.81×10^{-6}, 1.01×10^{-14}, 1.00×10^{-7} mol dm^3.

CHAPTER 14

14.1a copper; **14.1b** 0.740 volts; **14.3a** $K_{sp} \cong 5 \times 10^{-13}$; **14.3b** $[Ag^+] = 7 \times 10^{-7}$ There are no units because this is a ratio to a standard state; **14.5** $[Fe^{+3}(aq)] \approx 10^{-29}$ mol dm^3 (With these approximations, we can take this as an approximate concentration.) **14.7** pH = 5.6 (slightly acidic); **14.9** –2.12 volts.

CHAPTER 15

15.1 4.57×10^{14} Hz, 3.03×10^{-19} J; **15.5** $\mathrm{x} \cdot \mathrm{x} = \begin{pmatrix} 30 & 36 & 42.3 \\ 66 & 81 & 96.6 \\ 102.7 & 126.8 & 151.81 \end{pmatrix}$,

$$\mathrm{x} \cdot \mathrm{x}^{-1} = \begin{pmatrix} 1 & -3.553 \times 10^{-15} & 0 \\ 3.553 \times 10^{-15} & 1 & 0 \\ 4.441 \times 10^{-15} & -4.621 \times 10^{-15} & 1 \end{pmatrix}$$

CHAPTER 16

16.1 $y(x) = \int_{-\infty}^{\infty} x e^{-x^2} dx = 0$; **16.3** $\phi(x) = e^{-\lambda x}$; **16.5** $\Psi(x) = \sqrt{2} \sin \frac{2\pi x}{\lambda}$; **16.9** $\Psi(x) = \sqrt{2} \sin \frac{2\pi x}{\lambda}$.

CHAPTER 17

17.1 –2, 1, -1, 0, $\sin^2\theta - \cos^2\theta$; **17.3a** 22.22×10^{-6} cm; **17.3b** 22.22×10^{-8} m; **17.3c** 22.22×10^{-15} nm; **17.3d** 22.22×10^{-18} pm, 22.22×10^{-14}A, 1.349×10^{15} Hz, 8.938×10^{-19} J; **17.5** 1.00×10^{-28} J; **17.7** $\frac{1}{\sqrt{n}}$; **17.9** $E_{\text{He}} = -77.45$, about 2%.

CHAPTER 18

18.1 6.63×10^{-20} J; **18.5** 2142.5 cm^{-1}, 1856 N m^{-1}; **18.9** 2.82×10^{-29} C m = 8.47 D; **18.11** 15.9×10^{-24} cm^3 mol^{-1}, 1.44 D; **18.13** $CH_3C(=O)OCH_3$

CHAPTER 19

19.1a -168 kJ mol^{-1}; **19.1b** -7 kJ mol^{-1}; **19.3** 105.1°. The experimental value is 104.5°; **19.7** -31.7 kcal $mol^{-1} = -132.7$ kJ mol^{-1}. The experimental value is -32.60 ± 0.05 kcal $mol^{-1} = 136.4 \pm 0.2$ kJ mol^{-1}.

CHAPTER 20

20.3 the degeneracy pattern is 1,2,1,2, unlike the 1,2,2,1 pattern in the more common Huckel model; **20.7a** `UHF=-2.8874388`; **20.7b** `UHF=-4.8708157`; **20.7c** The molecule does not exist; **20.7d** The molecule is lower in energy than its atoms by 30.3 kcal mol^{-1} = 127 kJ mol^{-1}. LiH exists and is used in organic chemistry as a reducing agent.

CHAPTER 21

21.1 search for $PhCH_2CH_2Ph$ in the product mixture; **21.3** 100.2 kcal mol^{-1} and 74.2 kcal mol^{-1}; **21.7** 0.17.

INDEX

Printed in the United States of America
ED-01-05-12